Industrial IoT-Enabled Modeling and Optimization for the Process Industry

Industrial IoT-Enabled Modeling and Optimization for the Process Industry

Guest Editors

Gongzhuang Peng
Shenglong Jiang

MDPI

Basel • Beijing • Wuhan • Barcelona • Belgrade • Novi Sad • Cluj • Manchester

Guest Editors
Gongzhuang Peng
International Institute for
Interdisciplinary and
Frontiers
Beihang University
Beijing
China

Shenglong Jiang
School of Automation
Chongqing University
Chongqing
China

Editorial Office
MDPI AG
Grosspeteranlage 5
4052 Basel, Switzerland

This is a reprint of the Special Issue, published open access by the journal *Processes* (ISSN 2227-9717), freely accessible at: https://www.mdpi.com/journal/processes/special_issues/RF2112D4H3.

For citation purposes, cite each article independently as indicated on the article page online and as indicated below:

Lastname, A.A.; Lastname, B.B. Article Title. *Journal Name* **Year**, *Volume Number*, Page Range.

ISBN 978-3-7258-8131-4 (Hbk)
ISBN 978-3-7258-8132-1 (PDF)
https://doi.org/10.3390/books978-3-7258-8132-1

Contents

About the Editors vii

Shenglong Jiang and Gongzhuang Peng
Industrial IoT-Enabled Modeling and Optimization for the Process Industry
Reprinted from: *Processes* **2026**, *16*, 2331, https://doi.org/10.3390/pr16142331 1

Zhenkai Qin, Qining Luo, Xunyi Nong, Xiaolong Chen, Hongfeng Zhang and Cora Un In Wong
MAS-LSTM: A Multi-Agent LSTM-Based Approach for Scalable Anomaly Detection in IIoT Networks
Reprinted from: *Processes* **2025**, *13*, 753, https://doi.org/10.3390/pr13030753 4

Yan Zhang, Junsheng Wang, Jie Sun, Ruiqi Sun and Dawei Qin
Power Load Forecasting System of Iron and Steel Enterprises Based on Deep Kernel–Multiple Kernel Joint Learning
Reprinted from: *Processes* **2025**, *13*, 584, https://doi.org/10.3390/pr13020584 27

Cong Yu, Shuo Chen, Haiquan Yu, Yukun Zhu, Qiang Wang, Guangting Liao and Ling Shi
Quick Combustion Optimization for Utility Boilers Using a Novel Adaptive Hybrid Case Library
Reprinted from: *Processes* **2025**, *13*, 469, https://doi.org/10.3390/pr13020469 36

Aida Amini-Rankouhi, Abdurrafay Siddiqui and Yinlun Huang
Thermodynamic Model-Based Synthesis of Heat-Integrated Work Exchanger Networks
Reprinted from: *Processes* **2024**, *12*, 2293, https://doi.org/10.3390/pr12102293 64

Yonglin Guo, Di Zhou, Huimin Chen, Xiaoli Yue and Yuyu Cheng
Fault Intelligent Diagnosis for Distribution Box in Hot Rolling Based on Depthwise Separable Convolution and Bi-LSTM
Reprinted from: *Processes* **2024**, *12*, 1999, https://doi.org/10.3390/pr12091999 91

Fanghong Jian, Jiangfeng Li, Xiaomei Liu, Qiong Wu and Dan Zhong
Modified Deng's Grey Relational Analysis Model for Panel Data and Its Applications in Assessing the Water Environment of Poyang Lake
Reprinted from: *Processes* **2024**, *12*, 1935, https://doi.org/10.3390/pr12091935 108

Xianhua Ou, Zhongnan You and Xiongxiong He
Local Path Planner for Mobile Robot Considering Future Positions of Obstacles
Reprinted from: *Processes* **2024**, *12*, 984, https://doi.org/10.3390/pr12050984 121

Xingfan Zhang, Lanhui Ye, Cheng Zhang and Chun Wei
Data-Driven-Based Intelligent Alarm Method of Ultra-Supercritical Thermal Power Units
Reprinted from: *Processes* **2024**, *12*, 889, https://doi.org/10.3390/pr12050889 137

Xinnan Yu, Xiaowang Chen, Minggang Du, Yang Yang and Zhipeng Feng
Rotating Machinery Fault Diagnosis under Time–Varying Speed Conditions Based on Adaptive Identification of Order Structure
Reprinted from: *Processes* **2024**, *12*, 752, https://doi.org/10.3390/pr12040752 160

Bin Ren, Chunxi Yang, Xiufeng Zhang and Wenyuan Mao
Modeling and Switched Control of Modular Reconfigurable Flight Array for Faulty Redundancy
Reprinted from: *Processes* **2024**, *12*, 646, https://doi.org/10.3390/pr12040646 177

Shuzong Yan, Dong Xu, He Yan, Ziqiang Wang, Hainan He, Xiaochen Wang and Quan Yang
Measurement Method of Bar Unmanned Warehouse Area Based on Binocular Vision
Reprinted from: *Processes* **2024**, *12*, 466, https://doi.org/10.3390/pr12030466 . . . . **195**

Qiuna Wang, Jiquan Sun, Jiaxuan Yang, Haishen Wang, Lijie Dong, Yanlong Jiao, et al.
Online Partition-Cooling System of Hot-Rolled Electrical Steel for Thermal Roll Profile and Its Industrial Application
Reprinted from: *Processes* **2024**, *12*, 410, https://doi.org/10.3390/pr12020410 . . . . **212**

Xuejun Zhang, Zhimin Lv, Yang Liu, Xiong Xiao and Dong Xu
Novel Multi-Criteria Group Decision Making Method for Production Scheduling Based on Group AHP and Cloud Model Enhanced TOPSIS
Reprinted from: *Processes* **2024**, *12*, 305, https://doi.org/10.3390/pr12020305 . . . . **227**

About the Editors

Gongzhuang Peng

Gongzhuang Peng is an Associate Professor at the International Institute for Interdisciplinary and Frontiers, Beihang University. His research focuses on industrial big data, intelligent optimization, and decision-making. He received his B.S. degree from the School of Automation and Electrical Engineering at Beihang University in 2012 and his Ph.D. degree from the Department of Automation at Tsinghua University in 2018. He has led more than 10 research projects, including the National Key R&D Program and the National Natural Science Foundation of China (General Program).

Shenglong Jiang

Shenlong Jiang is an Associate Professor at the School of Automation, Chongqing University. He received his Ph.D. from the Department of Automation at Tsinghua University and has conducted academic visits and research exchanges at University College London (UCL). His research primarily focuses on system modeling and optimization algorithms under uncertainty, human-in-the-loop cyber–physical systems integrating perception, decision-making, and control, and their applications in intelligent manufacturing process optimization and smart transportation system management.

processes MDPI

Editorial

Industrial IoT-Enabled Modeling and Optimization for the Process Industry

Shenglong Jiang [1,*] and Gongzhuang Peng [2]

[1] School of Automation, Chongqing University, Chongqing 400044, China
[2] International Institute for Interdisciplinary and Frontiers, Beihang University, Beijing 100191, China; gzpeng@buaa.edu.cn
* Correspondence: jiang_shl@cqu.edu.cn

The process industry is undergoing a profound transformation driven by the rapid development of the Industrial Internet of Things (IIoT). Through pervasive sensing, real-time connectivity, and continuous data acquisition, IIoT supports advanced modeling, prediction, and optimization of complex industrial processes. These capabilities are particularly important in steel manufacturing, thermal power generation, energy conversion, materials processing, and warehousing, where operational safety, energy efficiency, equipment reliability, and environmental performance must be maintained under uncertain operating conditions.

Against this background, this Special Issue explores how IIoT-enabled modeling and optimization can support the transition from industrial sensing and predictive analytics to intelligent decision-making and autonomous operations. The contributions collectively highlight an emerging shift toward integrated, adaptive, and sustainable industrial systems.

The first group of studies advances secure and predictive industrial analytics. Qin et al. developed MAS-LSTM, a decentralized anomaly-detection framework that combined multi-agent systems (MASs) with long short-term memory (LSTM) networks to enhance the security and reliability of heterogeneous IIoT networks. Zhang et al. addressed power-load forecasting in iron and steel enterprises by proposing a joint learning method based on deep kernel learning (DKL) and multiple kernel learning (MKL), thereby improving the modeling of nonlinear and time-varying electricity demand patterns and supporting production planning, power purchasing, and energy cost reduction.

The second group addresses process and energy optimization across different system scales. Yu et al. proposed a rapid online combustion optimization framework for utility boilers based on an adaptive hybrid case library. The framework integrated data filtering, case-based reasoning, extreme gradient boosting (XGBoost), and multi-objective optimization to improve combustion efficiency and reduce nitrogen oxides (NO_x) emissions. Amini-Rankouhi et al. extended the optimization perspective from individual units to process networks through a thermodynamic-model-based methodology for synthesizing heat-integrated work exchanger networks, thereby enabling the coordinated recovery of thermal and mechanical energy while reducing the total annualized cost.

The third group focuses on intelligent condition monitoring, fault diagnosis, and alarm management. Guo et al. combined depthwise separable convolution with bidirectional LSTM (Bi-LSTM) to diagnose distribution-box faults in hot rolling, extracting both spatial and temporal features from vibration signals while maintaining a lightweight model structure. Yu et al. addressed rotating machinery fault diagnosis under time-varying speed conditions through the adaptive identification of order structures, improving nonstationary fault-feature extraction without requiring additional speed measurements. Zhang et al.

proposed a data-driven intelligent alarm method for ultra-supercritical thermal power units, improving threshold selection and reducing missed and nuisance alarms.

The final group broadens the application of IIoT-enabled modeling and optimization to autonomous operations, intelligent logistics, production decision-making, and sustainability assessment. Ou et al. improved mobile-robot local path planning by incorporating predicted future obstacle positions, while Ren et al. investigated switched control for modular reconfigurable flight arrays under actuator faults. Yan et al. developed a binocular-vision-based measurement method for unmanned bar warehouses, supporting three-dimensional perception for automated crane-hoisting operations. Within steel manufacturing, Wang et al. proposed an online partition-cooling system for hot-rolled electrical steel to regulate the thermal roll profile and improve strip-shape control, while Zhang et al. developed a multi-criteria group decision-making method based on Group analytic hierarchy process (AHP) and cloud-model-enhanced technique for order preference by similarity to ideal solution (TOPSIS) for evaluating production scheduling schemes. In addition, Jian et al. refined Deng's grey relational analysis model for panel data and applied it to assess the water environment of Poyang Lake, illustrating the broader relevance of industrial data-analysis methods to sustainability evaluation.

Taken together, the contributions to this Special Issue highlight three emerging research trends:

(1) Hybrid industrial modeling: Industrial modeling is increasingly integrating data-driven learning with domain knowledge, adaptive mechanisms, and distributed intelligence to address complex and nonstationary processes.
(2) Online and deployable optimization: Modeling and optimization methods are moving toward online implementation under practical constraints, supporting closer integration among process control, condition monitoring, energy management, logistics, and operational decision-making.
(3) Engineering-oriented validation: Greater emphasis is being placed on validating new methods using real industrial data, representative operating scenarios, pilot platforms, and field applications, thereby strengthening their practical relevance and deployment potential.

Building on these advances, future research should focus on translating increasingly sophisticated models and optimization methods into reliable, trustworthy, and adaptive solutions for real-world industrial operations. Key challenges remain in robust learning from limited labeled data, uncertainty-aware optimization, interpretable and deployable industrial AI, secure distributed analytics, and adaptation to changing operating conditions.

The Guest Editors would like to thank all the authors for their valuable contributions and all the reviewers for their constructive comments. The support of the editorial team of *Processes* is also gratefully acknowledged.

Conflicts of Interest: The authors declare no conflicts of interest.

List of Contributions:

1. Qin, Z.; Luo, Q.; Nong, X.; Chen, X.; Zhang, H.; Wong, C. MAS-LSTM: A Multi-Agent LSTM-Based Approach for Scalable Anomaly Detection in IIoT Networks. *Processes* **2025**, *13*, 753. https://doi.org/10.3390/pr13030753.
2. Zhang, Y.; Wang, J.; Sun, J.; Sun, R.; Qin, D. Power Load Forecasting System of Iron and Steel Enterprises Based on Deep Kernel–Multiple Kernel Joint Learning. *Processes* **2025**, *13*, 584. https://doi.org/10.3390/pr13020584.
3. Yu, C.; Chen, S.; Yu, H.; Zhu, Y.; Wang, Q.; Liao, G.; Shi, L. Quick Combustion Optimization for Utility Boilers Using a Novel Adaptive Hybrid Case Library. *Processes* **2025**, *13*, 469. https://doi.org/10.3390/pr13020469.

https://doi.org/10.3390/pr1614233

4. Amini-Rankouhi, A.; Siddiqui, A.; Huang, Y. Thermodynamic Model-Based Synthesis of Heat-Integrated Work Exchanger Networks. *Processes* **2024**, *12*, 2293. https://doi.org/10.3390/pr12102293.
5. Guo, Y.; Zhou, D.; Chen, H.; Yue, X.; Cheng, Y. Fault Intelligent Diagnosis for Distribution Box in Hot Rolling Based on Depthwise Separable Convolution and Bi-LSTM. *Processes* **2024**, *12*, 1999. https://doi.org/10.3390/pr12091999.
6. Yu, X.; Chen, X.; Du, M.; Yang, Y.; Feng, Z. Rotating Machinery Fault Diagnosis under Time–Varying Speed Conditions Based on Adaptive Identification of Order Structure. *Processes* **2024**, *12*, 752. https://doi.org/10.3390/pr12040752.
7. Zhang, X.; Ye, L.; Zhang, C.; Wei, C. Data-Driven-Based Intelligent Alarm Method of Ultra-Supercritical Thermal Power Units. *Processes* **2024**, *12*, 889. https://doi.org/10.3390/pr12050889.
8. Ou, X.; You, Z.; He, X. Local Path Planner for Mobile Robot Considering Future Positions of Obstacles. *Processes* **2024**, *12*, 984. https://doi.org/10.3390/pr12050984.
9. Ren, B.; Yang, C.; Zhang, X.; Mao, W. Modeling and Switched Control of Modular Reconfigurable Flight Array for Faulty Redundancy. *Processes* **2024**, *12*, 646. https://doi.org/10.3390/pr12040646.
10. Yan, S.; Xu, D.; Yan, H.; Wang, Z.; He, H.; Wang, X.; Yang, Q. Measurement Method of Bar Unmanned Warehouse Area Based on Binocular Vision. *Processes* **2024**, *12*, 466. https://doi.org/10.3390/pr12030466.
11. Wang, Q.; Sun, J.; Yang, J.; Wang, H.; Dong, L.; Jiao, Y.; Li, J.; Zhi, Z.; Yang, L. Online Partition-Cooling System of Hot-Rolled Electrical Steel for Thermal Roll Profile and Its Industrial Application. *Processes* **2024**, *12*, 410. https://doi.org/10.3390/pr12020410.
12. Zhang, X.; Lv, Z.; Liu, Y.; Xiao, X.; Xu, D. Novel Multi-Criteria Group Decision Making Method for Production Scheduling Based on Group AHP and Cloud Model Enhanced TOPSIS. *Processes* **2024**, *12*, 305. https://doi.org/10.3390/pr12020305.
13. Jian, F.; Li, J.; Liu, X.; Wu, Q.; Zhong, D. Modified Deng's Grey Relational Analysis Model for Panel Data and Its Applications in Assessing the Water Environment of Poyang Lake. *Processes* **2024**, *12*, 1935. https://doi.org/10.3390/pr12091935.

https://doi.org/10.3390/pr16142331

processes

MDPI

Article

MAS-LSTM: A Multi-Agent LSTM-Based Approach for Scalable Anomaly Detection in IIoT Networks

Zhenkai Qin [1,2], Qining Luo [1], Xunyi Nong [1], Xiaolong Chen [3,*], Hongfeng Zhang [3,*] and Cora Un In Wong [3]

1 Network Security Research Center, Guangxi Police College, Nanning 530028, China; qinzhenkai@gxjcxy.edu.cn (Z.Q.); luoqining22@gxjcxy.site (Q.L.); nongxunyi23@gxjcxy.site (X.N.)
2 School of Computer Science and Artificial Intelligence, Southwest Jiaotong University, Chengdu 611756, China
3 Faculty of Humanities and Social Sciences, Macao Polytechnic University, Macao, China; corawong@mpu.edu.mo
* Correspondence: osakacool@hait.edu.cn (X.C.); hfengzhang@mpu.edu.mo (H.Z.)

Abstract: The increasing complexity of interconnected systems in the Internet of Things (IoT) demands advanced methodologies for real-time security and management. This study presents MAS-LSTM, an anomaly-detection framework that combines multi-agent systems (MASs) with long short-term memory (LSTM) networks. By training agents on IoT traffic datasets (NF-ToN-IoT, NF-BoT-IoT, and their V2 versions), MAS-LSTM offers scalable, decentralized anomaly detection. The LSTM networks capture temporal dependencies, enhancing anomaly detection in time-series data. This framework overcomes key limitations of existing methods, such as scalability in heterogeneous traffic and computational efficiency in resource-constrained IIoT environments. Additionally, it leverages graph signal processing for adaptive and modular detection across diverse IoT scenarios. Experimental results demonstrate its effectiveness, achieving F1 scores of 0.9861 and 0.8413 on NF-BoT-IoT and NF-ToN-IoT, respectively. For V2 versions, MAS-LSTM achieves F1 scores of 0.9965 and 0.9678. These results highlight its robustness in handling large-scale IIoT traffic. Despite challenges in real-world deployment, such as adversarial attacks and communication overhead, future research could focus on self-supervised learning and lightweight architectures for resource-constrained environments.

Keywords: anomaly detection; Internet of Things; long short-term memory

1. Introduction

In early 2024, a significant disruption in one of Europe's largest manufacturing centers exposed critical vulnerabilities within industrial Internet of Things (IIoT) environments. A carefully orchestrated cyberattack infiltrated multiple smart factories, manipulating sensor data streams and halting production lines for over 48 hours. The incident caused widespread delays in global supply chains and substantial financial losses. Investigations revealed that the root cause was not simply the widespread adoption of IoT devices. Rather, the primary issue was the absence of robust anomaly-detection mechanisms within IIoT networks. Specifically, the inability to detect and respond promptly to irregular traffic and sensor data patterns led to catastrophic consequences.

This incident underscores the urgent need for robust anomaly-detection systems tailored to IIoT environments. Although previous research on IoT security explores vulnerabilities and mitigation strategies, significant gaps remain in accurately identifying subtle yet impactful anomalies. Researchers Hu et al. [1], Thamilarasu and Chawla [2], and Fang et al. [3] propose machine learning and deep learning-based methods for anomaly detection. However, these methods often fail to address the computational constraints and scalability challenges unique to industrial settings. Moreover, temporal anomalies—patterns that evolve over extended periods in sequential data—pose additional challenges that remain insufficiently addressed.

The Internet of Things (IoT) transforms industrial operations by interconnecting millions of devices. These devices form the backbone of critical infrastructures such as manufacturing facilities, smart factories, and energy grids [1,2]. While this connectivity revolutionizes automation and efficiency, it introduces significant security risks. IIoT systems are increasingly targeted by cyber threats such as Distributed Denial-of-Service (DDoS) attacks, botnet intrusions, and data manipulation, all of which threaten productivity and infrastructure integrity [3,4]. The scale, heterogeneity, and dynamic nature of IIoT networks pose formidable challenges to conventional security solutions. Traditional intrusion-detection systems (IDSs) are designed for standard network traffic and often struggle to handle the high volume, velocity, and diversity of industrial data [5,6]. Additionally, centralized anomaly-detection systems face serious scalability and adaptability issues in dynamic IIoT environments [7]. Furthermore, industrial systems require real-time threat detection that is not only accurate but also fast, adaptive, and scalable [8].

Deep learning techniques, such as recurrent neural networks (RNNs) and long short-term memory (LSTM) networks, show promise in time-series anomaly detection [9–11]. However, their centralized implementations often fail to scale in large and diverse networks [12–14]. Similarly, while Transformer models excel at parallel computation, their high computational demands make them unsuitable for resource-constrained IIoT environments [15].

To address these challenges, this paper proposes a novel anomaly-detection framework called MAS-LSTM, which integrates Multi-Agent Systems (MAS) with LSTM networks. MAS provides decentralized and modular decision-making, allowing independent agents to process specific datasets autonomously [16–18]. Meanwhile, LSTM networks are adept at capturing temporal dependencies, which is crucial for detecting time-series anomalies in industrial applications [19,20]. By tackling scalability and adaptability challenges in IIoT environments, MAS-LSTM provides a distributed yet efficient solution for anomaly detection.

Given these considerations, our research aims to answer the following key research question: "How can a decentralized anomaly-detection framework effectively address scalability, computational efficiency, and real-time adaptability challenges in IIoT networks?" This question guides the design and evaluation of MAS-LSTM, ensuring that it provides an effective and scalable solution tailored to the unique requirements of IIoT security.

Although MAS and LSTM offer strong advantages, alternative approaches have also been explored in the literature. Support Vector Machines (SVMs) and Random Forests demonstrate effectiveness in detecting known attack patterns, but they heavily depend on labeled datasets, which limit their adaptability in dynamic IIoT environments [21]. Similarly, unsupervised clustering techniques, such as k-means and DBSCAN, can identify outliers but often fail to generalize to unseen attack types due to their lack of sequential dependency modeling [22]. Graph Neural Networks (GNNs) have recently emerged as an alternative for anomaly detection in IIoT by capturing topological dependencies in network traffic [23]; however, their computational overhead makes them unsuitable for real-time industrial settings. Compared to these approaches, MAS-LSTM balances scalability, real-time adaptability, and lightweight computation, making it well-suited for IIoT applications.

- Scalable Multi-Agent Architecture: To tackle scalability issues in IIoT with diverse and high-volume traffic, we propose a decentralized MAS framework, where agents independently process traffic segments, minimizing single-point failures and enhancing network adaptability.

- Temporal Modeling with LSTM: IIoT systems are faced with evolving, stealthy attacks over extended periods. We integrate LSTM networks to capture these temporal patterns, enabling the detection of long-term anomalies that traditional methods often overlook.
- Resource-Constrained Adaptation: Traditional deep learning models struggle in resource-limited IIoT environments. Our lightweight, agent-based approach mitigates the computational burden through decentralized processing and efficient LSTM feature extraction, making it feasible for real-world deployment.

The remainder of this paper is structured as follows. Section 2 discusses related work on anomaly detection in IIoT, highlighting the strengths and limitations of existing approaches. Section 3 presents the MAS-LSTM framework, detailing its architecture and operational mechanisms. Section 4 describes the experimental setup, datasets, and evaluation metrics used to assess the model's performance. Section 5 analyzes the results and compares MAS-LSTM with alternative methods. Finally, Section 6 summarizes key findings and outlines future research directions.

2. Related Work

The domain of anomaly detection in Industrial Internet of Things (IIoT) has received considerable attention in recent years due to its critical role in industrial operations. The increasing prevalence of cyberattacks and device malfunctions gives rise to an urgent need for robust security mechanisms tailored to IIoT environments. Traditional intrusion-detection methods, while effective in conventional network contexts often fall short in addressing the unique demands of IIoT systems, such as heterogeneous traffic patterns, stringent real-time requirements, and diverse operational characteristics [24,25].

Machine learning-based approaches, such as Support Vector Machines (SVMs) and Random Forests, have demonstrated effectiveness in detecting known attack patterns [26,27]. However, these approaches rely heavily on labeled datasets, which are often labor-intensive to produce and scarce in industrial scenarios, thereby limiting their scalability and adaptability. Similarly, unsupervised methods like k-means clustering and density-based spatial clustering (DBSCAN) can identify outliers in unlabeled data [25]. However, these methods often fail to generalize to unseen attack types and lack the ability to capture temporal dependencies, which are critical in IIoT anomaly detection.

Deep learning techniques have significantly advanced anomaly-detection capabilities, particularly with Recurrent Neural Networks (RNNs) and Long Short-Term Memory (LSTM) networks. These models excel at capturing long-term temporal dependencies, making them highly effective for time-series anomaly-detection tasks [9,28]. However, centralized LSTM-based methods face challenges in handling large-scale and diverse IIoT traffic, particularly in high-volume networks with heterogeneous data distributions, which makes them less efficient and scalable compared to distributed solutions. Transformer-based models have also been explored for anomaly detection due to their strong parallel computation capabilities [15], but their high computational cost makes them less suitable for resource-constrained IIoT environments. Similarly, Graph Neural Networks (GNNs) have shown promise in capturing topological dependencies, but their integration with temporal modeling introduces additional complexity and computational overhead, which further challenges their application in real-time IIoT settings.

While Transformer models have demonstrated superior performance in capturing long-range dependencies within sequential data, their application in IIoT anomaly detection is limited by high computational complexity [29]. Compared to MAS-LSTM, Transformer-based methods require substantial computational resources, making them impractical for real-time and resource-constrained IIoT deployments. Additionally, Transformer models often require significantly more training data and hyperparameter tuning to achieve optimal accuracy, whereas LSTM-based methods, particularly in a decentralized MAS architecture, are more computationally efficient while maintaining competitive detection accuracy. Empirical studies have shown that MAS-LSTM achieves higher efficiency in detecting real-time anomalies with lower latency, whereas Transformer models may introduce computational bottlenecks when applied to high-volume industrial traffic.

Anomaly detection in Industrial Internet of Things (IIoT) networks has seen the development of various hybrid frameworks, each designed to address the unique challenges posed by industrial environments. Several such hybrid approaches combine the strengths of different techniques to enhance detection accuracy and scalability. Autoencoder-based methods, for instance, have been used in combination with LSTM networks to capture both feature representations and temporal dependencies in IIoT data, improving the robustness of anomaly detection. Similarly, reinforcement learning (RL)-enhanced anomaly-detection frameworks dynamically optimize model parameters and decision rules to adapt to evolving attack patterns, offering a flexible approach to address dynamic threats. Another emerging approach is the integration of federated learning (FL) with deep anomaly detection, enabling models to be trained across distributed IIoT nodes while preserving data privacy. These methods offer alternative solutions to improve scalability, adaptability, and efficiency in industrial anomaly detection, though they may introduce additional complexity or overhead.

Several studies have explored hybrid approaches similar to MAS-LSTM, illustrating the increasing interest in combining decentralized systems with advanced machine learning techniques for IIoT anomaly detection. For example, recent work by Zulfiqar [30] proposed a hybrid MAS and deep learning framework for anomaly detection in smart manufacturing, where the MAS was used to divide tasks across decentralized agents, and deep learning methods (including LSTM) were employed for temporal pattern recognition. While this method also uses a hybrid approach, it does not emphasize the distributed decision-making and computational efficiency that MAS-LSTM offers. Additionally, Sun [31] combined reinforcement learning with MAS to optimize decision-making in IIoT networks, with a focus on improving adaptability to dynamic traffic and attack patterns. This approach is similar to ours, but it lacks the temporal modeling capability of LSTM, which is a crucial component in detecting time-series anomalies in IIoT environments. These studies provide valuable context for the effectiveness of hybrid models in IIoT anomaly detection, while also highlighting key differences in model structure and the application of temporal dependencies.

In light of these challenges, recent research has increasingly focused on distributed learning architectures such as Multi-Agent Systems (MAS) [16]. MAS-based frameworks enable independent agents to process specific datasets autonomously, improving network efficiency and reducing single points of failure [32–34]. These hybrid models, which combine LSTM's temporal modeling capabilities with MAS's decentralized decision-making, have shown promise in enhancing adaptability while minimizing centralized processing overhead [35]. This integration of MAS with deep learning methods, particularly LSTM, provides a more scalable and adaptive solution than centralized methods alone, offering improved performance in IIoT environments with high data variability.

Furthermore, research in adjacent domains, such as distributed learning and reinforcement learning, provides valuable insights. Reinforcement learning techniques, when integrated into MAS, have demonstrated potential in dynamically optimizing decision-making processes and improving adaptability in evolving environments [32,34]. These approaches highlight the potential of decentralized architectures to manage complex and dynamic IIoT systems. However, many existing MAS-based frameworks lack robust temporal modeling capabilities, which limits their effectiveness in capturing the sequential nature of anomalies.

MAS-LSTM addresses these gaps by combining the strengths of MAS and LSTM networks, offering a scalable, adaptive, and computationally efficient solution tailored to the complex requirements of IIoT environments. Unlike Transformer-based approaches, which are computationally intensive, MAS-LSTM provides a lightweight alternative that maintains high detection performance. Furthermore, the modular nature of the MAS enables the seamless integration of additional agents, allowing the system to dynamically adapt to new traffic patterns and evolving attack scenarios. This unique combination of distributed decision-making, computational efficiency, and temporal modeling positions the MAS-LSTM framework as a robust solution for anomaly detection in IIoT networks.

3. Methodology

3.1. Problem Definition and Background

The Industrial Internet of Things (IIoT) consists of a vast network of interconnected devices in industrial environments, generating large-scale and highly heterogeneous data. Formally, let $\mathcal{D} = \{\mathbf{d}_1, \mathbf{d}_2, \ldots, \mathbf{d}_N\}$ represent the dataset of traffic information collected from N devices, where each $\mathbf{d}_i \in \mathbb{R}^m$ is a data vector of dimension m corresponding to the traffic features from device i. The complexity of IIoT systems, due to the diverse nature of devices and traffic patterns, makes them vulnerable to various cyberattacks and anomalies, such as Distributed Denial-of-Service (DDoS) attacks, botnet intrusions, and sensor malfunctions [6,36].

Traditional anomaly-detection methods face several challenges in IIoT settings, including scalability issues when handling large data volumes, the need for real-time detection of evolving attack patterns, and the limitations imposed by resource-constrained devices [20,37]. Let $\mathcal{A} = \{a_1, a_2, \ldots, a_K\}$ represent the set of attack types, where $a_k \in \mathbb{R}^m$ denotes the feature representation of attack type k. In many cases, the detection of a_k requires models capable of learning temporal dependencies within the data, denoted by:

$$\mathbf{X}_t = f(\mathbf{X}_{t-1}, \mathbf{X}_{t-2}, \ldots, \mathbf{X}_{t-n}; \theta) \tag{1}$$

where $\mathbf{X}_t$ is the data vector at time t, and θ represents the model parameters. This temporal dependency makes it necessary to use methods that can model such sequences effectively, as shown in studies of time-series anomaly detection [38,39].

Existing centralized approaches to anomaly detection often struggle with the heterogeneity and dynamic nature of IIoT traffic, leading to high false positives or delayed detection [40]. These methods also require substantial computational resources, which are impractical in IIoT environments where devices may have limited processing power [41]. Therefore, there is a need for scalable, efficient, and adaptive solutions capable of detecting anomalies while accommodating the constraints of IIoT systems. Let $\mathbf{y}_t$ represent the anomaly label at time t, where:

$$y_t = \begin{cases} 0, & \text{if normal traffic at time } t, \\ 1, & \text{if anomalous traffic at time } t. \end{cases} \tag{2}$$

To address these challenges, we propose the MAS-LSTM framework, which integrates Multi-Agent Systems (MAS) for decentralized learning and decision-making with Long Short-Term Memory (LSTM) networks for capturing temporal dependencies in IIoT data. Multi-Agent Systems (MASs) are collections of autonomous agents that interact with each other and the environment to achieve individual or collective goals [42]. MASs have been widely used in decentralized systems to improve scalability and efficiency, particularly in environments like IIoT where traditional centralized systems often fail to meet real-time and resource limitations [43,44]. Each agent in an MAS can process local data, make decisions, and communicate with others to achieve a broader goal. In the context of IIoT, an MAS helps distribute the computational load and allows the system to scale efficiently.

LSTM is a type of Recurrent Neural Network (RNN) designed to model sequential data by capturing long-term dependencies. In IIoT anomaly detection, LSTM networks are especially effective for identifying abnormal patterns in time-series data generated by devices. The LSTM-based temporal modeling can be formally expressed as:

$$h_t = LSTM(\mathbf{X}_t, h_{t-1}; \theta_{LSTM}), \tag{3}$$

where h_t is the hidden state at time t, and θ_{LSTM} are the parameters of the LSTM model. This allows LSTM to capture the temporal dependencies in data, which are crucial for detecting anomalies that evolve over time.

The combination of MAS and LSTM in the MAS-LSTM framework enables decentralized anomaly detection. Each agent uses LSTM to analyze its local data and detect anomalies. The agents can then share their findings, allowing the system to coordinate and improve the accuracy of anomaly detection. This collaboration improves scalability, real-time detection, and efficiency in large-scale IIoT environments. By leveraging both MAS for decentralization and LSTM for temporal modeling, the MAS-LSTM framework is better equipped to handle the complexity and dynamic nature of IIoT data.

3.2. System Architecture

This section presents the proposed anomaly-detection framework, which integrates a multi-agent system (MAS) with long short-term memory (LSTM) networks. MAS-LSTM is designed to address the scalability, adaptability, and real-time detection challenges in large-scale Industrial Internet of Things (IIoT) networks. To improve the model's generalization and prevent overfitting, we introduce an adaptive dropout strategy during training. This strategy dynamically adjusts the dropout rate throughout the training process, starting with a higher rate and decreasing over time. By doing so, we can prevent the model from overfitting in the initial stages while allowing it to capture long-term dependencies in the later stages. This adaptive dropout mechanism enhances the robustness of the model without compromising the accuracy of anomaly detection. By leveraging distributed learning and temporal modeling, the system effectively identifies anomalies in IIoT traffic. The modular structure, as shown in Figure 1, ensures each component contributes to the overall robustness and efficiency of MAS-LSTM.

The integration of MAS and LSTM within MAS-LSTM addresses critical challenges in IIoT anomaly detection. We chose MAS due to its ability to offer decentralized decision-making, allowing agents to work autonomously on distinct subsets of the data. This design reduces the risk of single-point failures, which is crucial for real-time applications in dynamic IIoT environments. Moreover, LSTM networks within each agent capture temporal dependencies in IIoT traffic, making MAS-LSTM highly effective for detecting sequential anomalies, such as stealthy botnet behaviors. The decision to use LSTM was based on its capability to detect time-series anomalies, which are common in IIoT systems. The combination of MAS's adaptability and LSTM's precision enables MAS-LSTM to outperform traditional centralized methods in dynamic and heterogeneous environments. Furthermore, the modular design of MAS-LSTM allows the seamless integration of new agents and supports scalability without compromising detection accuracy.

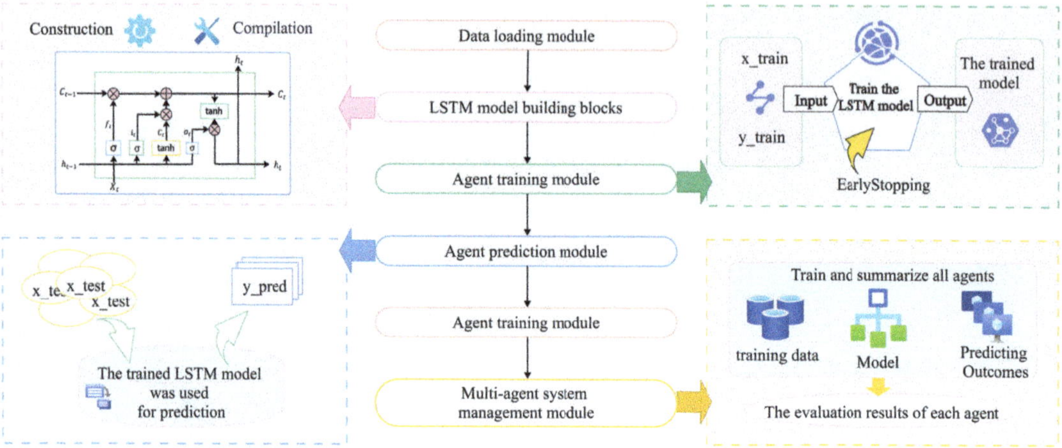

Figure 1. Framework for anomaly detection in IIoT networks, covering data loading, LSTM model construction, agent training, and system-level decision-making.

3.3. Multi-Agent System (MAS)

In the design of MAS-LSTM, the decision to incorporate a Multi-Agent System (MAS) was made to address the need for scalability in heterogeneous IIoT networks. Each agent is assigned a specific traffic dataset, ensuring that data processing is distributed across multiple nodes, rather than being handled centrally. This approach enables parallelism and facilitates more efficient learning from the traffic data. The backbone of MAS-LSTM relies on MAS, which offers a decentralized and scalable solution to managing the diverse and heterogeneous traffic patterns in IIoT networks [16]. Each agent operates independently, analyzing specific datasets to ensure localized learning and decision-making. This modular design enables parallelism, significantly enhancing MAS-LSTM's scalability and adaptability.

The agents in the MAS are designed to independently process and analyze specific portions of the IIoT data. However, when it comes to decision-making, agents share their anomaly-detection insights with each other through a communication mechanism. This enables agents to exchange information about potential anomalies, which helps improve the overall accuracy of the system. In the case of conflicting anomaly classifications (e.g., one agent classifies traffic as normal while another classifies it as anomalous), a majority voting mechanism is employed to resolve conflicts. This ensures that the final decision reflects the consensus of the majority of agents, reducing the impact of individual conflicting classifications.

Let $D = \{D_1, D_2, \ldots, D_N\}$, where D_i represents the traffic dataset for agent A_i. Each dataset is segmented into time-series sequences $\mathbf{X}_i = \{\mathbf{x}_{i1}, \mathbf{x}_{i2}, \ldots, \mathbf{x}_{iT}\}$ with corresponding labels $\mathbf{Y}_i = \{y_{i1}, y_{i2}, \ldots, y_{iT}\}$ [25,26].

Before being fed into the LSTM network, IIoT traffic data undergoes preprocessing steps to make it suitable for time-series analysis. Raw traffic data is cleaned by removing irrelevant data, handling missing values, and filtering out noise. Time windows are then defined to segment the traffic data into smaller chunks, allowing the LSTM model to process sequential patterns. These time-series sequences are created for each agent to analyze, and normalization is applied to ensure consistent feature scaling.

The network traffic metrics used for anomaly detection include packet arrival rates, which indicate the rate at which packets are received and can help identify congestion or unusual traffic behavior. Transmission delays are also monitored, as they represent the time taken for packets to reach their destination and can be useful in identifying issues like DDoS attacks or general network performance problems. Additionally, the total number of bytes and packets transmitted, known as byte and packet counts, are tracked to detect traffic surges or sudden changes in the network, which may signify an anomaly. Protocol types, such as TCP, UDP, and ICMP, are analyzed as they can reveal certain attack types or abnormal system behavior based on their usage patterns. Flow duration and flow counts are also important, as the number of active flows and their duration can help detect botnet activities or patterns indicative of DDoS attacks. Finally, round-trip time (RTT) is monitored to identify network performance issues or detect behaviors associated with slow-rate flooding attacks. These metrics together form the foundation for anomaly detection in the MAS-LSTM framework.

By breaking the datasets into smaller time-series sequences, agents can focus on more manageable portions of the data, reducing the processing load and speeding up detection. Here, $y_t = 0$ denotes normal traffic, and $y_t = 1$ denotes anomalous traffic. Each agent processes its assigned dataset independently, eliminating dependencies on a centralized controller and reducing the risk of a single point of failure.

This decentralized architecture ensures system robustness and allows seamless integration of new agents, enabling the system to adapt to evolving traffic patterns.

3.4. LSTM-Based Temporal Modeling

One of the key decisions in the design of MAS-LSTM was to integrate LSTM networks for capturing temporal dependencies in IIoT traffic. This was motivated by the need to model long-term patterns that traditional methods could not detect. To capture the temporal dependencies in IIoT traffic, each agent leverages LSTM networks. LSTM's proven ability to model sequential data makes it ideal for detecting complex anomalies, such as Distributed Denial-of-Service (DDoS) and botnet behaviors [9]. While Transformer models excel in parallel processing, their high computational cost limits their applicability in resource-constrained IIoT environments. Therefore, LSTM was chosen for its efficiency in time-series anomaly detection, without compromising the system's real-time capabilities. Similarly, Graph Neural Networks (GNNs), though effective in capturing topological dependencies, introduce unnecessary complexity for time-series modeling.

The LSTM network used in MAS-LSTM consists of two LSTM layers with 128 hidden units per layer, which have been shown to balance model complexity and computational efficiency. We use a batch size of 32 and train the model for 50 epochs to ensure that it captures sufficient temporal dependencies while avoiding overfitting. The Adam optimizer is used for training, as it is well-suited for large-scale datasets and adaptive learning rates. Hyperparameters such as the learning rate and batch size were tuned using a grid search to optimize performance.

The LSTM network's architecture consists of three gates: input (i_t), forget (f_t), and output (o_t), enabling it to retain long-term dependencies while mitigating the vanishing gradient problem. The internal equations are:

$$i_t = \sigma(W_i x_t + U_i h_{t-1} + b_i), \quad f_t = \sigma(W_f x_t + U_f h_{t-1} + b_f), \tag{4}$$

$$o_t = \sigma(W_o x_t + U_o h_{t-1} + b_o), \quad c_t = f_t \cdot c_{t-1} + i_t \cdot \tanh(W_c x_t + U_c h_{t-1} + b_c), \tag{5}$$

$$h_t = o_t \cdot \tanh(c_t), \tag{6}$$

where h_t is the hidden state, and c_t is the cell state at time t. These mechanisms enable LSTM to model sequential traffic patterns effectively.

The final hidden state h_T is passed through a dense layer with a sigmoid activation to predict anomaly probabilities:

$$\hat{y}_T = \sigma(W_h h_T + b_h), \tag{7}$$

where $\hat{y}_T > 0.5$ indicates anomalous traffic, and $\hat{y}_T \leq 0.5$ indicates normal traffic.

3.5. Anomaly Detection and Decision-Making

The decision-making process in MAS-LSTM involves aggregating the individual predictions from each agent. By employing a majority voting system, the framework ensures that the final anomaly-detection decision remains robust, even in cases where individual agents may detect anomalies with slight variations. If two or more agents disagree on the classification of a particular traffic instance, the final decision is determined by the majority vote. This voting mechanism helps resolve conflicts and enhances the reliability of the overall system.

During inference, each agent predicts anomaly probabilities for its assigned dataset. A majority voting mechanism is then used to aggregate the individual decisions, flagging an anomaly if the majority of agents detect abnormal behavior:

$$\hat{y}_t = \begin{cases} 1 & \text{if } \hat{y}_t > 0.5, \\ 0 & \text{if } \hat{y}_t \leq 0.5. \end{cases} \tag{8}$$

This ensemble approach reduces false positives and ensures robust detection. By decentralizing decision-making, the system maintains high adaptability and reliability, even in dynamic IIoT environments.

3.6. Scalability and Real-Time Detection

MAS-LSTM excels in scalability due to its decentralized MAS architecture, where agents process traffic independently. Formally, let $\mathcal{D}_i = \{\mathbf{d}_i^{(1)}, \mathbf{d}_i^{(2)}, \ldots, \mathbf{d}_i^{(T)}\}$ represent the traffic dataset assigned to agent A_i, where T is the number of time steps in the traffic sequence. Each agent A_i processes its own data independently, and the global model decision is made by aggregating local decisions, as described by the majority voting mechanism:

$$\hat{y}_t = \frac{1}{N} \sum_{i=1}^{N} \hat{y}_t^{(i)} \quad \text{for } t = 1, 2, \ldots, T, \tag{9}$$

where $\hat{y}_t^{(i)}$ is the prediction of agent A_i at time t, and N is the total number of agents. This decentralized decision-making reduces the bottlenecks associated with centralized systems, enabling faster real-time anomaly detection.

The modular nature of MAS-LSTM allows seamless integration of additional agents as the network grows or new devices are added. Mathematically, this can be represented by the dynamic addition of agents to the system, where each new agent A_{i+1} is assigned a subset of the data and the model is updated as follows:

$$\mathcal{D}_{i+1} = \{\mathbf{d}_{i+1}^{(1)}, \mathbf{d}_{i+1}^{(2)}, \ldots, \mathbf{d}_{i+1}^{(T)}\}, \tag{10}$$

and the model is adapted to incorporate new data from $\mathcal{D}_{i+1}$ without retraining the entire system. This ensures scalability and adaptability as the system expands.

Additionally, the use of a distributed learning approach enables MAS-LSTM to process large-scale IIoT traffic in real-time. Each agent updates its model parameters locally based on the data it receives, using the following update rule:

$$\theta_i^{(t)} = \theta_i^{(t-1)} - \eta \nabla_{\theta_i} \mathcal{L}_i(\mathbf{d}_i^{(t)}, \hat{y}_i^{(t)}), \tag{11}$$

where $\theta_i^{(t)}$ represents the model parameters for agent A_i at time t, η is the learning rate, and $\mathcal{L}_i$ is the loss function for agent A_i. The parameters are updated independently, which helps distribute the computational load and ensures real-time processing capabilities.

The combination of MAS and LSTM provides a robust solution for complex IIoT networks. The LSTM component models the temporal dependencies in traffic data, ensuring accurate anomaly detection over time, while the MAS component ensures scalability and adaptability by distributing the computation across multiple agents. This results in a system that is efficient, scalable, and capable of handling the dynamic nature of IIoT environments without compromising detection accuracy [45,46].

4. Experiment

The goal of this experiment is to evaluate the effectiveness and flexibility of the proposed MAS-LSTM framework in detecting anomalies across different IoT environments. The experiment focuses on testing the model's ability to handle various attack scenarios in both traditional and modern IoT traffic, using both the original and v2 versions of the datasets. It also compares MAS-LSTM's performance with other machine learning and deep learning methods. To understand the contribution of different components of the architecture, ablation studies are conducted. The experiment uses four IoT traffic datasets—NF-ToN-IoT, NF-BoT-IoT, and their v2 versions—to simulate real-world conditions. We measure performance based on detection accuracy, precision-recall trade-offs, and computational efficiency, while also using tools like 3D UMAP projections and radar charts to visualize feature distributions and decision boundaries. This approach helps address key challenges in IoT security, such as class imbalance, temporal dependencies, and scalability.

4.1. Dataset and Preprocessing

This study utilizes four publicly available IoT traffic datasets: NF-ToN-IoT, NF-BoT-IoT, and their respective versions v2 [47–54]. These datasets were chosen for their ability to comprehensively represent typical IoT traffic patterns, encompassing both benign and malicious behaviors, and providing a robust basis for evaluating the proposed anomaly-detection framework.

The NF-ToN-IoT dataset captures traffic generated by diverse IoT devices, such as sensors and controllers, and includes multiple attack scenarios, including Distributed Denial-of-Service (DDoS) attacks, password guessing, and ransomware. Its heterogeneous traffic patterns and realistic device diversity make it a valuable resource for training and testing anomaly-detection models.

In contrast, NF-BoT-IoT focuses on botnet-generated traffic, a critical threat in IoT environments. This dataset provides detailed records of benign and botnet-induced anomalous behaviors, offering insights into the characteristics of botnet attacks and their impact on network traffic.

The v2 versions of these datasets, NF-ToN-IoT-v2 and NF-BoT-IoT-v2, extend their predecessors by incorporating additional attack types, higher traffic volumes, and more complex traffic patterns. NF-ToN-IoT-v2 simulates diverse and realistic IoT environments, capturing a wider range of benign and malicious interactions. Meanwhile, NF-BoT-IoT-v2 reflects the evolving sophistication of botnet attacks, including high-frequency, large-scale traffic anomalies, which are essential for evaluating the scalability and robustness of anomaly-detection systems in modern IoT settings.

Table 1 summarize the distribution of benign and attack samples across these datasets. For instance, the v2 datasets exhibit significantly larger volumes and a broader variety of attack types compared to their predecessors. NF-ToN-IoT-v2 contains over 10 million attack samples (63.99% of total traffic), while NF-BoT-IoT-v2 exhibits nearly 99.64% malicious traffic. This diversity not only highlights the relevance of these datasets for modeling modern IoT network anomalies but also ensures a rigorous evaluation of MAS-LSTM's adaptability and effectiveness.

Table 1. NF-ToN-IoT, NF-BoT-IoT, NF-ToN-IoT-v2, and NF-BoT-IoT-v2 dataset overview.

Dataset	NF-ToN-IoT		NF-BoT-IoT		NF-ToN-IoT-v2		NF-BoT-IoT-v2	
	Count	%	Count	%	Count	%	Count	%
Benign	270,279	19.6	13,859	2.31	6,099,469	36.01	135,037	0.36
Backdoor	17,247	1.25	-	-	16,809	0.10	-	-
DoS	17,717	1.28	56,833	9.47	712,609	4.21	16,673,183	44.11
DDoS	326,345	23.6	56,844	9.47	2,026,234	11.97	18,331,847	48.56
Injection	468,539	33.9	-	-	684,465	4.05	-	-
MITM	1295	0.09	-	-	7723	0.05	-	-
Password	156,299	11.3	-	-	1,153,323	6.81	-	-
Ransomware	142	0.01	-	-	3425	0.02	-	-
Scanning	21,467	1.56	-	-	3,781,419	22.31	-	-
XSS	99,944	7.25	-	-	2,455,020	14.50	-	-
Attack Samples	1,108,995	80.4	586,241	97.69	10,841,027	63.99	37,628,460	99.64

In terms of data preprocessing, several critical steps were performed to ensure the dataset's quality for model training and evaluation. Initially, irrelevant features such as IPV4_SRC_ADDR, L4_SRC_PORT, IPV4_DST_ADDR, L4_DST_PORT, Attack, and Label were removed to prevent the model from relying on potentially leaking information. This was achieved using the remove_features function, which excluded these specific columns. Missing values were then handled by removing rows containing null entries, ensuring a complete dataset free from gaps that could affect model training.

To further enhance model performance, data normalization was applied, scaling the feature values into the range [0, 1]. This step is particularly important for models like LSTM, as it helps speed up convergence and prevents issues related to features with disproportionate value ranges. The dataset was then split into training and testing sets in an 80/20 ratio, with stratified sampling to maintain the class distribution of benign and malicious traffic. Additionally, the attack labels from the test set were saved separately to facilitate error analysis and assess the performance of the anomaly-detection framework.

These preprocessing steps ensure that the dataset is well-prepared for training and testing, allowing for a more accurate evaluation of the anomaly-detection models, and addressing the complexities of IoT traffic patterns across different datasets.

4.2. Experimental Setup

The proposed multi-agent system was evaluated using the four datasets described earlier. For each dataset, we trained a separate agent within the system, allowing each agent to independently process and learn from its assigned dataset. This setup ensures that the multi-agent framework can handle diverse and heterogeneous IoT traffic patterns effectively. The experimental configuration and hyperparameters are summarized in Table 2.

Table 2. Experiment setup overview.

Component	Description
Batch Size	128
Random Seed	42
Model Type	LSTM
Input Dimension	Dynamic (based on data)
LSTM Units	150
Activation Function	ReLU
Regularization	L2 (lambda = 0.001)
Dropout Rate	0.3
Optimizer	Adam
Loss Function	Binary Crossentropy
Metrics	Accuracy
Epochs	10
Validation Split	20%
Early Stopping	3 epochs without improvement
Prediction Threshold	1
Threshold	0.7
CPU	Intel(R) Core(TM) i7-10870H
GPU	NVIDIA GeForce RTX 3060 Laptop GPU

In our experiments, we explored how adjusting the decision threshold can affect the performance of the model, specifically the trade-off between precision and recall. A higher threshold can improve precision by reducing the number of false positives, though it may slightly reduce recall. In our setup, the default threshold was set to 1, which is optimal for our scenario where minimizing false positives is crucial. This setting is particularly important in real-world industrial IoT applications, where excessive false alarms could lead to operational disruptions, increased costs, and unnecessary security interventions.

However, in certain scenarios where higher recall is required—such as early-stage threat-detection systems or environments where missing an attack is more critical than occasional false positives—adjusting the threshold to a lower value can enhance sensitivity. To analyze how the model behaves with different thresholds, we also evaluated it with a threshold of 0.7, as shown in the confusion matrices in Figure 2. This allows for a flexible trade-off between precision and recall, ensuring the model can be tuned based on specific deployment requirements.

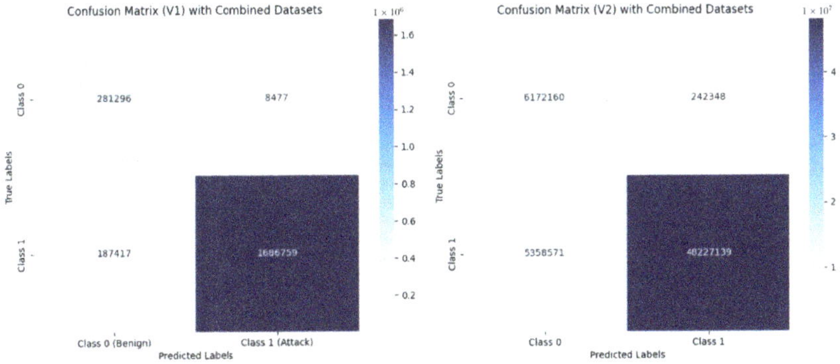

Figure 2. Confusion matrices of the model with different thresholds (threshold = 0.7).

4.3. Results and Analysis

The experimental results, summarized in Tables 3 and 4, provide a comprehensive evaluation of the proposed system compared to traditional machine learning and deep learning methods. These results highlight the system's ability to handle diverse IoT datasets and detect complex anomalies, addressing scalability and adaptability challenges in IIoT environments.

Table 3. Performance metrics for NF-BoT-IoT and NF-ToN-IoT.

Model	**NF-BoT-IoT**				**NF-ToN-IoT**			
	F1-Score	**Accuracy**	**Precision**	**Recall**	**F1-Score**	**Accuracy**	**Precision**	**Recall**
EFC	0.9649	0.9336	0.9988	0.9331	0.8087	0.7385	0.9821	0.6873
Naive Bayes	0.9652	0.9334	0.9871	0.9442	0.8937	0.8089	0.8084	0.9992
Logreg	0.9932	0.9866	0.9870	0.9995	0.8958	0.8151	0.8190	0.9886
RBF	0.9986	0.9811	0.9975	0.9998	0.9906	0.9822	0.9827	0.9987
GRU-FCN	0.9617	0.8830	0.9975	0.8657	0.8730	0.7892	0.8612	0.8734
LSTM	0.9270	0.7649	0.7600	0.7597	0.8897	0.8423	0.7893	0.8932
GMS-IDS	0.9816	0.9686	0.9483	0.9570	0.8400	0.7965	0.9067	0.9012
MAS-LSTM	0.9861	0.9863	0.9996	0.9929	0.8413	0.8405	0.9906	0.9094

Table 4. Performance metrics for NF-BoT-IoT-v2 and NF-ToN-IoT-v2.

el	**NF-BoT-IoT-v2**				**NF-ToN-IoT-v2**			
	F1-Score	**Accuracy**	**Precision**	**Recall**	**F1-Score**	**Accuracy**	**Precision**	**Recall**
	0.9730	0.9477	0.9997	0.9478	0.9455	0.9601	0.9012	0.9945
Bayes	0.9960	0.9931	0.9988	0.9943	0.3911	0.2631	0.2441	0.9822
g	0.9993	0.9986	0.9989	0.9998	0.9766	0.9888	0.9840	0.9693
	0.9994	0.9988	0.9992	0.9996	0.9875	0.9913	0.9905	0.9844
FCN	0.9902	0.8830	0.9975	0.8657	0.8730	0.7892	0.8612	0.8734
	0.9922	0.7649	0.7600	0.7597	0.8897	0.8423	0.7893	0.8932
IDS	0.9820	0.8016	0.9532	0.9453	0.9860	0.8901	0.9011	0.9321
LSTM	0.9965	1.0000	0.9982	0.9809	0.9678	0.9522	0.9599	0.9599

The proposed multi-agent system consistently outperforms or remains highly competitive with baseline methods across key metrics, including F1-score, accuracy, precision, and recall. For instance, on the NF-BoT-IoT-v2 dataset, it achieves an F1-score of 0.9965 with 1.0000 accuracy, performing comparably to RBF networks (F1-score 0.9994). Similarly, on the NF-ToN-IoT-v2 dataset, it achieves an F1-score of 0.9678, while Deep Autoencoders (DeepAE) underperform significantly due to their limited ability to model temporal dependencies in IoT traffic. This robust performance demonstrates the proposed system's capacity to detect rare and complex anomalies—a critical requirement for IoT anomaly detection.

The superior results are attributed to MAS-LSTM's modular and decentralized architecture, where each agent specializes in a subset of the data. This design not only enhances scalability but also ensures adaptability to heterogeneous traffic patterns. Additionally, the use of LSTM networks enables effective modeling of long-term temporal dependencies, allowing the detection of sequential patterns characteristic of advanced attack types, such as Distributed Denial-of-Service (DDoS) and stealthy botnet behaviors.

To ensure a fair comparison, baseline models were selected based on their relevance and popularity in anomaly-detection tasks. Logistic Regression and Naive Bayes serve as foundational benchmarks for binary classification problems [55,56], while DeepAE represents unsupervised feature learning approaches [57]. RBF networks, widely recognized for their success in non-linear classification, provide a strong benchmark in network intrusion detection [58]. Together, these models establish a comprehensive baseline for evaluating the proposed system's effectiveness.

In contrast, baseline methods often exhibit a trade-off between precision and recall, limiting their reliability for real-world anomaly detection. For example, while RBF networks may achieve high accuracy or F1-score on certain datasets, they do not consistently generalize across all scenarios. Similarly, DeepAE struggles with temporal modeling, resulting in suboptimal detection of sequential anomalies.

The scalability and adaptability of the proposed system are further underscored by its multi-agent architecture. By delegating specific datasets to independent agents, the system efficiently handles heterogeneous IoT traffic. Decentralized decision-making reduces computational overhead and eliminates single points of failure. These capabilities are particularly beneficial in real-world deployments where IoT networks generate diverse and dynamic traffic.

The Receiver Operating Characteristic (ROC) curve presented in Figure 3 visualizes the performance of the proposed MAS-LSTM model on both the V1 and V2 combined datasets. For the V1 dataset, MAS-LSTM achieves an AUC of 0.95 on the NF-BoT-IoT dataset, significantly outperforming the Silo NF-ToN-IoT model with an AUC of 0.82. On the V2 dataset, MAS-LSTM maintains strong performance, achieving an AUC of 0.94 for the NF-BoT-IoT dataset and outperforms the Silo NF-ToN-IoT model, which achieves an AUC of 0.78.

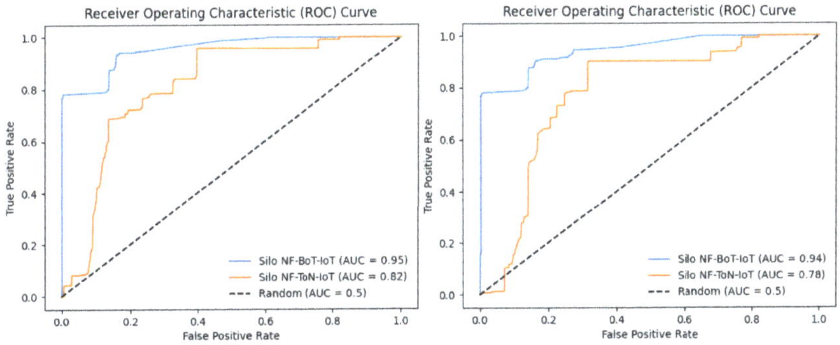

Figure 3. Receiver Operating Characteristic (ROC) curve for MAS-LSTM on V1 and V2 datasets.

These results demonstrate the superior anomaly-detection performance of MAS-LSTM, particularly in distinguishing between normal and anomalous traffic across both V1 and V2 datasets. The AUC scores for both datasets show a clear distinction in performance, with MAS-LSTM consistently outperforming traditional models, reinforcing its effectiveness in IoT environments.

In summary, the proposed multi-agent system, through its integration of LSTM networks and decentralized learning, achieves superior or at least highly competitive performance across diverse IoT datasets. Its balance across precision, recall, and F1-score ensures reliable anomaly detection, while its scalability and adaptability address critical challenges in IIoT environments. These results validate MAS-LSTM as a robust and practical solution for real-world IoT anomaly detection.

4.4. Visual Analysis

To better understand the data characteristics and model performance, we conducted an in-depth analysis using UMAP 3D dimensionality-reduction visualizations and radar charts. The following images illustrate the results.

In Figure 4, the UMAP algorithm was used to visualize the high-dimensional traffic features in 3D. The clustered points of different colors and positions indicate significant distribution differences between normal and attack traffic in the datasets. For instance, in the NF-BoT-IoT-v2 dataset, attack traffic forms distinct and compact clustered regions, while normal traffic is more scattered. This distinct separation provides strong evidence of unique high-dimensional patterns in attack traffic, which our model can effectively learn. Additionally, the scattered nature of normal traffic suggests greater diversity in feature representation, potentially requiring a more flexible model architecture to achieve high recall.

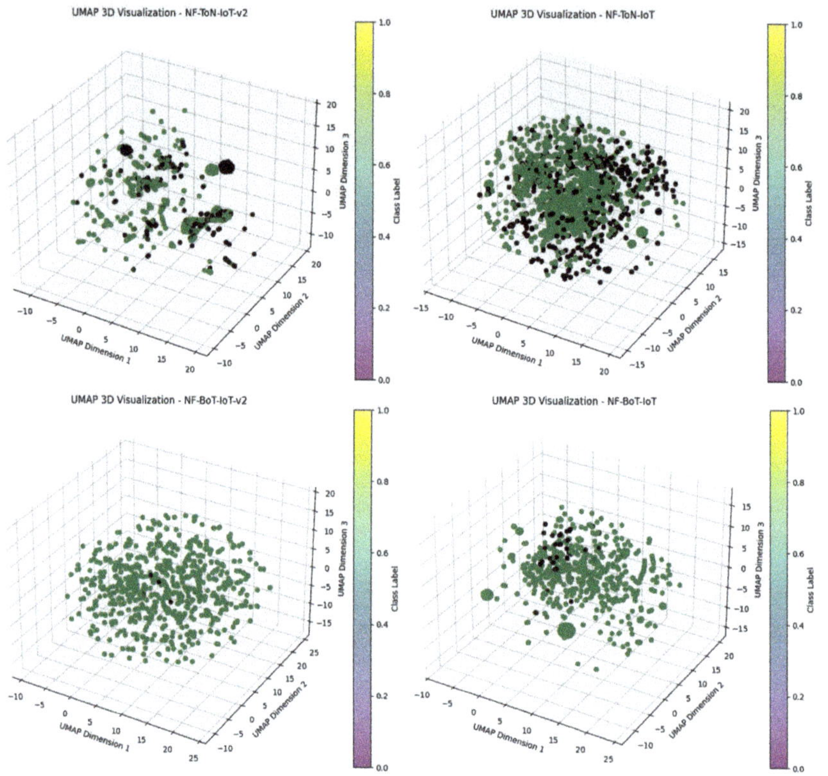

Figure 4. UMAP 3D visualization: dimensionality-reduction distribution of NF-BoT-IoT, NF-ToN-IoT, NF-BoT-IoT-v2, and NF-ToN-IoT-v2 datasets.

Cross-comparing datasets reveals that v2 versions exhibit more compact clusters for attack traffic compared to their original versions. This could indicate enhanced labeling or preprocessing that highlights distinguishing features. Enhanced labeling may involve better tagging of attack instances or more careful data cleaning, which leads to clearer distinctions between attack and normal traffic. Preprocessing steps such as feature scaling, noise reduction, or advanced data augmentation techniques might have contributed to this clearer separation, enabling the model to better capture the attack patterns. Analyzing such differences in feature distributions offers deeper insights into how the datasets influence the learning process and model generalization.

Figure 5 presents a radar chart comparing our model's performance with baseline methods. Notably, our model achieves the highest F1-score, accuracy, precision, and recall, particularly on the v2 datasets. The superior recall demonstrates the model's ability to detect diverse and complex attack types effectively. However, precision, while competitive, shows smaller margins of improvement. This suggests potential false positives, possibly due to overlapping feature spaces of attack and normal traffic. Such overlaps could be further analyzed by visualizing feature importance or model decision boundaries. Compared to traditional models, our method shows exceptional adaptability in handling imbalanced datasets. For example, the significant improvement on NF-BoT-IoT-v2 highlights the robustness of the proposed architecture in capturing nuanced patterns in highly imbalanced traffic, where normal samples dominate.

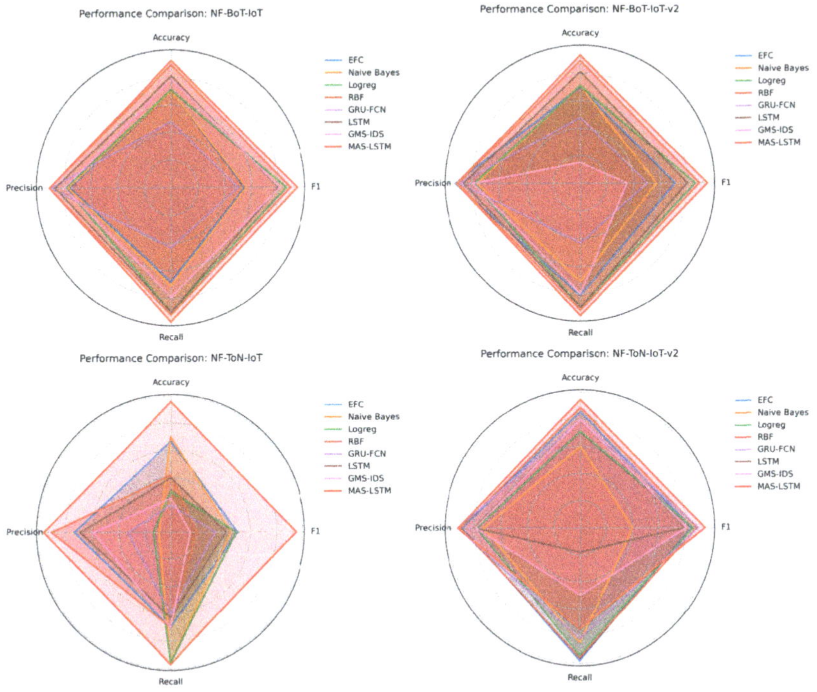

Figure 5. Radar chart: performance comparison of various models on NF-ToN-IoT and NF-BoT-IoT datasets and their v2 versions.

To gain further insights into model performance, we evaluated the impact of different hyperparameter settings, specifically the dropout rates and the number of LSTM hidden units. The combined results are shown in Figure 6.

The first experiment focused on the effect of dropout rates (0.1, 0.3, 0.5, and 0.7). The results reveal that the F1-score peaked at a dropout rate of 0.3, while higher rates (e.g., 0.7) led to a decline in performance. A deeper investigation indicates that excessive dropout may remove critical features required for accurate classification. By analyzing feature activation maps, we observed that lower dropout rates preserve key patterns in traffic data, contributing to improved recall and F1-score.

The second experiment evaluated the impact of the number of LSTM hidden units (50, 100, 150, and 250). Increasing the number of hidden units consistently improved recall and F1-score. However, the marginal gains diminished for larger hidden units (e.g., 250). This suggests that while larger hidden layers capture more complex patterns, they also introduce the risk of overfitting and increased computational costs. Analysis of model-training dynamics further revealed that larger hidden layers required more epochs to converge, underscoring the importance of balanced hyperparameter settings.

From these analyses, it is evident that the distinct distribution differences between normal and attack samples enable deep learning-based models to effectively extract critical features. The proposed multi-agent system leverages its modular architecture to better adapt to the heterogeneity of the datasets. The significant improvements on v2 datasets highlight the model's robustness and its ability to generalize across complex and high-dimensional traffic scenarios. The comparison between v1 and v2 datasets reveals that the enhancements in labeling and preprocessing have improved the clarity of the attack traffic patterns, which in turn supports the learning process and boosts the model's performance on v2 datasets.

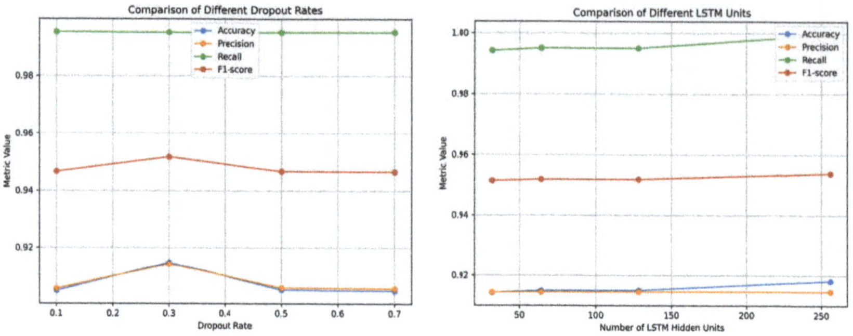

Figure 6. Impact of different dropout rates (**left**) and LSTM hidden units (**right**) on model performance.

4.5. Ablation Study

This section presents ablation experiments on two key components of our system: the LSTM module and the Agent module. We evaluate their individual contributions by examining configurations where only one module is active. The datasets used for this study, v1 and v2, represent integrated versions of the NF-BoT-IoT and NF-ToN-IoT datasets, respectively, with v2 incorporating more diverse and complex traffic patterns. Table 5 summarizes the results of these experiments.

Table 5. Ablation study comparing only lstm, only agent, and no voting on v1 and v2 datasets.

Model	**v1**				**v2**			
	Accuracy	**Precision**	**Recall**	**F1-Score**	**Accuracy**	**Precision**	**Recall**	**F1-Score**
MAS-LSTM	0.9823	0.9823	0.9996	0.9994	0.9904	0.9945	1.0000	0.9972
-LSTM	0.8246	0.8201	0.9804	0.8935	0.6382	0.5298	0.5431	0.5364
-Agent	0.8104	0.8145	0.9763	0.8873	0.8153	0.7623	0.7915	0.7767
-Voting	0.9051	0.9049	0.9923	0.9472	0.7894	0.4756	0.4798	0.4777

First, we examine the impact of the LSTM module in the absence of the Agent module (only lstm). On the simpler v1 dataset, the only lstm configuration achieves an F1-score of 0.8935, with a precision of 0.8201 and a recall of 0.9804, indicating that the LSTM alone can capture temporal dependencies reasonably well. However, on the more complex v2 dataset, the F1-score drops to 0.5364, highlighting that relying solely on LSTM for time-series modeling is insufficient when facing a broader variety of IoT traffic patterns and advanced attacks.

Next, we focus on retaining only the Agent module (only agent) while removing the LSTM. On v1, the performance is comparable to only lstm with an F1-score of 0.8873, suggesting that the Agent module alone can also learn certain relevant features. On v2, the only agent configuration achieves a slightly lower F1-score of 0.7767 compared to only lstm, implying that the agent-driven approach is less effective than LSTM in handling more diverse and complex traffic patterns. However, there remains a noticeable gap between only agent and the full system configuration, suggesting that Agent without LSTM-based temporal modeling is still suboptimal in handling the intricacies of high-dimensional network attacks.

We also include a configuration without the voting mechanism (no voting). This setup achieves an F1-score of 0.9472 on v1, reflecting its ability to integrate predictions effectively in simpler scenarios. However, on the more complex v2 dataset, its performance drops significantly to an F1-score of 0.4777, indicating that the absence of voting hinders its ability to generalize to diverse traffic patterns.

Finally, when both the LSTM and Agent modules are integrated, the system achieves near-perfect performance on v1 with an F1-score of 0.9994 and maintains an impressive F1-score of 0.9972 on v2. These results underscore that the synergy between the LSTM module's ability to capture temporal dependencies and the Agent module's specialized decision-making process is critical for robust detection in both simpler (v1) and more advanced (v2) IoT traffic scenarios.

5. Discussion

5.1. Bridging Anomaly-Detection Challenges

This study tackles the critical challenge of anomaly detection in Industrial Internet of Things (IIoT) networks, focusing on scalability, adaptability, and computational efficiency. By leveraging a multi-agent system (MAS) integrated with Long Short-Term Memory (LSTM) networks, MAS-LSTM demonstrates a robust capability to identify both rare and complex attack patterns. The evaluation across diverse datasets, such as NF-ToN-IoT and NF-BoT-IoT, underscores the system's practical applicability in industrial scenarios. These datasets, rich in attack vectors and traffic behaviors, enable MAS-LSTM to reflect real-world IIoT environments, such as smart factories and energy grids.

The findings resonate with prior research demonstrating the efficacy of LSTM models in modeling sequential data for IoT anomaly detection [59]. While traditional centralized approaches have been effective in detecting temporal anomalies, the proposed system's decentralized MAS architecture offers a distinct advantage by enhancing scalability and robustness. This modular design enables seamless handling of heterogeneous and large-scale traffic patterns.

Unlike conventional anomaly-detection techniques, which often struggle with diverse traffic behaviors, the proposed system's adaptability aligns with recent research advocating scalable and modular frameworks [60]. Moreover, our results complement findings from [61], which highlight the importance of diverse datasets in evaluating anomaly-detection systems. However, the reliance on labeled training data presents a notable limitation. While semi-supervised methods have mitigated this issue in related studies, the current framework's dependence on labeled data limits its generalizability to unseen attack patterns. Addressing this gap could pave the way for further advancements in scenarios with sparse labeled data availability.

5.2. Key Strengths of the MAS-LSTM

The integration of LSTM networks in the MAS architecture enables the system to effectively model the temporal dynamics of IoT traffic. This design is particularly advantageous for detecting sequential anomalies, such as coordinated DDoS attacks. Additionally, the MAS structure enhances scalability, allowing the system to process traffic from heterogeneous IIoT environments while maintaining high detection accuracy. This flexibility in scaling is critical for handling the diverse and growing traffic patterns in IIoT networks, where the volume and complexity of data can vary greatly.

The modular design is another significant strength, as it balances computational efficiency with performance. The ability to dynamically add or remove agents enhances adaptability, making MAS-LSTM suitable for rapidly evolving IoT ecosystems. As new devices are deployed, agents can be quickly integrated without disrupting the overall system's performance, which is especially valuable in large-scale IIoT applications. Moreover, the diversity of datasets used in this study ensures the system's evaluation under realistic conditions, providing robust evidence for its application in practical industrial contexts, such as smart factories, energy grids, and healthcare systems. These contexts often involve heterogeneous devices with varying data characteristics, further proving the system's robustness and generalizability across different industries.

In terms of computational efficiency, MAS-LSTM demonstrates strong performance. The total training time for the system was 460.45 seconds, with an average training time of 0.000291 seconds per sample. Prediction times are also efficient, with a total prediction time of just 16.67 seconds and an average prediction time of 0.000042 seconds per sample. These results indicate that MAS-LSTM is capable of processing large-scale IoT data in real time, making it highly suitable for real-world applications that require both scalability and fast detection. Such efficiency ensures that MAS-LSTM can be deployed in real-time systems with stringent time requirements, like automated traffic monitoring in smart cities or real-time cybersecurity in critical infrastructure.

5.3. Limitations

Despite its strengths, MAS-LSTM faces several limitations that need to be addressed for more robust deployment in real-world IIoT environments.

The reliance on labeled training data limits the model's ability to generalize to unseen attack patterns, a challenge that has been frequently discussed in the literature [62]. In industrial settings, datasets often lack comprehensive labeling, especially for novel attack types, which can negatively affect the model's performance in real-world deployments. One potential solution to this limitation is to explore semi-supervised or unsupervised learning approaches, which are less dependent on large amounts of labeled data, or anomaly-detection methods that can identify new patterns without predefined labels. Moreover, incorporating techniques like few-shot learning or transfer learning could further alleviate the dependency on large labeled datasets by leveraging knowledge from related domains or tasks.

While the decentralized MAS architecture offers improved scalability, excessive partitioning of data among agents can diminish the global context available for anomaly detection. This issue becomes more pronounced when the attack behavior involves multiple agents, as in the case of complex, multi-stage attacks. This may result in reduced performance, particularly in the detection of complex attack scenarios, such as advanced persistent threats (APTs), where the attack activities span across multiple agents. Future research could focus on investigating collaborative learning techniques or cross-agent communication protocols to strike an optimal balance between scalability and global context awareness, thereby enhancing the detection capabilities for sophisticated attack types.

5.4. Future Directions

To address the current limitations, integrating semi-supervised and self-supervised learning methods presents a promising research avenue. These approaches can leverage unlabeled traffic data to enhance MAS-LSTM's anomaly-detection capabilities, particularly in scenarios with limited labeled data. Additionally, reinforcement learning could improve the system's adaptability by enabling agents to dynamically optimize decision-making processes. For instance, agents could learn to adjust detection thresholds or allocate resources in response to emerging attack patterns. Moreover, meta-learning techniques could enable the system to quickly adapt to new attack types with minimal retraining, improving its generalization across unseen scenarios.

The integration of graph neural networks (GNNs) with LSTM models offers another compelling direction, allowing the system to capture both spatial and temporal dependencies in IIoT traffic. This hybrid architecture could be effective in identifying distributed or coordinated attacks. Additionally, GNNs could enhance the system's understanding of network topologies, improving anomaly detection by considering device relationships. Federated learning could further enhance scalability and generalization by enabling agents to share insights while preserving data privacy, making the system more robust and adaptable across diverse industrial settings.

In terms of deployment, the integration of MAS-LSTM within existing IIoT infrastructures poses challenges, such as hardware constraints, legacy system compatibility, and real-time processing efficiency. Future research could explore optimizations like model pruning or edge AI techniques to reduce computational overhead, making MAS-LSTM more suitable for resource-constrained devices. Exploring low-latency processing strategies, such as edge computing, would also help improve anomaly-detection speed in environments requiring immediate responses. Compatibility with industrial communication protocols (e.g., MQTT, OPC UA) would facilitate easier integration with IIoT ecosystems.

Empirical validation through real-world IIoT testbeds or industry collaborations would help assess MAS-LSTM's performance in dynamic operational environments. Testing under varied network conditions—such as fluctuating data loads, cyber–physical interactions, and different levels of attack sophistication—would provide deeper insights into its robustness and adaptability. As IoT attacks continue to evolve, static datasets may no longer capture emerging threats. Future research should focus on dynamic datasets that are regularly updated to reflect new attack patterns, ensuring that MAS-LSTM remains effective against evolving cyber threats. Addressing these deployment challenges will facilitate the practical adoption of MAS-LSTM in real-world industrial applications.

6. Conclusions

This study presents a novel multi-agent anomaly-detection system tailored for Industrial Internet of Things (IIoT) networks, leveraging the temporal modeling capabilities of Long Short-Term Memory (LSTM) networks. The system was rigorously evaluated on multiple benchmark datasets, including NF-ToN-IoT, NF-BoT-IoT, and their v2 versions. The experimental results consistently demonstrate the system's superiority over existing methods, achieving strong performance across metrics such as accuracy, precision, recall, and F1-score.

The multi-agent architecture plays a key role in enhancing the system's scalability and robustness. By decentralizing the learning process, each agent specializes in different aspects of the data, effectively managing the heterogeneity inherent in IIoT traffic. The integration of LSTM networks further strengthens the system, enabling it to detect temporal patterns that characterize long-duration and subtle anomalies—patterns often missed by traditional models. This combination proves particularly effective in identifying complex attack scenarios, such as Distributed Denial-of-Service (DDoS) attacks and botnet behaviors.

The findings also highlight the significant contributions of the system's core components. The multi-agent framework ensures modularity and adaptability, while the LSTM-based temporal modeling provides the precision required to detect sequential anomalies. Together, these features form a robust foundation for anomaly detection in IIoT environments. However, challenges remain. The dependency on labeled data limits the system's applicability in real-world scenarios, where annotations are often scarce or incomplete. Addressing this limitation requires innovative approaches, such as leveraging semi-supervised learning and reinforcement learning, to enable the system to learn effectively from partially labeled or unlabeled data.

In conclusion, the proposed system represents a significant advancement in IIoT security, offering a scalable, adaptable, and effective solution for anomaly detection. Its ability to evolve with emerging attack strategies ensures its relevance in protecting increasingly complex and interconnected IIoT environments. This work lays a strong foundation for future research, paving the way for more intelligent, secure, and resilient IoT networks.

Author Contributions: Conceptualization, Z.Q., X.C. and H.Z.; methodology, Z.Q., Q.L. and X.N.; software, Q.L.; validation, H.Z. and C.U.I.W.; formal analysis, Z.Q.; investigation, Z.Q., X.C., Q.L. and X.N.; resources, X.C., H.Z. and C.U.I.W.; data curation, Q.L.; writing—original draft preparation, Z.Q., Q.L. and X.N.; writing—review and editing, X.C., H.Z. and C.U.I.W.; visualization, Q.L. and X.N.; supervision, H.Z. and X.C.; project administration, X.C. and H.Z. All authors have read and agreed to the published version of the manuscript.

Funding: This research received no external funding.

Data Availability Statement: The implementation of the reinforcement learning algo-rithms discussed in this paper can be found in this repository: https://github.com/bitbitlemon/MAS-LSTM (accessed on 7 February 2025). The data set can be obtained from the following connections: https://staff.itee.uq.edu.au/marius/NIDS_datasets/ (accessed on 7 February 2025).

Acknowledgments: This paper is supported by Macao Polytechnic University (RP/FCHS-02/2022).

Conflicts of Interest: The authors declare no conflicts of interest.

References

1. Hu, C.; Wang, W.; He, Q. A survey on security and privacy issues in edge computing-assisted Internet of Things. *IEEE Internet Things J.* **2019**, *6*, 8115–8137.
2. Thamilarasu, G.; Chawla, S. Towards deep-learning-driven intrusion detection for the Internet of Things. *Sensors* **2019**, *19*, 1977. [CrossRef] [PubMed]
3. Fang, Y.; Xu, Y.; Pan, S.; Zhao, Y. IoT security: Review, blockchain solutions, and open challenges. *Future Gener. Comput. Syst.* **2020**, *107*, 164–174.
4. Inayat, U.; Zia, M.F.; Mahmood, S.; Khalid, H.M.; Benbouzid, M. Learning-based methods for cyber attacks detection in IoT systems: A survey on methods, analysis, and future prospects. *Electronics* **2022**, *11*, 1502. [CrossRef]
5. Tsiknas, K.; Taketzis, D.; Demertzis, K.; Skianis, C. Cyber threats to industrial IoT: A survey on attacks and countermeasures. *IoT* **2021**, *2*, 163–186. [CrossRef]
6. Dhirani, L.L.; Armstrong, E.; Newe, T. Industrial IoT, cyber threats, and standards landscape: Evaluation and roadmap. *Sensors* **2021**, *21*, 3901. [CrossRef]
7. Qu, X.; Liu, Z.; Wu, C.Q.; Hou, A.; Yin, X.; Chen, Z. MFGAN: Multimodal Fusion for Industrial Anomaly Detection Using Attention-Based Autoencoder and Generative Adversarial Network. *Sensors* **2024**, *24*, 637. [CrossRef]
8. Liu, L.; Vollmer, T.; Manic, M. Neural network based intrusion detection system for critical infrastructures. In Proceedings of the 2009 International Joint Conference on Neural Networks, Atlanta, GA, USA, 14–19 June 2009; pp. 1827–1834.
9. Malhotra, P.; Vig, L.; Shroff, G.; Agarwal, P. Long short term memory networks for anomaly detection in time series. In *European Symposium on Artificial Neural Networks, Computational Intelligence and Machine Learning (ESANN)*; 2015; pp. 89–94. Available online: https://paperswithcode.com/paper/long-short-term-memory-networks-for-anomaly (accessed on 26 February 2025).
10. Nguyen, V.Q.; Van Ma, L.; Kim, J. LSTM-based anomaly detection on big data for smart factory monitoring. *J. Digit. Contents Soc.* **2018**, *19*, 789–799.
11. Hsieh, R.-J.; Chou, J.; Ho, C.-H. Unsupervised online anomaly detection on multivariate sensing time series data for smart manufacturing. In Proceedings of the 2019 IEEE 12th Conference on Service-Oriented Computing and Applications (SOCA), Kaohsiung, Taiwan, 18–21 November 2019; pp. 90–97.
12. Dong, H.; Kotenko, I. Multi-task learning for IoT traffic classification: A comparative analysis of deep autoencoders. *Future Gener. Comput. Syst.* **2024**, *158*, 242–254. [CrossRef]
13. Liu, Q.; Wang, D.; Jia, Y.; Luo, S.; Wang, C. A multi-task based deep learning approach for intrusion detection. *Knowl.-Based Syst.* **2022**, *238*, 107852. [CrossRef]
14. Dong, H.; Kotenko, I. Multi-Task Learning Approach for Network Traffic Classification: A Comparative Analysis for Deep Auto Encoders. 2023. Available online: https://papers.ssrn.com/sol3/papers.cfm?abstract_id=4542930 (accessed on 5 February 2025)

15. Munir, M.; Chattha, M.A.; Dengel, A.; Ahmed, S. A comparative analysis of traditional and deep learning-based anomaly detection methods for streaming data. In Proceedings of the 2019 18th IEEE International Conference on Machine Learning and Applications (ICMLA), Boca Raton, FL, USA, 16–19 December 2019; pp. 561–566.
16. Stone, P.; Veloso, M. Multiagent systems: A survey from a machine learning perspective. *Auton. Robot.* **2000**, *8*, 345–383. [CrossRef]
17. García, N.M. Multi-agent system for anomaly detection in Industry 4.0 using Machine Learning techniques. *ADCAIJ Adv. Distrib. Comput. Artif. Intell. J.* **2019**, *8*, 33.
18. Rosenberger, J.; Urlaub, M.; Rauterberg, F.; Lutz, T.; Selig, A.; Bühren, M.; Schramm, D. Deep reinforcement learning multi-agent system for resource allocation in industrial Internet of Things. *Sensors* **2022**, *22*, 4099. [CrossRef] [PubMed]
19. Ullah, I.; Mahmoud, Q.H. Design and development of RNN anomaly detection model for IoT networks. *IEEE Access* **2022**, *10*, 62722–62750. [CrossRef]
20. Chen, Z.; Li, Z.; Huang, J.; Liu, S.; Long, H. An effective method for anomaly detection in industrial Internet of Things using XGBoost and LSTM. *Sci. Rep.* **2024**, *14*, 23969. [CrossRef]
21. Park, J.K.; Baek, Y. Real-Time Adaptive and Lightweight Anomaly Detection Based on a Chaotic System in Cyber–Physical Systems. *Electronics* **2025**, *14*, 598. [CrossRef]
22. Belay, M.A.; Blakseth, S.S.; Rasheed, A.; Salvo Rossi, P. Unsupervised Anomaly Detection for IoT-Based Multivariate Time Series: Existing Solutions, Performance Analysis and Future Directions. *Sensors* **2023**, *23*, 2844. [CrossRef]
23. Wu, Y.; Dai, H.-N.; Tang, H. Graph Neural Networks for Anomaly Detection in Industrial Internet of Things. *IEEE Internet Things J.* **2022**, *9*, 9214–9231. [CrossRef]
24. Sommer, R.; Paxson, V. Outside the closed world: On using machine learning for network intrusion detection. In Proceedings of the 2010 IEEE Symposium on Security and Privacy, Oakland, CA, USA, 16–19 May 2010; pp. 305–316.
25. Chandola, V.; Banerjee, A.; Kumar, V. Anomaly detection: A survey. *ACM Comput. Surv.* **2009**, *41*, 15. [CrossRef]
26. Ahmed, M.M.; Mahmood, A.N.; Hu, J. A survey of network anomaly detection techniques. *J. Netw. Comput. Appl.* **2016**, *60*, 19–31. [CrossRef]
27. Shone, N.; Nguyen, T.N.; Phai, V.D.; Shi, Q. A deep learning approach to network intrusion detection. *IEEE Trans. Emerg. Top. Comput. Intell.* **2018**, *2*, 41–50. [CrossRef]
28. He, Y.; Zhao, J. Temporal convolutional networks for anomaly detection in time series. *J. Phys. Conf. Ser.* **2019**, *1213*, 042050. [CrossRef]
29. Yassine, M.; Théo, F. Anomaly detection for industrial sensors using transformers. In Proceedings of the 2023 10th International Conference on Future Internet of Things and Cloud (FiCloud), Marrakesh, Morocco, 14–16 August 2023; pp. 167–174.
30. Zulfiqar, Z.; Malik, S.U.R.; Moqurrab, S.A.; Zulfiqar, Z.; Yaseen, U.; Srivastava, G. DeepDetect: An innovative hybrid deep learning framework for anomaly detection in IoT networks. *J. Comput. Sci.* **2024**, *83*, 102426. [CrossRef]
31. Sun, Q.; Chen, Z.; Liu, H. Application and Optimization of Multi-agent Reinforcement Learning in Collaborative Decision-Making. In *Cognitive Computing—ICCC 2024 (ICCC 2024), Proceedings of the 8th International Conference, Held as Part of the Services Conference Federation, SCF 2024, Bangkok, Thailand, 16–19 November 2024*; Xu, R., Chen, H., Wu, Y., Zhang, L.J., Eds.; Lecture Notes in Computer Science; Springer: Cham, Switzerland, 2024; Volume 15426.
32. Belhadi, A.; Djenouri, Y.; Srivastava, G.; Lin, J.C.-W. Reinforcement learning multi-agent system for faults diagnosis of microservices in industrial settings. *Comput. Commun.* **2021**, *177*, 213–219. [CrossRef]
33. Ren, K.; Zeng, Y.; Zhong, Y.; Sheng, B.; Zhang, Y. MAFSIDS: A reinforcement learning-based intrusion detection model for multi-agent feature selection networks. *J. Big Data* **2023**, *10*, 137. [CrossRef]
34. Kheddar, H.; Dawoud, D.W.; Awad, A.I.; Himeur, Y.; Khan, M.K. Reinforcement-learning-based intrusion detection in communication networks: A review. *IEEE Commun. Surv. Tutor.* **2024** . [CrossRef]
35. Meshram, A.; Haas, C. Anomaly detection in industrial networks using machine learning: A roadmap. In *Machine Learning for Cyber-Physical Systems: Selected Papers from the International Conference ML4CPS 2016, Karlsruhe, 29 September 2016*; Springer: Berlin/Heidelberg, Germany, 2017; pp. 65–72.
36. Wang, X.; Garg, S.; Lin, H.; Hu, J.; Kaddoum, G.; Piran, M.J.; Hossain, M.S. Toward Accurate Anomaly Detection in Industrial Internet of Things Using Hierarchical Federated Learning. *IEEE Internet Things J.* **2022**, *9*, 7110–7119. [CrossRef]
37. Sharma, P.; Sharma, S.K. A Survey on Anomaly Detection Techniques in IoT. In *Proceedings of Second Doctoral Symposium on Computational Intelligence*; Gupta, D., Khanna, A., Kansal, V., Fortino, G., Hassanien, A.E., Eds.; Advances in Intelligent Systems and Computing; Springer: Singapore, 2022; Volume 1374.
38. Ergen, T.; Kozat, S.S. Unsupervised Anomaly Detection With LSTM Neural Networks. *IEEE Trans. Neural Netw. Learn. Syst.* **2020**, *31*, 3127–3141. [CrossRef]
39. Lee, G.; Yoon, Y.; Lee, K. Anomaly Detection Using an Ensemble of Multi-Point LSTMs. *Entropy* **2023**, *25*, 1480. [CrossRef]
40. Lian, Z.; Su, C. Decentralized Federated Learning for Internet of Things Anomaly Detection. In Proceedings of the 2022 ACM on Asia Conference on Computer and Communications Security, Nagasaki, Japan, 30 May–3 June 2022; ACM: New York, NY, USA, 2022; pp. 1249–1251.

41. Haldikar, S.V.; Kader, O.F.M.A.; Yekollu, R.K. Edge Computing and Federated Learning for Real-Time Anomaly Detection in Industrial Internet of Things (IIoT). In Proceedings of the 2024 International Conference on Inventive Computation Technologies (ICICT), Lalitpur, Nepal, 24–26 April 2024; pp. 1699–1703.
42. Panait, L.; Luke, S. Cooperative multi-agent learning: The state of the art. *Auton. Agents Multi-Agent Syst.* **2005**, *11*, 387–434. [CrossRef]
43. Stefanova-Stoyanova, V.; Stankov, I. Multi-agent systems (MAS) in the area of IoT and using a model with Distributed Shared Memory system (DSM). In Proceedings of the 2020 XXIX International Scientific Conference Electronics (ET), Sozopol, Bulgaria, 16–18 September 2020; pp. 1–4.
44. Zhang, W.; Yang, D.; Wu, W.; Peng, H.; Zhang, N.; Zhang, H.; Shen, X. Optimizing Federated Learning in Distributed Industrial IoT: A Multi-Agent Approach. *IEEE J. Sel. Areas Commun.* **2021**, *39*, 3688–3703. [CrossRef]
45. Hao, W.; Yang, T.; Yang, Q. Hybrid statistical-machine learning for real-time anomaly detection in industrial cyber-physical systems. *IEEE Trans. Autom. Sci. Eng.* **2021**, *20*, 32–46. [CrossRef]
46. Eiteneuer, B.; Niggemann, O. LSTM for model-based anomaly detection in cyber-physical systems. *arXiv* **2020**, arXiv:2010.15680.
47. Moustafa, N. A new distributed architecture for evaluating AI-based security systems at the edge: Network TON_IoT datasets *Sustain. Cities Soc.* **2021**, *72*, 102994. [CrossRef]
48. Booij, T.M.; Chiscop, I.; Meeuwissen, E.; Moustafa, N.; den Hartog, F.T.H. ToN IoT—The role of heterogeneity and the need for standardization of features and attack types in IoT network intrusion datasets. *IEEE Internet Things J.* **2021**, *9*, 485–496. [CrossRef]
49. Alsaedi, A.; Moustafa, N.; Tari, Z.; Mahmood, A.; Anwar, A. TON_IoT telemetry dataset: A new generation dataset of IoT and IIoT for data-driven Intrusion Detection Systems. *IEEE Access* **2020**, *8*, 165130–165150. [CrossRef]
50. Moustafa, N.; Keshk, M.; Debie, E.; Janicke, H. Federated TON_IoT Windows Datasets for Evaluating AI-Based Security Applications. In Proceedings of the 2020 IEEE 19th International Conference on Trust, Security and Privacy in Computing and Communications (TrustCom), Guangzhou, China, 29 December 2020–1 January 2021; pp. 848–855.
51. Moustafa, N.; Ahmed, M.; Ahmed, S. Data Analytics-Enabled Intrusion Detection: Evaluations of ToN_IoT Linux Datasets. In Proceedings of the 2020 IEEE 19th International Conference on Trust, Security and Privacy in Computing and Communications (TrustCom), Guangzhou, China, 29 December 2020–1 January 2021; pp. 727–735.
52. Moustafa, N. New Generations of Internet of Things Datasets for Cybersecurity Applications based Machine Learning: TON_IoT Datasets. In Proceedings of the eResearch Australasia Conference, Brisbane, Australia, 21–25 October 2019.
53. Moustafa, N. A systemic IoT-Fog-Cloud architecture for big-data analytics and cyber security systems: A review of fog computing *arXiv* **2019**, arXiv:1906.01055.
54. Ashraf, J.; Keshk, M.; Moustafa, N.; Abdel-Basset, M.; Khurshid, H.; Bakhshi, A.D.; Mostafa, R.R. IoTBoT-IDS: A Novel Statistical Learning-enabled Botnet Detection Framework for Protecting Networks of Smart Cities. *Sustain. Cities Soc.* **2021**, *72*, 103041 [CrossRef]
55. Cox, D.R. The regression analysis of binary sequences. *J. R. Stat. Soc. Ser. B (Methodol.)* **1958**, *20*, 215–232. [CrossRef]
56. Lewis, D.D. Naive (Bayes) at forty: The independence assumption in information retrieval. In *Machine Learning: ECML-98 Proceedings of the 10th European Conference on Machine Learning, Chemnitz, Germany, 21–23 April 1998*; Springer: Berlin/Heidelberg, Germany, 1998; pp. 4–15.
57. Hinton, G.E.; Salakhutdinov, R.R. Reducing the dimensionality of data with neural networks. *Science* **2006**, *313*, 504–507 [CrossRef]
58. Broomhead, D.S.; Lowe, D. *Radial Basis Functions, Multi-Variable Functional Interpolation and Adaptive Networks*; Royal Signals and Radar Establishment: Malvern, UK, 1988; Volume 4148.
59. Doe, J.; Smith, J. Sequential Anomaly Detection in IoT Networks Using LSTM. *J. IoT Res.* **2022**, *15*, 321–335.
60. Lee, M.; Zhang, W. Modular and Scalable Anomaly Detection for Industrial IoT. *IEEE Trans. Ind. Inform.* **2023**, *18*, 1234–1245.
61. Ahmed, A.; Wang, Y. Benchmarking Anomaly Detection Systems Using Diverse IoT Datasets. *ACM Trans. IoT* **2021**, *12*, 98–112.
62. Brown, E.; Gupta, R. Overcoming Labeled Data Limitations in Anomaly Detection. *Mach. Learn. IoT* **2020**, *10*, 45–60.

processes

MDPI

Article

Power Load Forecasting System of Iron and Steel Enterprises Based on Deep Kernel–Multiple Kernel Joint Learning

Yan Zhang [1,*], Junsheng Wang [1], Jie Sun [2], Ruiqi Sun [1] and Dawei Qin [1]

[1] Beijing Research Institute of Ansteel Co., Ltd., Beijing 102200, China; phdwjs@sina.com (J.W.); qindaweiemail@163.com (D.Q.)
[2] The State Key Laboratory of Rolling and Automation, Northeastern University, Shenyang 110819, China
* Correspondence: wu_kunkui@163.com

Abstract: The traditional power load forecasting learning method has problems such as overfitting and incomplete learning of time series information when dealing with complex nonlinear data, which affects the accuracy of short–medium term power load forecasting. A joint learning method, LSVM-MKL, was proposed based on the bidirectional promotion of deep kernel learning (DKL) and multiple kernel learning (MKL). The multi-kernel method was combined with the input layer, the highest coding layer, and the highest encoding layer to model the network of the stack autoencoder (SAE) to obtain more comprehensive information. At the same time, the deep kernel was integrated into the optimization training of Gaussian multi-kernel by means of the nonlinear product to form the nonlinear composite kernel. Through a large number of reference datasets and actual industrial data experiments, it was shown that compared with the Elman and LSTM-Seq2Seq methods, the proposed method achieved a higher prediction accuracy of 4.32%, which verified its adaptability to complex time-varying power load forecasting processes and greatly improved the accuracy of power load forecasting.

Keywords: multiple kernel learning; deep learning; data driven; power load forecasting

1. Introduction

Electricity is very important in the production processes of steel enterprises, accounting for about 90% of the total energy loss [1]. Power load forecasting is based on historical load data to predict the total electricity consumption in future periods. Accurate load forecasting can guide steel enterprises' production planning, rational power purchase, and the reasonable arrangement of equipment maintenance and repair, and improve their economic efficiency. Because of the non-stationarity, non-linearity, complexity, and dynamic nature of load data, it is challenging to accurately forecast the short- and medium-term load. The traditional load forecasting methods are mainly based on statistical models, which cannot fully explore the inherent laws and dynamic changes in power loads [2]. In recent years, deep learning has achieved significant results in the field of temporal prediction, and kernel learning-based pattern analysis is able to efficiently handle nonlinear and temporal features in systems [3]. Multiple new data-driven methods have been proposed, and reference [4] proposed a deep learning-based LSTM (Long Short-Term Memory) Seq2Seq load forecasting model. The algorithm is prone to gradient vanishing and exploding as the number of layers and training iterations increases. A mid-term load forecasting model based on the Elman neural network was proposed [5], which used an improved BP algorithm to predict the load values of the regional power grid for the next 12 months of the year. However, the algorithm had problems with a slow rate of convergence and local minima [6].

In contrast, deep kernels based on deep neural networks have nonlinear mapping expressions. They only use the highest hidden layer of neural networks to construct kernels, requiring a large amount of data to ensure the information quality of the highest hidden layer, and there is an overfitting problem [7–9]. This paper proposes a bidirectional promotion joint learning method that improves the performance of deep kernels using multi-kernel learning, while using deep kernels to improve the training process of Gaussian multi-kernels [10]. A deep kernel prediction model with adaptive time-varying characteristics is established, and a special correlation layer is introduced into the multi-layer feedforward network to form a network with a "memory" ability [11], which can map the nonlinear and dynamic characteristics of the power load forecasting system. This combined model can exploit more abstract potential features from multiple angles of massive data, reflect the nonlinear and dynamic characteristics of the system, and overcome the problems of slow convergence and local minimal points [12]. The power load forecasting system of the steel enterprise adopts a mixed structure of B/S and C/S and implements deep learning algorithm functions in Python. The experimental results show that the method proposed not only effectively improves the accuracy of kernel learning, but also has a stronger universality to the dataset. It can meet the higher accuracy and adaptability requirements of production data, and can deal with the challenges of nonlinearity and the randomness of power energy forecasting, considering the incomplete learning of time-sequence information [13]. According to the entire process from theory to practical application, it can be divided into fundamental research methods and the deep multi-kernel joint learning model of experimental data.

2. Deep Multi-Kernel Joint Learning Model

The kernel method is one of the methods that transforms nonlinear inseparable problems in low-dimensional spaces into linearly separable problems in high-dimensional spaces [14]. Multi-kernel learning is a linear or nonlinear combination of basic kernels, which takes the form of a convex combination of base kernels, that is, a linear non-negative combination of weight vectors with a fixed 1-norm:

$$\begin{aligned} &k_\mu(x_i, x_j) = \sum_{r=1}^{M} u_r k_r(x_i, x_j) = \sum_{r=1}^{M} u_r \langle \varphi_r(x_i), \varphi_r(x_j) \rangle \\ &u_r \geq 0, \|u\|_1 = 1 \end{aligned} \tag{1}$$

where k_μ is the combination of base kernels, φ_r is the feature map from the original space to the high-dimensional Hilbert space, k_r is the r-th kernel (the Gaussian kernel or the polynomial kernel), u_r is the r-th kernel weight vector, and M is the number of multiple kernels. The multi-kernel can capture different data similarity information, effectively combine them to adapt to different modalities of information, and reduce the workload of selecting kernel hyperparameters by directly learning the base kernel weights from the data.

The deep neural network relies on a stacked series of nonlinear mappings to learn representations through the increasing complexity of feature levels. The mathematical model can be described by a nonlinear function φ_{net} from the input space to the output space. In typical feedforward neural networks, it is defined as a combination of basic mapping functions:

$$\varphi_{net}(x) = \psi_{out} \circ \psi_l \circ \cdots \circ \psi_2 \circ \psi_1(x) \tag{2}$$

where Ψ_1 is the mapping from the input layer to the first layer, Ψ_{out} is the mapping from the highest layer to the output, and $\circ$ is the iteration between mappings. The nonlinear mapping from the hidden layer of $r-1$th to the hidden layer of rth is

$$\psi_r(x) = \sigma(W_r x + b) \tag{3}$$

where W_r is the weight matrix, b is the bias, and σ is the activation function. The kernel function constructed based on the deep neural network approximation kernel mapping idea has an explicit function expression form. Kernel mapping transforms the original input space into a feature space using neural networks, and the deep kernel mapping function is expressed as:

$$K_d = \psi_n(K_n \cdots (\psi_1(K_1))) = \sigma(W_n(K_n \cdots (\sigma(W_1 K_1 + b_1))) + b_n) \tag{4}$$

As shown in Figure 1, in response to the problems of incomplete information and overfitting in deep kernels, a multi-kernel method is adopted to combine the input layer, highest encoding layer, and highest decoding layer of the SAE kernel, which improves the expressive ability of the deep kernel. The obtained deep kernel is integrated into the optimization training of Gaussian multi-kernels in a nonlinear multiplication manner, and they are coordinated with each other to form a nonlinear combination kernel [15]. NLk1. M means the deep information of nonlinear kernels.

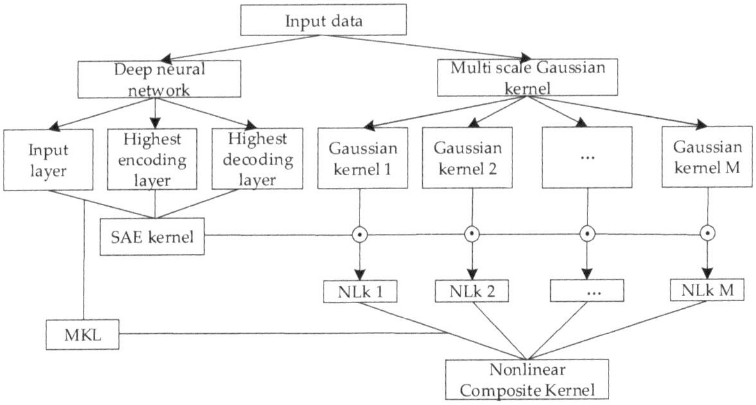

Figure 1. Joint learning framework.

As shown in Figure 2, the network representations of the input layer, the highest encoding layer, and the highest decoding layer of the SAE core are combined and reconstructed [16]. The number of layers in the deep autoencoder are set to $2k + 1$, with the 0th layer as the input layer. The first to k th layers are set as the encoding layers. The $k + 1$th to $2k$th layers are set as the decoding layers. The nonlinear mapping function between each layer of the encoding layer is $f_r\ (x)$, and the nonlinear mapping function between the decoding layers is $g_r\ (x)$,

$$\psi_r(x) = x \circ f_0 \circ f_1 \circ \cdots \circ f_r \circ g_r \circ \cdots \circ g_1 = \psi_{r-1}(x) \circ f_r \circ g_r, 0 \le r \le k \tag{5}$$

Among them, $\psi_0(x) = x$ as the input, and the internal representation of the hidden layer were determined through calculations performed by continuously applying the basic mapping ψ_0, ψ_1, ψ_2. Among them, $f_1, \ldots f_k$ are the nonlinear mapping functions from the input layer to the encoding layer, and $g_1, \ldots, g_k$ are the nonlinear mapping functions from the encoding layer to the decoding layer.

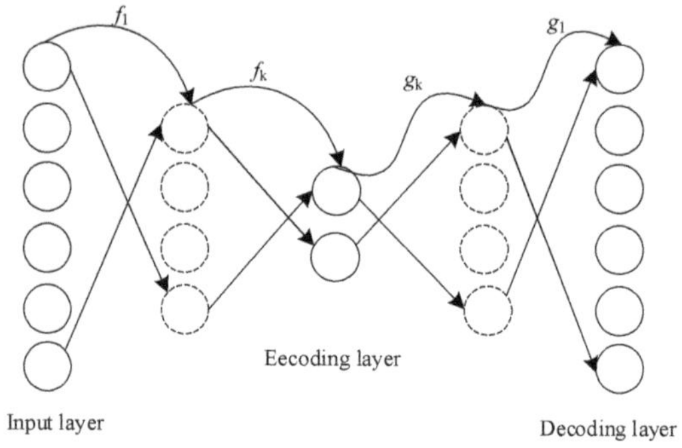

Figure 2. Stack autoencoder structure.

Given that the encoder layer can automatically and effectively extract high-dimensional data and map it to low-dimensional hidden data:

$$\begin{aligned} k_0(x_i, x_j) &= \langle x_i, x_j \rangle \\ k_1(x_i, x_j) &= \langle \varphi_k(x_i), \varphi_k(x_j) \rangle \\ k_2(x_i, x_j) &= \langle \phi_k(x_i), \phi_k(x_j) \rangle \end{aligned} \tag{6}$$

Three kernels containing different depth information can be obtained, and the SAE core can be integrated through a multi-kernel method.

The organic combination of deep kernel learning and multi-kernel learning achieves the coordinated training and joint learning between the deep kernel and multi-kernel, improving the flexibility of the model. The algorithm process is as follows:

Step 1: Predefine the hyperparameters such as network layers, number of neurons, training algorithm, activation function, etc., and train the deep autoencoder using unsupervised pretraining and fine-tuning methods.

Step 2: Extract the highest encoding layer representation $\varphi_k(x)$, the highest decoding layer representation $\phi_k(x)$, and the input layer representation $\phi_0(x)$ from the network, and kernel the Equation (4).

Step 3: Integrate the three types of kernelling layers from Step 2 using a multi-kernel, and determine the hyperparameters using empirical and experimental methods.

Step 4: Predefine M Gaussian basis kernels $g_1, \ldots, g_M$. Each kernel is multiplied with the deep autoencoder kernel obtained in Step 3 to obtain a set of nonlinear kernels $k_1, \ldots, k_M$.

Step 5: Use the same multi-kernel method and kernel parameters as in Step 3 to optimize the $k_1, \ldots, k_M$ combinatorial optimization to form a nonlinear combinatorial kernel.

Step 6: Apply the obtained nonlinear combination and the kernel machine parameters (Lagrange multipliers and biases) from Step 5 for classification.

3. Experimental Results and Analysis

3.1. Benchmark Dataset

In order to ensure data timeliness, the data are divided into a sequence every ten days on a daily basis, and the total load value for the next ten days is predicted based on the process losses and power load consumption values of the ten days. Experiments are conducted on 15 benchmark datasets with different features and sample sizes. The data have been preprocessed and normalized, and the prediction accuracy is used to evaluate the performance of the model. The information of the dataset is compared before and after optimization, and the accuracy results are shown in Table 1. The prediction accuracy is the average of 20 random experiments, with 70% of the samples randomly selected as the training set and 30% as the testing set.

Table 1. Dataset information.

Serial	Dataset	Sample Size	Dimension	LSVM-MKL/ Before Optimization	ElmanNeural Network/ Before Optimization	Seq2Seqmodel/ BeforeOptimization
1	MON1	124 + 432	6	87.73 ± 0.01/66.46 ± 0.00	86.04 ± 0.17/67.25 ± 0.62	87.18 ± 0.28/66.65 ± 0.01
2	MON2	169 + 432	6	86.66 ± 0.28/65.60 ± 0.01	86.60 ± 0.24/73.38 ± 0.01	84.19 ± 0.36/65.19 ± 1.10
3	MON3	122 + 432	6	95.37 ± 0.01/82.28 ± 0.10	95.14 ± 0.01/82.12 ± 0.90	95.08 ± 0.11/81.94 ± 0.00
4	HOR	300 + 68	25	83.82 ± 0.01/80.81 ± 1.09	83.75 ± 0.73/80.88 ± 0.00	87.13 ± 0.64/81.32 ± 1.05
5	OOC1	1022	40	79.24 ± 2.62/76.67 ± 1.26	80.19 ± 1.47/78.58 ± 1.20	77.14 ± 1.35/75.05 ± 1.79
6	OOC2	912	25	81.24 ± 1.28/79.84 ± 1.51	78.86 ± 2.02/77.16 ± 1.76	82.49 ± 1.41/80.23 ± 1.84
7	ION	351	34	95.06 ± 1.35/87.64 ± 1.31	94.23 ± 1.17/87.70 ± 2.29	94.97 ± 1.60/88.01 ± 2.06
8	SON	208	60	83.08 ± 4.01/76.88 ± 3.02	84.38 ± 3.84/75.29 ± 1.93	83.41 ± 3.44/77.21 ± 3.75
9	SPL	1535	60	92.17 ± 1.91/90.05 ± 1.96	92.02 ± 0.95/89.86 ± 1.45	91.35 ± 1.65/90.23 ± 1.43
10	BRE1	699	9	96.79 ± 0.81/86.43 ± 0.81	96.63 ± 2.66/85.06 ± 1.06	96.91 ± 0.78/86.73 ± 0.66
11	BRE2	569	30	95.63 ± 2.01/84.86 ± 3.02	93.72 ± 2.61/83.72 ± 2.66	95.37 ± 2.35/85.21 ± 3.41
12	PRO	106	55	76.23 ± 7.49/74.53 ± 4.83	77.08 ± 4.75/58.87 ± 2.66	76.89 ± 6.48/75.57 ± 5.04
13	CLI	540	18	91.96 ± 1.27/81.96 ± 1.28	92.28 ± 1.22/82.28 ± 1.22	92.15 ± 1.46/82.15 ± 1.05
14	BLO	748	5	78.03 ± 2.42/75.91 ± 1.34	74.10 ± 2.94/72.35 ± 3.02	78.58 ± 1.88/77.47 ± 1.28
15	BAN	512	35	71.13 ± 2.27/69.94 ± 2.73	68.28 ± 4.42/66.15 ± 3.47	72.79 ± 1.79/70.88 ± 2.10

Three data-driven methods for instantiation are used, such as LSVM-MKL, Elman, and LSTM-Seq2Seq [17]. Based on this, comparative experiments are conducted using SAE (AE), Gaussian multi-kernel (Gau), and joint methods (G-A). The kernel learning machine and hyperparameter settings are the same. Elman uses a three hidden layer regression neural network and trains the network by an improved BP algorithm (an adaptive learning rate momentum gradient descent backpropagation algorithm) [18]. The encoder part of the LSTM-Seq2Seq model consists of 10 sets of LSTM modules stacked in a two-layer LSTM network [19]. The number of hidden layers is set to 24, and the initial states are all zero. The weight initialization is a normal distribution with a mean of 0 and a variance of 1. The LSVM-MKL penalty parameter is 1, and the maximum number of iterations is 10,000 using hard spaced SVM. The hyperparameter is 0. The multi-scale Gaussian kernel set has kernel widths of $\{5^{-2}, 5^{-1}, 5^{0}, 5^{1}, 5^{2}, 5^{3}, 5^{4}\}$, and the SAE kernel network has seven layers. Adam is used for training, and the activation function is ReLU For nonlinear functions, the kernel width and ReLU have a stronger expressive power, especially in deep networks. ReLU has a constant gradient in the non-negative interval, so there is no gradient vanishing problem, which maintains the convergence speed of the model in a stable state.

As shown in Figure 3, after the comprehensive comparison of the three methods, the joint method achieved excellent results, with the highest classification accuracy on datasets 13, 10, and 12. Compared with the Gaussian multi-kernel method, it improved almost all of the 15 datasets and had significant effects on some datasets. Under LSVM-MKL, the accuracy improvements of the HOR, OOC1, OOC2, SON, PRO, BAN, and SPL datasets were 2.94%, 2.57%, 3.15%, 3.66%, 5.19%, 4.37% and 2.58% respectively. Under Elman, the improvement levels of the MON1, MON2, SON, and PRO datasets were 2.94%, 2.57%, 3.15%, and 3.66%, respectively. Under LSTM-Seq2Seq, the improvement levels of the OOC1, OOC2, and PRO datasets were 2.09%, 2.26%, and 4.44%, respectively. Overall, in the high-dimensional datasets of HOR, OOC2, SON, PRO, and BAN, the SAE kernel has a strong feature extraction ability and integrates prior information into Gaussian multi-kernel training.

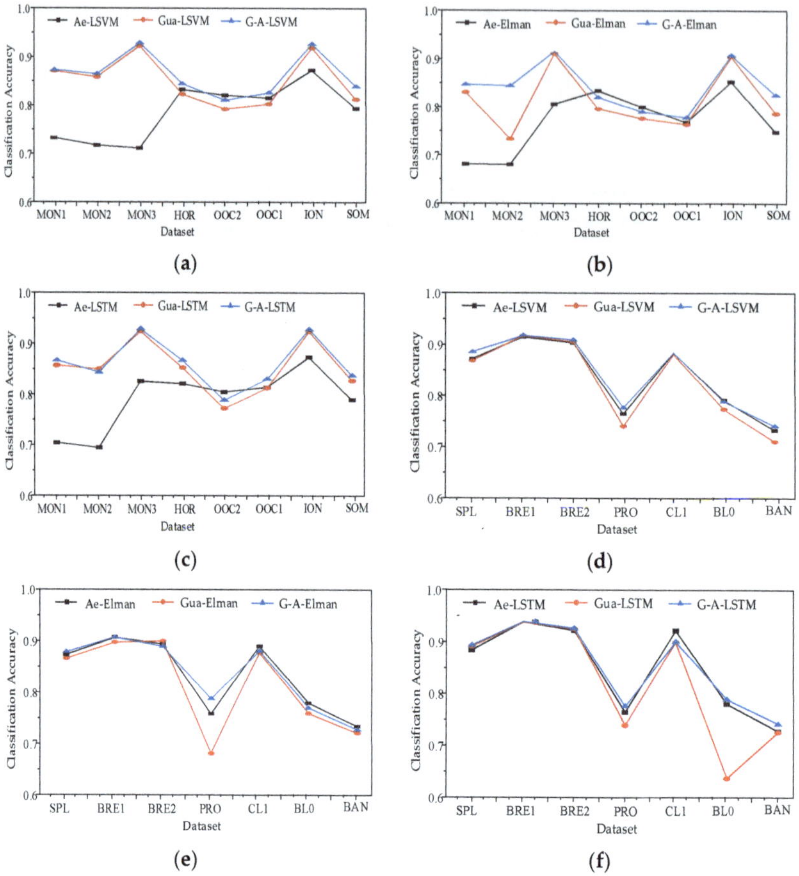

Figure 3. Classification accuracy under three kinds of multi-kernel instances. (**a**) LSVM-MKL1; (**b**) Elman1; (**c**) LSTM-Seq2Seq1; (**d**) LSVM-MKL2; (**e**) Elman2; and (**f**) LSTM-Seq2Seq2.

The good results in datasets with lower dimensions, such as MON1, MON2, MON3, and BLO, further demonstrate that our method has good universality and stability, and can be applied to multiple types of data. On the other hand, the joint approach is not simply about neutralizing the kernel under two systems, but rather integrating the advantages of both. In the ION, SON, and MON1 datasets, the method can still achieve high-precision prediction, indicating that it can eliminate the influence of adverse factors and prefers to select the best one to achieve the optimal combination state of the two kernel systems.

3.2. Power Load Forecasting

This prediction system is developed in the Micro Visual Studio 2010 environment. Net technology is used to design the web interface, and the C # language is used to complete the layout of the system pages. SQL Server 2019 is selected as the database to complete data access functions. The Python 3.1 uses TensorFlow, and the neural network structure adopts an unsupervised model autoencoder network. As shown in Figure 4, it is a partial screenshot of the dataset. The historical load data and environmental collection information (such as temperature, humidity, air pressure, etc.) obtained from the power system will be processed for missing/NaN values, seasonal adjustments, noise removal, outlier handling, and stability checks to obtain the preprocessed load data.

Figure 4. Datasets and timing databases.

To address this issue, a power load forecasting model is established, and the prediction target is set as the active power of electricity. The input data includes historical day parameters, holiday definitions, scheduling maintenance plans, loading load data (source data), and modifying source data. Load forecasting includes setting forecast dates, conducting medium- to short-term load forecasting, compiling forecast results, and adjusting forecast results. Data output includes adding curves, deleting curves, clearing the screen, refreshing, comparing curves, displaying all of the algorithm results, and conducting assessment statistics. As shown in Figure 5a, the dataset comes from the active power consumption of 800–1400 MW in a certain steel enterprise in the first 11 months of 2023, and the active power in the 12th month is predicted, with a total of 2880 records. Considering the real-time requirements of the actual production process, the efficient LSVM-MKL was selected for the experiment, with a multi-scale Gaussian set kernel width of $\{5^{-1}, 5^{0}, 5^{1}, 5^{2}, 5^{3}, 5^{4}\}$.

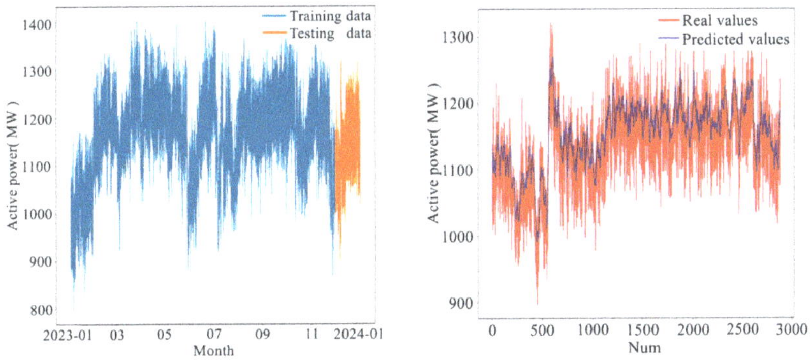

(a) (b)

Figure 5. Power load forecasting results. (**a**) Training values and actual values; and (**b**) predicted values and real values.

As shown in Figure 5b, compared with the actual value, the predicted load for December 2023 is basically in the middle of the actual value, with an average relative error of 4.32%. Using the LSVM-MKL model for prediction, the neural networks can better fit the relationship between the input and output. With the increase in information sources, this method achieves the highest prediction accuracy. Therefore, this method has been proven to be adaptable and effective for complex time-varying industrial problems, and can meet the working conditions of different processes. Overall, the average relative error of the electricity load forecasting system for steel enterprises in 2024 is within 5% to avoid the penalty price settlement of electricity bills. The maximum demand is accurately predicted based on the dynamic load, and the peak demand is reduced by the load dispatch measures, which directly reduces the expenditure for purchased electricity.

4. Conclusions

Aimed at the problems of poor stability and unsatisfactory prediction accuracy caused by the nonlinear and temporal characteristics of power load forecasting, a joint learning model based on the multi-kernel is proposed for steel enterprises. The effectiveness of the model is ultimately verified through experiments. It constructs a more flexible kernel learning framework by the joint bidirectional learning between the data-driven SAE kernel and Gaussian multi-kernel, which improves the expression ability of the kernel.

In the experiment, the prediction accuracies of three prediction models were compared, and prediction accuracy showed a different performance on different data types. The LSVM-MKL model had a higher prediction accuracy, training iteration speed, and stability. Comparison with the Elman and LSTM-Seq2Seq, the feasibility and effectiveness of the joint learning methodit was further demonstrated.

Due to the poor learning ability of the Elman and LSTM-Seq2Seq models for power load temporal information, the joint models outperformed the other two models in terms of training iteration speed and stability, and the average relative error was improved to 4.32%. In the model, only historical load data was considered as input, without taking into account the influence of other factors. In future studies, temperature and weather conditions can be included to further improve the prediction accuracy.

Author Contributions: Conceptualization, Y.Z., J.W., J.S., R.S. and D.Q.; methodology, Y.Z.; software, Y.Z. and J.W.; validation, Y.Z., J.W., J.S. and D.Q.; formal analysis, Y.Z. and R.S.; investigation, D.Q.; resources, J.S.; data curation, Y.Z., J.W. and R.S.; writing—original draft preparation, Y.Z. and J.W.; writing—review and editing, Y.Z. and J.S.; visualization, Y.Z. and D.Q.; supervision, Y.Z.; project administration, Y.Z. and J.W. All authors have read and agreed to the published version of the manuscript.

Funding: This research was funded by the National Key Research and Development Program Funding Project (Grant No. 2022YFB3304800).

Data Availability Statement: The data that support the findings of this study are available from the corresponding author, Yan Zhang, upon reasonable request.

Conflicts of Interest: Authors Yan Zhang, Junsheng Wang, Ruiqi Sun and Dawei Qin were employed by Beijing Research Institute of Ansteel Co., Ltd. The remaining authors declare that the research was conducted in the absence of any commercial or financial relationships that could be construed as a potential conflict of interest.

References

1. Luo, X.; Chen, F.; Lu, Y.; Zhao, Z.; Peng, Y.; Pang, J. Intelligent Fault Diagnosis Method for Transformer Driven by Multiple Vibration Data. *J. Phys. Conf. Ser.* **2024**, *2774*, 4255–4262. [CrossRef]
2. Li, H.; Li, Y.; Huang, J.; Shen, C.; Wang, C.; Jing, T.; Liu, Z.; Xu, W. Physical Metallurgy Guided Industrial Big Data Analysis System with Data Classification and Property Prediction. *Steel Res. Int.* **2022**, *93*, 3029–3043. [CrossRef]
3. Gonen, M.; Alpaydin, E. Multiple kernel learning algorithms. *J. Mach. Res.* **2011**, *12*, 2211–2268.
4. Anan, M.; Kanaan, K.; Benhaddou, D.; Nasser, N.; Qolomany, B.; Talei, H.; Sawalmeh, A. Occupant-Aware Energy Consumption Prediction in Smart Buildings Using a LSTM Model and Time Series Data. *Energies* **2024**, *17*, 6451. [CrossRef]

5. He, Z.; Li, C.; Shen, Y.; He, A. A Hybrid Model Equipped with the Minimum Cycle Decomposition Concept for Short-Term Forecasting of Electrical Load Time Series. *Neural Process. Lett.* **2017**, *46*, 1059–1081. [CrossRef]
6. Arvanitidis, A.I.; Bargiotas, D.; Daskalopulu, A.; Laitsos, V.M.; Tsoukalas, L.H. Enhanced Short-term Load Forecasting using Artificial Neural Networks. *Energies* **2021**, *14*, 7788. [CrossRef]
7. Li, Y.; Zhang, T.; Hu, H. Deep Kernel Mapping Support Vector Machine based on Multilayer Perceptron. *J. Beijing Univ. Technol.* **2016**, *42*, 1652–1661.
8. Wang, Q.; Lv, Z.; Wang, L.; Wang, W. Long Term Prediction Model based on Deep Denoising Kernel Mapping. *Control Decis.* **2019**, *34*, 989–996.
9. Ma, L.; Dong, J.; Peng, K. A Novel Hierarchical Detection and Isolation Framework for Quality-related Multiple Faults in Large-scale Processes. *IEEE Trans. Ind. Electron.* **2020**, *67*, 1316–1327. [CrossRef]
10. Jiao, J.F.; Yu, H.; Wang, G. A Quality-related Fault Detection Approach Based on Dynamic Least Squares for Process Monitoring. *IEEE Trans. Ind. Inform.* **2016**, *63*, 2625–2632. [CrossRef]
11. Han, M.; Zhang, H. Multiple Kernel Learning for Label Relation and Class Imbalance in Multi-label Learning. *Inf. Sci.* **2022**, *613*, 344–356. [CrossRef]
12. Geng, Z.; Li, S.; Yu, F. Ulta-short-term and Short-term Power Load Single-step and Multi-step Prediction Considering Spatial Association. *Comput. Eng.* **2024**, *7*, 22–25.
13. Aiolli, F.; Donini, M. Easymkl. A Scalable Multiple Kernel Learning Algorithm. *Neurocomputing* **2015**, *169*, 215–224. [CrossRef]
14. Wang, G. Research on Theory and Algorithm of Support Vector Machine. Ph.D. Thesis, Beijing University of Posts and Telecommunications, Beijing, China, 2008; pp. 1–30.
15. Hassanzadeh, S.; Danyali, H.; Karami, A.; Helfroush, M.S. A Novel Graph-Based Multiple Kernel Learning Framework for Hyperspectral Image Classification. *Int. J. Remote Sens.* **2024**, *45*, 3075–3103. [CrossRef]
16. Wang, T.; Zhang, L.; Hu, W. Bridging Deep and Multiple Kernel Learning: A Review. *Inf. Fusion* **2021**, *23*, 18. [CrossRef]
17. Uğurel, E.; Huang, S.; Chen, C. Learning to Generate Synthetic Human Mobility Data: A Physics-Regularized Gaussian Process Approach based on Multiple Kernel Learning. *Transp. Res. Part B* **2024**, *189*, 103064. [CrossRef]
18. Mori, H.; Kurata, E. An Efficient Kernel Machine Technique for short-term Load Forecasting under Smart Grid Environment. In Proceedings of the 2012 IEEE Power and Energy Society General Meeting, San Diego, CA, USA, 22–26 July 2012.
19. Hernandez, L.; Baladron, C.; Aguiar, J.M.; Carro, B.; Sanchez-Esguevillas, A.; Lloret, J. Artifical Neural Networks for Short-term Load Forecasting in Microgrids Environment. *Energy* **2014**, *75*, 252–264. [CrossRef]

processes

MDPI

Article

Quick Combustion Optimization for Utility Boilers Using a Novel Adaptive Hybrid Case Library

Cong Yu [1,2], Shuo Chen [2], Haiquan Yu [3], Yukun Zhu [2], Qiang Wang [2], Guangting Liao [2] and Ling Shi [1,*]

1 Hubei Key Laboratory of Industrial Fume and Dust Pollution Control, Jianghan University, Wuhan 430056, China; congy@jhun.edu.cn
2 School of Intelligent Manufacturing, Jianghan University, Wuhan 430056, China
3 School of Electrical Engineering, Nanjing Vocational University of Industry Technology, Nanjing 210023, China; hqyu@niit.edu.cn
* Correspondence: ling_s_h@163.com

Abstract: To achieve carbon neutrality, thermal power plants must undertake heavier peak shaving tasks; consequently, utility boilers will be required to operate under frequently changing operating conditions. In light of this new circumstance, a combustion optimization decision that is executed more rapidly is necessary. A novel online combustion optimization framework is proposed for the combustion system of utility boilers. First, a robust filter for extracting high-quality steady-state data samples is designed and executed. Then, the K-means algorithm is used to divide the cleaned sample space and construct the working condition case library. Based on the constructed library, the boiler combustion model is constructed using the XGBoost algorithm. Therefore, the corresponding optimization case library can be established using the multiobjective optimization algorithm and working condition case library. To further capture the phenomenon of data distribution migrating as the operating conditions change, an adaptive update strategy for the combustion system is proposed, including online querying and data and model updates. The findings of this study conducted on a 660 MW utility boiler show that the proposed online optimization method can effectively decrease NO_x emissions and improve combustion efficiency in approximately 2 milliseconds.

Keywords: utility boiler; data filter; case-based reasoning; online combustion optimization; adaptive update strategy

1. Introduction

Coal is currently utilized as a fuel in many thermal power plants worldwide. However, the combustion process of coal generates nitrogen oxides (NO_x), which can directly impact human respiratory systems, damage the ozone layer, and contribute to greenhouse effects [1]. With growing concerns about energy utilization and environmental emissions [2], coal-fired power plants must not only enhance boiler combustion efficiency but also reduce pollutant emissions [3,4]. In recent years, data-driven modeling has grown vigorously and has been widely applied to combustion optimization in coal-fired utility boilers. However, due to the lengthy calculation time, the applications of traditional optimization methods in actual engineering are limited [5]. Moreover, the nonlinearity between combustion system parameters and the optimization objectives poses serious challenges to improving operational performance [6]. Therefore, studying more advanced methods for online optimizing the combustion process is worthwhile.

 https://doi.org/10.3390/pr13020469

Before combustion optimization, characteristic models are required to be established. A mechanism model is one of the typical methods. It can obtain the distributions of velocity, chemical components, temperature, and other furnace information like boiler thermal efficiency and pollutant emission concentrations [7]. However, this method often involves complex multiphysics coupled iterative calculations, requiring high calculation costs [8], so combining mechanism modeling with intelligent optimization algorithms and applying it to actual boiler online optimization is difficult. In contrast to a mechanism model, a data-driven model does not require a deep understanding of the mechanism or much designer experience. It mainly reflects the internal laws through the data themselves; furthermore, the excellent solution speed of a data-driven model also aids boiler online optimization applications [9].

Data-driven boiler combustion optimization has been widely recognized by both academia and industry. To meet the needs of on-site operation optimization of coal-fired power plants, two main optimization strategies exist: (a) obtaining the quantitative relationship between operational variables and target variables through data mining algorithms and then expressing the quantitative relationship as association rules for combustion optimization; (b) constructing a data-driven boiler combustion system model and using intelligent optimization algorithms to search for the optimal operating solutions for this model.

Strategy (a) uses boiler historical operating data to create optimization rules for combustion and then optimizes boiler parameters based on these rules. This approach offers fast solving capabilities and is appropriate for online optimization. Yang et al. [10] introduced fuzzy set theory in the association mining process to identify important optimization parameters in the combustion process. Using boiler historical data, the optimized values can be obtained quickly during operation. Zhao et al. [11] established an optimal rule base for historical operation data by using the fuzzy association rule mining method. Because the rule base is the result of a global search, it is suitable for online use and can be easily and quickly updated when the working conditions of the boiler combustion system change. In addition, Kusiak et al. [12–14] studied the relationships between uncontrollable variables, controllable variables, and target variables through cluster analysis. By adjusting controllable variables, the target variable can be optimized. The abovementioned optimization scheme based on data mining can quickly obtain a good optimization scheme under any boiler condition, but it cannot guarantee the best optimization scheme will be obtained.

To search for better optimization solutions that are different from historical operating solutions, strategy (b) including system modeling and multiobjective optimization can be used. First, the characteristic model of the boiler combustion system is established using historical data, and then the combustion process is optimized by intelligent optimization methods according to the model. Zhou et al. [15] constructed an artificial neural network (ANN) model with NO_x emissions and thermal efficiency as the output and used a genetic algorithm (GA) to obtain a good NO_x emission concentration and boiler combustion efficiency. Wang et al. [16] employed Gaussian process regression (GPR) to establish the relationship between boiler operational parameters and NO_x emission concentrations. Under specific combustion conditions, they utilized a GA to identify optimal boiler operational parameters that result in lower NO_x emissions. Song et al. [17] initially designed an enhanced generalized regression neural network (EGRNN) using Gaussian adaptive resonance theory (GART) learning and polynomial extrapolation. They then formulated a cost function that accounted for potential coal ash recovery. Finally, they utilized an improved version of the artificial bee colony algorithm (ABC) to determine the optimal parameters for the boiler. This comprehensive approach involves using advanced techniques such as neural networks, optimization algorithms, and cost considerations to optimize boiler combustion performance, accounting for coal ash recycling potential. In the discussed literature, researchers often convert multiobjective optimization problems into single-objective optimization problems to simplify calculations. However, this approach is subjective and cannot handle nonconvex sets. To comprehensively consider boiler combustion efficiency and environmental protection, a better solution is to compromise among multiple objectives and search for a set of Pareto optimal solutions so that their objective values are distributed on the Pareto front. Ma et al. [18] employed an improved extreme learning machine (ELM) to construct models for the efficiency, NO_x concentration, and SO_2 concentration of a circulating fluidized bed boiler. They then utilized a multiobjective modified teaching–learning-based optimization (MMTLBO) algorithm to optimize the combustion parameters of the boiler. Rahat et al. [19] used the nonlinear regression of the Gaussian process to establish models for unburned coal content and NO_x emissions. Then, they used the evolutionary multiobjective search algorithm (EMOSA) to search for the ideal combustion operating parameters for boiler combustion optimization. The above scheme based on a data-driven model and intelligent optimization algorithm can effectively search for the best optimization scheme. However, compared with the optimization method based on the case library, the optimization method based on the data-driven model takes a longer amount of time to obtain a series of optimization schemes and has certain defects in the real-time response of the boiler online combustion optimization.

In summary, strategy (b) usually leads to better optimization results than strategy (a). However, the optimization calculations involved in strategy (b) typically take several minutes (or longer) to complete, whereas strategy (a) can be accomplished within a short time frame (less than 1 s) and is based on actual operational data. This makes strategy (a) a simple, rapid, and secure approach. In addition, case-based reasoning (CBR) theory is widely used to achieve the online combustion optimization of boilers. CBR quickly solves current problems by retrieving similar cases from historical experience or optimization results [20,21]. Considering the pros and cons of the aforementioned optimization methods, a novel method that combines data-driven algorithms with CBR to construct an optimization case library for online boiler combustion optimization is proposed in this paper. This method aims to enhance boiler efficiency while reducing NO_x emissions, achieving real-time boiler combustion optimization. The proposed approach provides operational guidance based on real-time data streams, presenting significant research and practical value. Integrating data-driven algorithms and CBR offers a balanced solution, leveraging both accuracy and real-time responsiveness for boiler online combustion optimization.

2. Boiler Combustion System

2.1. Description of the Coal-Fired Boiler

Figure 1 illustrates a schematic of the boiler and the sensor locations. The combustion system of this boiler employs low-NO_x concentric firing technology, comprising six groups of primary air (PA) nozzles and six groups of secondary air (SA) nozzles. The secondary air nozzles include peripheral air nozzles (A, B, C, D, E, and F), concentric firing systems (CFS; AI/AII–FI/FII), auxiliary air nozzles (AA, AB, BC, CD, DE, and EF), close-coupled overfire air nozzles (CCOFA-I and CCOFA-II), and separated overfire air nozzles (SOFA-I–SOFA-V). During the actual operation of the boiler, the damper openings and vertical tilting angles of these nozzles are adjustable, whereas the horizontal yaw angles remain fixed. As shown in Figure 1, the oxygen content and NO_x concentration sensors are positioned at the entrance of the selective catalytic reduction (SCR) system, located before the ammonia injection grid (AIG). This location ensures that the NO_x concentration measurements represent the NO_x concentration produced by combustion within the boiler.

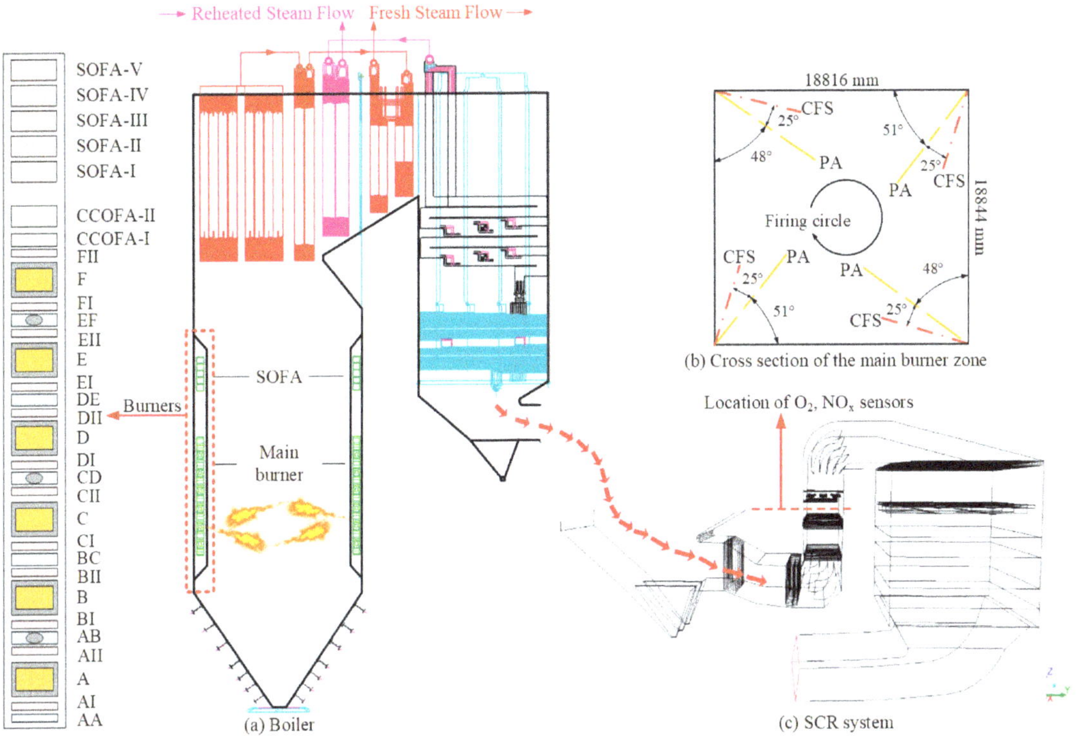

Figure 1. Schematic diagram of the supercritical coal-fired boiler.

2.2. Variable Analysis

Two boiler combustion models are constructed with boiler efficiency and NO_X emission concentration as the target variables. The influencing factors are primarily the boiler structure, coal characteristics, and operating parameters. Since the boiler structure is fixed, the main effects considered are the fuel characteristics and adjustable operating parameters. To obtain real-time information on coal composition during boiler combustion, the coal quality coefficient (Cqc) is used to represent coal composition [22]. In this paper, the coal quality coefficient is calculated by Equation (1):

$$C_{qc} = \frac{UL}{TC} \tag{1}$$

where UL represents the unit load (MW) and TC represents the total coal rate (t/h).

A total of 72,000 operational data points are collected with a sampling interval of 30 s. The preliminary selections for the boiler combustion process parameters and their respective ranges based on the provided analysis are presented in Tables 1–3. Table 1 lists the 3 condition variables. Table 2 lists the 27 operating variables. Table 3 lists the 2 boiler performance variables. The data presented in Tables 1–3 are statistically derived from our experimental data. Additional details about the experiments, including measurement point locations, and sensor accuracy and range, as well as the raw operational data collected, can be found in our previously published papers [23,24].

Table 1. List of the condition variables.

Index	Parameter Description	Symbol	Unit	Lower Limit	Upper Limit
x_1	Unit Load	L	MW	336.62	668.61
x_2	Coal quality coefficient	Cqc	MW·h/t	1.97	3.11
x_3	Primary air pressure	P_a	kPa	7.88	12.04

Table 2. List of the operating variables.

Index	Parameter Description	Symbol	Unit	Lower Limit	Upper Limit
x_4	Oxygen content	O_2	%	0.65	6.39
x_5	SOFA tilt angle	S_a	%	0	94.84
x_6	SOFA damper opening	I	%	0	100
x_7		II		1.36	99.29
x_8		III		0.62	98.59
x_9		IV		0	99.38
x_{10}		V		0	94.02
x_{11}	CCOFA damper opening	I	%	1.05	97.34
x_{12}		II		0	98.79
x_{13}	Secondary damper opening	A	%	11.09	74.70
x_{14}		AA		1.23	97.21
x_{15}		AI/AII		3.95	68.40
x_{16}		B		9.67	74.43
x_{17}		AB		5.97	79.02
x_{18}		BI/BII	%	5.43	66.40
x_{19}		C		17.06	69.50
x_{20}		BC		11.43	91.65
x_{21}		CI/CII		0.53	71.44
x_{22}		D		11.93	75.92
x_{23}		CD		10.45	70.82
x_{24}		DI/DII		8.29	69.73
x_{25}		E		1.90	71.38
x_{26}		DE		8.83	87.83
x_{27}		EI/EII		0	68.02
x_{28}		F		2.31	73.40
x_{29}		EF		0.72	89.18
x_{30}		FI/FII		2.04	74.13

Table 3. List of the performance variables.

Index	Parameter Description	Symbol	Unit	Lower Limit	Upper Limit
y_1	Boiler efficiency	η	%	92.37	94.07
y_2	NO_x concentration	NO_x	mg/m^3	115.14	599.87

2.3. Combustion Optimization Method

In this paper, online combustion optimization for boilers is examined. First, a filter for mining credible high-quality data samples from historical operating data is designed. Next, an initial boiler working condition case library is constructed by a clustering algorithm. Finally, a set of online combustion optimization methods, including offline optimization and online optimization, for boilers based on a data-driven hybrid strategy is proposed, as shown in Figure 2.

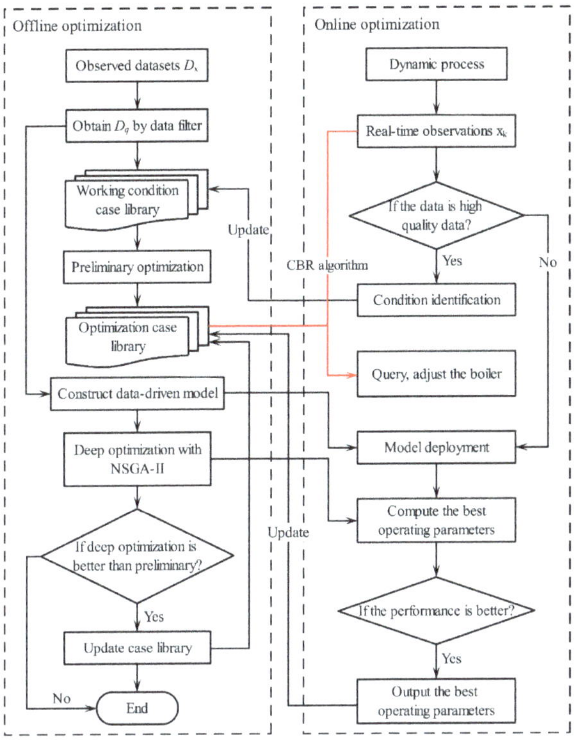

Figure 2. Structural diagram of the boiler online combustion optimization method. The red arrow in the figure indicates the application of optimization case library for online optimization.

3. Data Filter Design

Due to the need to establish a case library that accurately reflects the mechanism relationship examined in this study, high-quality data corresponding to the steady-state operating conditions of the unit must be selected. Therefore, a robust filter framework is designed to filter out nonsteady-state and abnormal data. The data filtering process is shown in Figure 3.

First, original data are collected from the plant's supervisory information system (SIS) database. To obtain high-quality data samples that accurately reflect the static mechanism of the boiler combustion system, steady-state diagnosis and outlier detection are performed on the original observed data. Aiming at this process, a filter algorithm framework integrating steady-state diagnosis and outlier detection using a steady-state detection method based on kernel-PCA is proposed by comparing the joint steady index (JSI) with a given threshold and filtering the appropriate steady-state samples from the original observed data. In addition, the VMD algorithm is used to determine the data flow trend, and the outliers are determined by calculating the interquartile range to obtain the confidence interval. Finally, the data samples that passed the steady-state diagnosis and outlier detection are aggregated to form the high-quality steady-state dataset containing combustion system information.

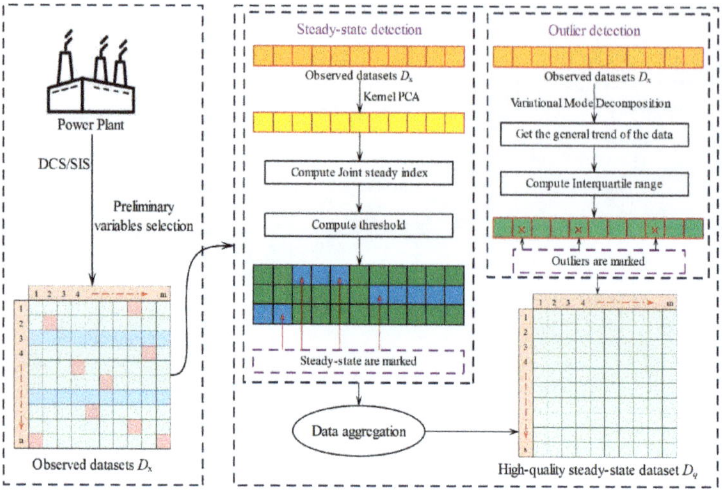

Figure 3. Flow chart of the filter designed in this paper. The blue square represents unsteady-state data. The green square represents steady-state data. The red cross represents outlier data.

3.1. Steady-State Identification

A sliding time window strategy is employed with the kernel-PCA algorithm for steady-state identification to determine the boiler's operational status [24]. Given a matrix $X= [x_1, x_2, \cdots, x_m]^T$ with m variables and n observations and considering the need to eliminate the influence of dimensions, the matrix X is first normalized by the mean value to obtain a new matrix X'. For a specific variable x_i, the vector $d_{i,t} = [x_{i,t-T+1}, \cdots, x_{i,t}]$ represents the subset corresponding to a certain sliding time window T of the matrix X'. Therefore, the data under a sliding window can be expressed as $D_t = [d_{1,t}, \cdots, d_{m,t}]$, and the vector $d_{i,t}(i \in [1, m])$ can be mapped to the high-dimensional space by the kernel function $\varnothing$ to obtain $\chi = [\varnothing_1, \varnothing_2, \cdots, \varnothing_m]$. Let S be the covariance matrix of the data samples in the feature space; then, the following formula is obtained:

$$(m-1)S = \chi^T \chi = \sum_{i=1}^{m} \varnothing_i \varnothing_i^T \quad (2)$$

In this paper, the Gaussian kernel function is used:

$$K(d_i, d_j) = e^{-\gamma \|d_i - d_j\|^2} \quad (3)$$

where $K(\cdot,\cdot)$ represents the kernel function and γ is a free parameter.

The calculation steps of the multivariate steady-state detection algorithm are as follows:

Step 1: Mean-normalize the original data to obtain a new matrix X'.

Step 2: Obtain the data D_t of the corresponding sliding time window, map it to the high-dimensional space through Equations (2) and (3), and obtain the result.

Step 3: Compute the covariance matrix S of χ.

Step 4: Compute the l principal eigenvalues $\Lambda = [\lambda_1, \cdots, \lambda_l]^T$ and eigenvectors $\Lambda = [\lambda_1, \cdots, \lambda_l]^T$ of S, and arrange them in descending order.

Step 5: Compute the steady-state index vector $C = Q\Lambda = [c_{t-T+1}, \cdots, c_t]$.

Step 6: Compute the joint steady index (*JSI*) according to the steady-state index vector C and Equation (4).

$$JSI = \sqrt{\frac{1}{T-1}\sum_{k=t-T+1}^{t}(c_k - \bar{c})^2} \tag{4}$$

Step 7: Compute the threshold ζ according to Equation (5).

$$\zeta = \lambda\sqrt{\frac{1}{n}\sum_{i=1}^{n}(JSI - \overline{JSI})^2} \tag{5}$$

where λ is a positive adjustment coefficient and $\overline{JSI}$ represents the average value of JSI.

Step 8: Judge whether $JSI < \zeta$ is true; if it is true, keep the data sample; otherwise, delete it.

3.2. Outlier Detection

Variational Modal Decomposition (VMD) is a signal decomposition algorithm [25] that divides the original signal into multiple signals from the perspective of the frequency domain and can be used to distinguish between normal and abnormal data [26]. To reasonably select the parameters of the VMD algorithm, the minimum envelope entropy [27] is used as the objective function, and the genetic algorithm is used to identify the best parameter combination for achieving the best signal decomposition effect and ensuring trend extraction accuracy. The envelope entropy is mathematically defined as follows:

$$E_p = -\sum_{j=1}^{m} p_j \lg p_j \tag{6}$$

$$p_j = a(j)/\sum_{j=1}^{m} a(j) \tag{7}$$

$$a(j) = \sqrt{[x(j)]^2 + \{H[x(j)]\}^2} \tag{8}$$

where $a(j)$ represents the envelope signal sequence obtained after the signal $x(j)$ $(j = 1, 2, \cdots, m)$ is demodulated by Hilbert; H represents the Hilbert transform of the signal; and p_j is the normalized form of $a(j)$.

The Pearson coefficient method is used to measure the degree of correlation between two variables; it is mathematically defined in Equation (9). In this paper, the effective signal is determined by judging the threshold in Equation (10), and the effective signal is reconstructed to extract the trend line. Herein, the decomposed signal with a Pearson coefficient greater than the threshold ξ is effective.

$$C = \frac{Cov(x_i, y_i)}{\sigma_x \sigma_y} \tag{9}$$

$$\xi = \frac{C_{max}}{10C_{max} - 3} \tag{10}$$

where $Cov(x_i, y_i)$ and σ_x, σ_y are the covariance and standard deviation of x_i and y_i, respectively, and C_{max} is the maximum Pearson coefficient of each component and the original signal.

The trend extraction algorithm is shown in Figure 4. The specific process is as follows:

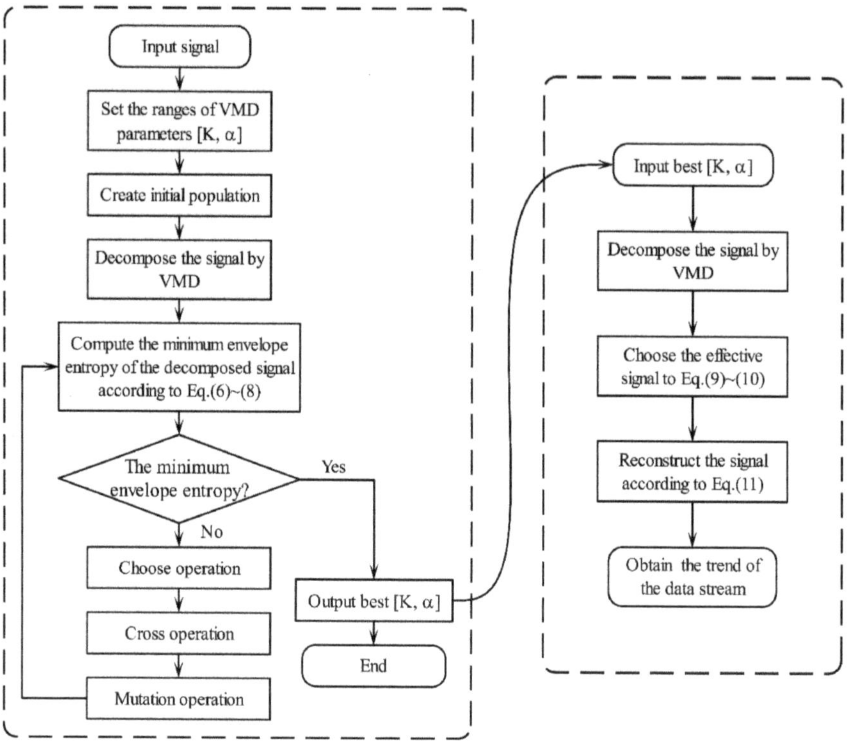

Figure 4. Flow chart of trend extraction based on GA-VMD.

Step 1: Input the original signal f, set the range of K to [2, 9], and set the range of α to [1000, 3000].

Step 2: Use the GA to search [K, α]. VMD is input, the original signal is decomposed, and KIMF components are obtained.

Step 3: Calculate the minimum envelope entropy of the decomposed signal according to Equations (6)–(8).

Step 4: Choose the effective signal according to Equations (9) and (10).

Step 5: Repeat steps 2~4 to identify the minimum envelope entropy of the decomposed signal and its corresponding best parameters [K, α].

Step 6: Decompose the original signal with the best parameters [K, α] by VMD and choose the effective signal.

Step 7: Reconstruct the signal according to Equation (11) to obtain the best trend line.

$$f_{rec} = \sum_{i=1}^{N} IMF_i \tag{11}$$

where f_{rec} is the reconstruction signal and N is the number of effective signals.

For the data point $x(t)$ on the extracted trend curve, the confidence interval is calculated using the interquartile range method. For the original signal $f(t)$, if a data point is distributed outside the confidence interval, it is considered an outlier; otherwise, it is considered a normal point. The absolute error is first calculated according to Equation (12), and the confidence interval is calculated by Equation (13):

$$Error = |f(t) - x(t)| \tag{12}$$

$$\begin{cases} IQR = Q3 - Q1 \\ Bandwidth = Q3 + 1.5 \times IQR \\ CI = [x(t) - \text{Bandwidth}, x(t) + \text{Bandwidth}] \end{cases} \tag{13}$$

where $Q1$ and $Q3$ are the smaller and larger quartiles in the absolute error, respectively.

4. Offline and Online Optimization

4.1. Working Condition Case Library

K-means++ [28] is used to cluster the three parameters that can reflect the operating state of the boiler, unit load, coal quality coefficient, and primary air pressure to construct an initial working condition case library. To consider the impact of data distribution changes on the case library, a reclustering strategy for the online adaptive calculation of working condition clusters is proposed.

According to engineering experience, when the working condition parameter range of data samples in a working condition cluster is kept within 2~3% of the parameter range in Table 1, the working condition division performs well. We thus use the maximum intracluster parameter range $\Delta x_i (i = 1, 2, 3)$ determined in the previous round of working condition clustering as the judgment condition of the strategy to determine the number of clusters k in the new round of working conditions. The process of a new working condition clustering round is shown in Figure 5.

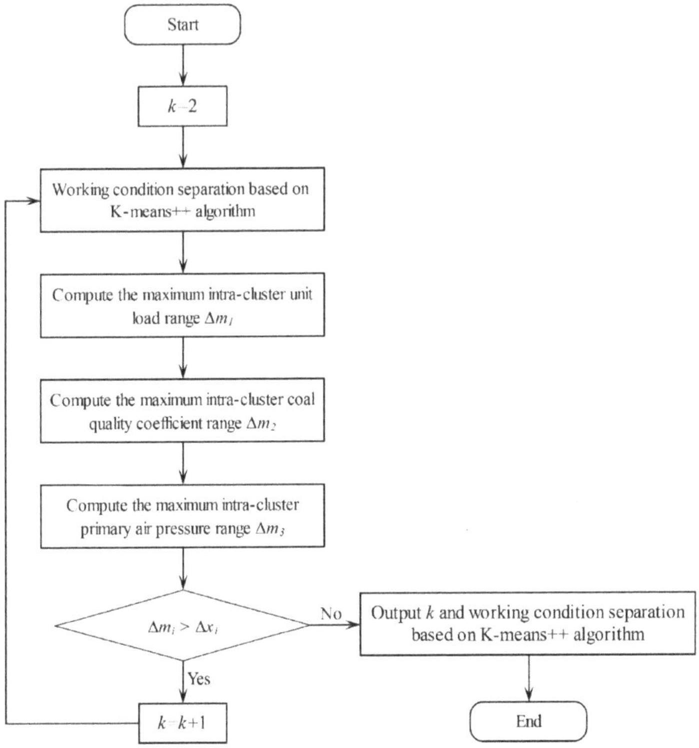

Figure 5. Flow chart of the reclustering strategy.

4.2. Preliminary Optimization

Based on the working condition case library, the initial boiler optimization case library is constructed by querying the two operating modes of the lowest NO_x concentration and the highest boiler efficiency for each working condition in the working condition case library. To find the case most similar to the target case, an online case retrieval strategy is presented for determining the optimization case.

First, the optimization case library is defined as $O_n = (x_{1,n}, x_{2,n}, \ldots, x_{m,n}, y_{1,n}, y_{2,n})$, $n = 1, 2, \ldots, p$, where p represents the total number of optimization cases; O_n represents the optimization case of the n-th working condition; $x_{m,n}$ represents the value of the m-th parameter of the n-th working condition optimization case; and $y_{1,n}$ and $y_{2,n}$ represent the boiler efficiency and NO_x concentration in the optimization case, respectively.

Second, the Euclidean distance between two points in space is the proximity, denoted as d, as shown in Equation (14). Then, the maximum proximity of all data samples of each operating condition in the operating condition case library to the center of that operating condition can be represented as $D_n = (d_1, d_2, \cdots, d_n)$. Next, this section defines the confidence distance (CD) as $CD = \max(D_n)$.

$$Dis(x_i, x_j) = \sqrt{\sum_{i=1}^{m} (x_i - x_j)^2} \tag{14}$$

The online retrieval strategy based on the confidence distance is depicted in Figure 6. We provide the calculation steps of the store and update operation as follows:

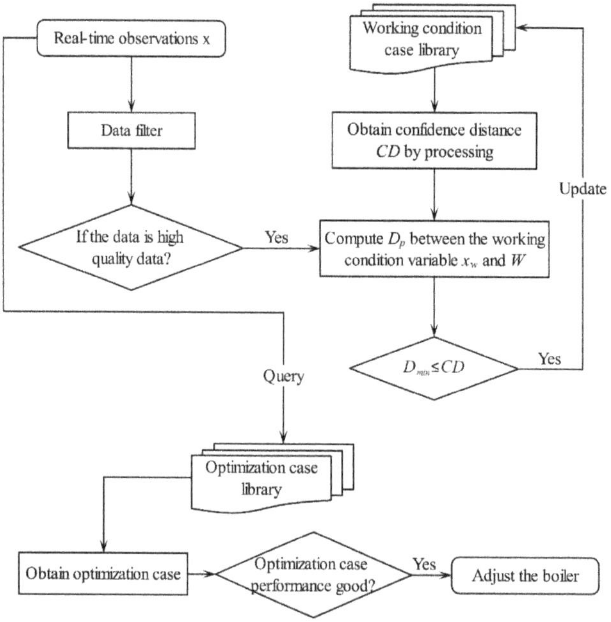

Figure 6. Flow chart of the online retrieval strategy.

Step 1: When observing a new data sample x, determine whether it passes the filter judgment; if it passes, the working condition variable of the data sample x is taken out as x_w; otherwise, the retrieval is stopped.

Step 2: The working condition variables and cluster centers in the working condition case library are normalized to obtain the normalized cluster center set $W = (w_1, w_2, \cdots, w_n)$. CD is calculated from the normalized cluster center set.

Step 3: Normalize the current operating condition parameters x_w to obtain x'_w, and calculate their proximity with the normalized operating condition parameters in the library to obtain $D_p = (d_1, d_2, \cdots, d_n)$.

Step 4: Calculate the minimum proximity $D_{\min} = \min(D_p)$ and judge whether $D_{\min} \leq CD$ is true; if it is true, store it in the corresponding working condition case library.

4.3. Deep Optimization

Model-based optimization is conducted to improve the depth of the boiler optimization case library.

4.3.1. System Modeling

The boiler efficiency and NO_x emission model is constructed using XGBoost (Version 0.90). Its predicted value $\hat{y}_i$ can be expressed as the sum of the results of multiple base models:

$$\hat{y}_i = \sum_{k=1}^{K} f_k(x_i),\ f_k \in N \tag{15}$$

where N is all the value spaces of the regression tree; K is the number of regression trees in the model; x_i is the input feature of the i-th sample; and $f_k(x_i)$ is the prediction result of the k-th tree. More details of the algorithm can be found in Ref. [29].

To evaluate performance, four evaluation metrics are used: mean absolute error (MAE), root mean square error (RMSE), mean absolute percentage error (MAPE), and R-squared (R^2). Smaller MAE, RMSE, and MAPE values indicate a better predictive performance. The closer the R^2 value is to 1, the more accurate the model's predictive performance is. The indices are calculated as follows:

$$\text{MAE} = \frac{1}{n}\sum_{i=1}^{n}|\hat{y}_i - y_i| \tag{16}$$

$$\text{RMSE} = \sqrt{\frac{1}{n}\sum_{i=1}^{n}(\hat{y}_i - y_i)^2} \tag{17}$$

$$\text{MAPE} = \frac{1}{n}\sum_{i=1}^{n}\frac{|\hat{y}_i - y_i|}{\hat{y}_i} \times 100\% \tag{18}$$

$$R^2 = 1 - \frac{\Sigma_{i=1}^{n}(y_i - \hat{y}_i)^2}{\Sigma_{i=1}^{n}(y_i - \overline{y}_i)^2} \tag{19}$$

where y_i, $\hat{y}_i$, and $\overline{y}_i$ denote the measured value, predicted value, and average value of the prediction target, respectively.

4.3.2. Multiobjective Optimization

NSGA-II [30] is used for combustion optimization, yielding a set of optimal optimization schemes, the Pareto front solution set. To quickly determine the only optimal optimization scheme in the equilibrium mode, an inflection point attribution strategy is proposed. The obtained Pareto solution set exhibits a distribution of an approximate concave function in the plane space, which has a maximum point, also known as an inflection point. In actual boiler combustion optimization, the inflection point indicates that when the boiler efficiency rises to a certain level, continuing to increase the small range efficiency will greatly increase the NO_x concentration in the boiler. Therefore, in the equilibrium mode, choosing the Pareto solution at the inflection point is the best optimization scheme. However, the distribution of the data samples of the Pareto solution set actually calculated by the NSGA-II algorithm is not uniform, making it difficult to identify the actual solution of the inflection point. To address this problem, a curve is first fit on the Pareto solution set space using polynomial fitting, and the curve is then discretized with a certain sampling interval to calculate the inflection point X_m on the fitted curve, and the closest Pareto solution to X_m is identified as the best optimization solution in the equilibrium mode.

4.4. Adaptive Update Strategy

To consider the variation in the working condition case library over time, an online adaptive update strategy is proposed. The strategy is shown in Figure 7, and the steps are as follows:

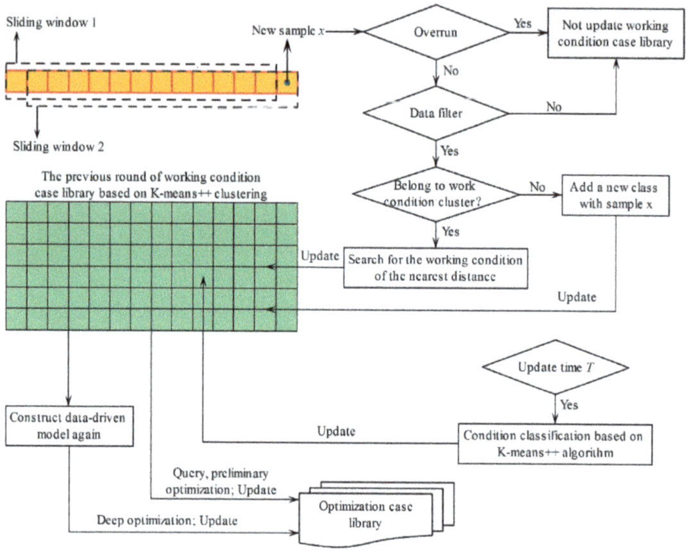

Figure 7. Flow chart of the adaptive update strategy. The orange square represents the operating data and the green square represents the working condition case library.

Step 1: The data filter is used to extract high-quality data samples D_q from the original data, and then the K-means++ algorithm is used to divide the working conditions and construct the initial boiler working condition case library; the online adaptive update period is defined as T.

Step 2: When a new data sample x is observed in real time, whether it is overrun is evaluated; if it is, the working condition case library is not updated.

Step 3: The data filter is used to determine whether x is a high-quality data sample; if it is, Step 4 is carried out; otherwise, the working condition case library is not updated.

Step 4: Using the online retrieval strategy, whether x belongs to the existing working condition cluster is judged; if it is, x will be stored in the nearest working condition cluster; otherwise, a new category of working conditions is added with the data sample as the cluster center and updated synchronously in the optimization case library.

Step 5: Steps 2~4 are repeated until the adaptive update period T is reached; at this time, the working conditions are reclassified based on the K-means++ algorithm to construct the working condition case library and the optimization case library. The data-driven model is trained based on the reconstructed working condition case library. After training, the model is updated online, and the optimization case library is updated again.

Step 6: The timing is repeated until the adaptive update period T is reached, and a new round of data and model updates is conducted.

5. Results and Discussion

5.1. Data Filtering Results

Figure 8 shows the steady-state detection results of the 24 h sampling points of the utility boiler. The boiler has a total of five steady-state ranges (SRs) in the 24 h. The results show that the proposed filter successfully retains the steady-state samples, performing well.

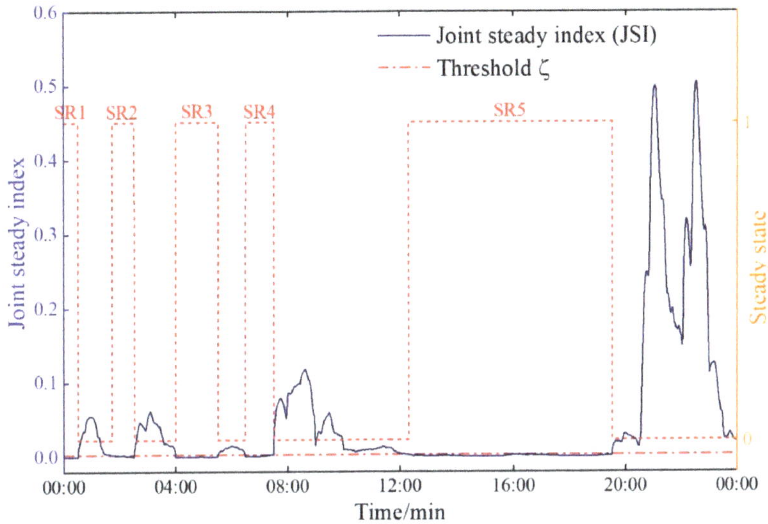

(a)

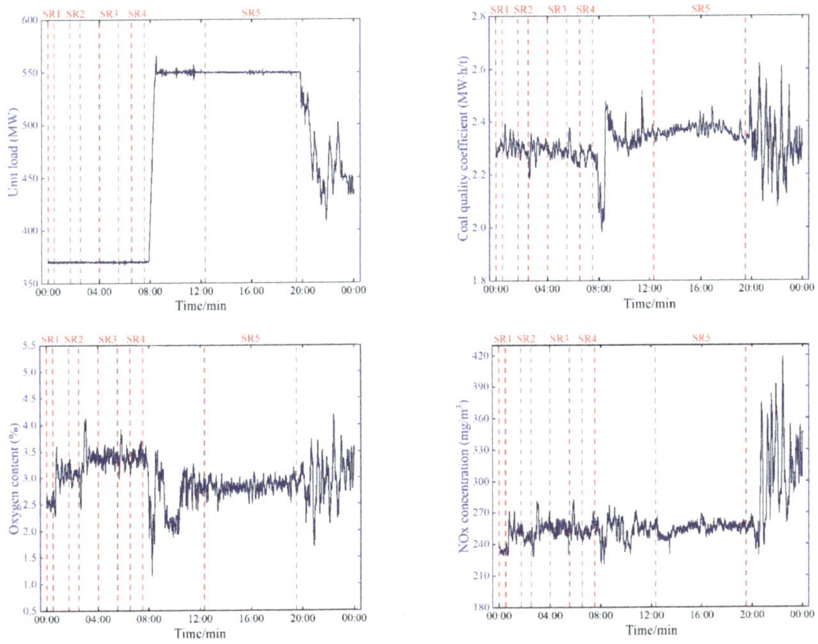

(b)

Figure 8. An example of steady-state identification in a typical 24 h. (**a**) Joint steady index, threshold, and original data. (**b**) Steady-state identification results for four variables.

Figure 9 shows the outlier detection results for the sampling points of the utility boiler on the same day, where original data are represented by black dots, trends are represented by dashed yellow lines, confidence intervals are represented by blue areas, and outliers are represented by red hollow circles. The results indicate that the designed filter has good outlier detection ability. The filter is conducted for 72,000 observation data points of utility boilers over 25 days. After the above filtering operations, 6392 high-quality steady-state operating data are maintained as the initial samples for constructing the working condition case library, data-driven model, and optimization case library.

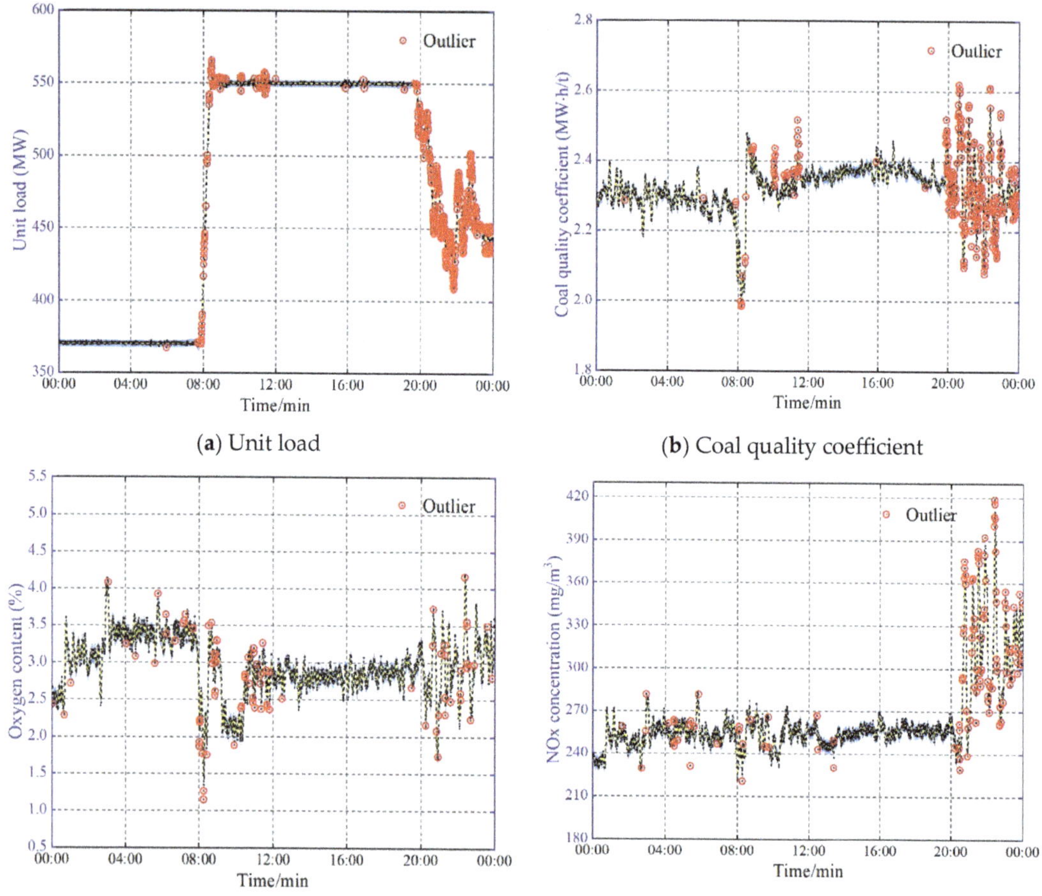

(**a**) Unit load

(**b**) Coal quality coefficient

(**c**) Oxygen content

(**d**) NO_x concentration

Figure 9. An example of outlier detection in a typical 24 h. The red circle represents the outliers, the blue represents the confidence interval, and the yellow represents the trend line of the data stream.

5.2. Construction of the Working Condition Case Library

Clustering and corresponding analysis are performed on the obtained 6392 high-quality data samples. First, the range of the hyperparameter k for the K-means++ algorithm is set to [2, 360], and the minimum intracluster sample number and the maximum intra-cluster working condition parameter range are calculated. Then, $k = 113$ is determined to be the number of working condition clusters for the initial clustering. Finally, Figure 10 illustrates that when $k = 113$, there are at least 13 high-quality data samples in each working cluster in the working condition case library to reflect the mechanism relationship under this working condition. In each working cluster, the range of working condition parameters can be kept below 3% of the overall variation range, which is sufficient for showing that this method has a good working condition division effect.

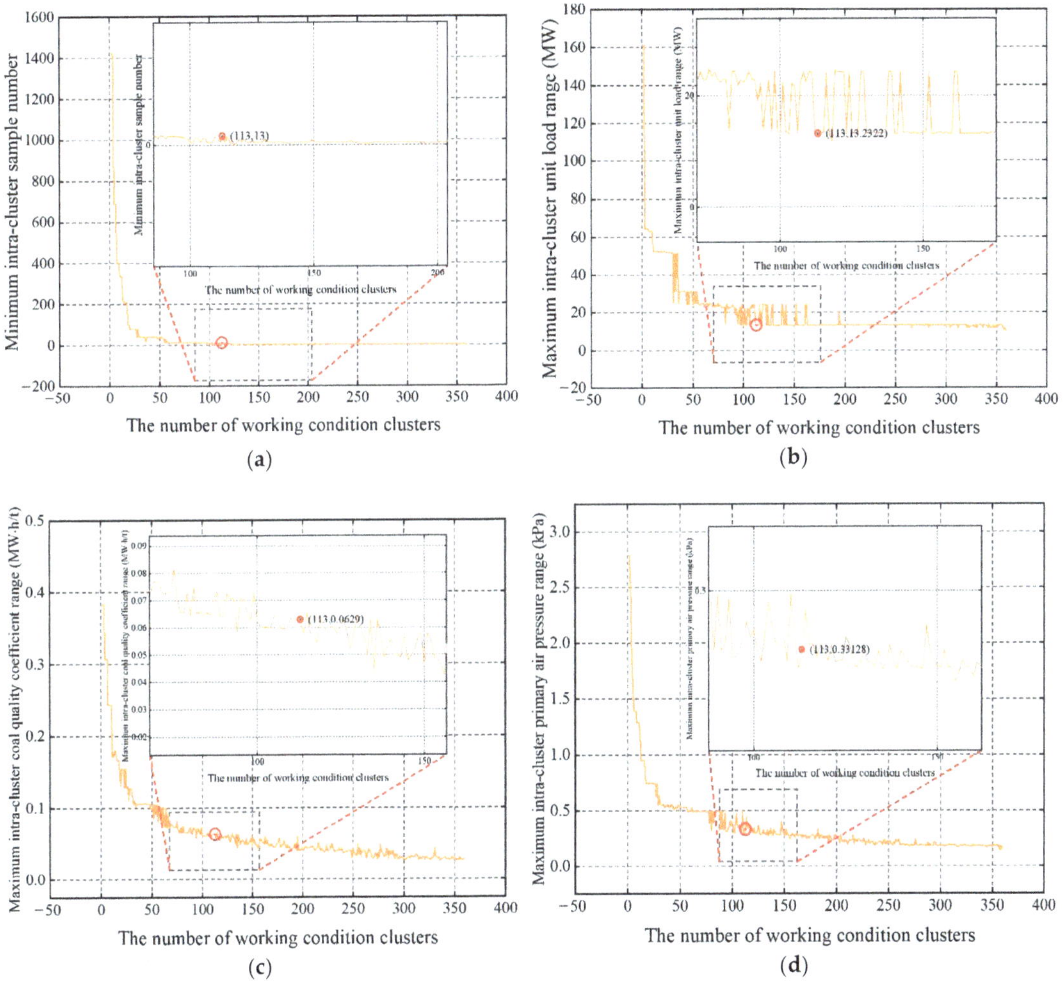

Figure 10. Statistical charts for choosing *k*. (**a**) The minimum intracluster sample number changes with the number of clusters. (**b**) The maximum intracluster unit load range changes with the number of clusters. (**c**) The maximum intracluster coal quality coefficient range changes with the number of clusters. (**d**) The maximum intracluster primary air pressure range changes with the number of clusters.

According to the above analysis, 6392 high-quality data samples are preliminarily divided into 113 working conditions. In Figure 11, (a) shows the spatial distribution of the centers of all condition clusters, with each sphere representing a class of boiler operating conditions, (b) represents the distribution of condition cluster samples for the working condition [659.7 MW, 2.44 MW·h/t, 9.01 kPa], and (c) represents the projection of the data samples in (b) onto the NO_x concentration and boiler efficiency dimensions.

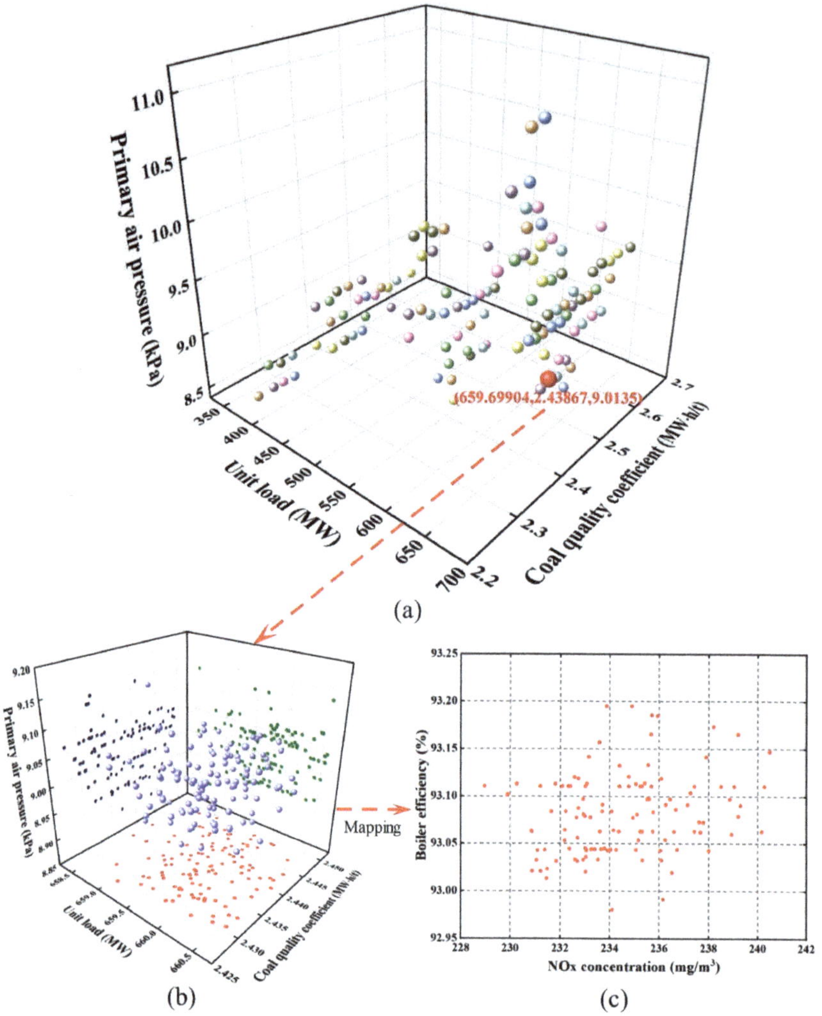

Figure 11. Case clustering and projection plots. (**a**) The distribution of cases in the case library. Dots of different colors represent the cases at different regions. (**b**) The distribution of data points in a certain case. The blue dots represent the projections of the data points on the plane of primary air pressure and coal quality coefficient. The red dots represent the projections of the data points on the plane of unit load and coal quality coefficient. The green dots represent the projections of the data points on the plane of unit load and primary air pressure. (**c**) The projections of the data points on the performance index plane (i.e., boiler efficiency and NOx concentration).

This figure illustrates that the distribution of a single working condition cluster in the sample space is relatively uniform, and the projection figure illustrates that although the optimization method based on the case library for directly adjusting the boiler is real, simple, and fast, the case library samples often do not present an equilibrium solution like the Pareto front. Therefore, deep optimization is essential for the online boiler optimization.

5.3. Construction of the Data-Driven Model

5.3.1. Model Training

From the 6392 data samples of the above 113 types of working conditions, 70% of the data samples of each working condition are selected to form the model training set, and the remaining 30% of the data are used as the testing set for evaluating the model performance. After the dataset is reconstructed, the operating condition parameters and performance parameter ranges of the training set and testing set are shown in Table 4.

Table 4. Dataset reconstruction results.

Description	Training Set	Testing Set
Unit load	[368.64, 668.61]	[368.46, 662.07]
Coal quality coefficient	[2.22, 2.65]	[2.23, 2.64]
Primary air pressure	[8.37, 11.20]	[8.31, 11.10]
Boiler efficiency	[92.39, 94.07]	[92.38, 94.07]
NO_x concentration	[207.69, 329.90]	[213.84, 329.83]
Count	4426	1966

First, the parameters n_estimators, subsample, and colsample_bytree of the XGBoost model are preliminarily determined to be 365, 0.3, and 0.7, respectively. In addition, the model is trained using a GA and a five-fold cross-validation method to determine the model hyperparameters. The optimal hyperparameters of the boiler efficiency model and the NO_x emission model are provided in Table 5.

Table 5. Result of hyperparameter optimization of the XGBoost model.

Description	Range	Boiler Efficiency Model	NO_x Emission Model
Learning_rate	$[1 \times 10^{-8}, 0.1]$	0.0810	0.0319
Max_depth	[3, 25]	19	24
Min_child_weight	[1, 10]	7.0224	1
gamma	$[1 \times 10^{-8}, 10]$	1×10^{-8}	1.8735

5.3.2. Model Evaluation

Figures 12 and 13 show the predictive performance of the boiler efficiency model and NO_x emission model on the training and testing datasets, respectively. The training data ($R^2 = 0.9998$) and testing data ($R^2 = 0.9982$) of the boiler efficiency model and the training data ($R^2 = 0.9982$) and testing data ($R^2 = 0.9831$) of the NO_x emission model are obviously distributed near the standard line. The results show that the characteristic model of the boiler combustion system has good predictive performance.

The specific prediction performances of the two combustion characteristic models are provided in Table 6.

Table 6. Result of model evaluation.

	Boiler Efficiency Model		NO_x Emission Model	
	Training Set	Testing Set	Training Set	Testing Set
MAE	0.0031	0.0104	0.7677	2.3672
RMSE	0.0049	0.0154	1.0225	3.1429
MAPE (%)	0.0034	0.0111	0.2923	0.8997
R^2	0.9998	0.9982	0.9982	0.9831

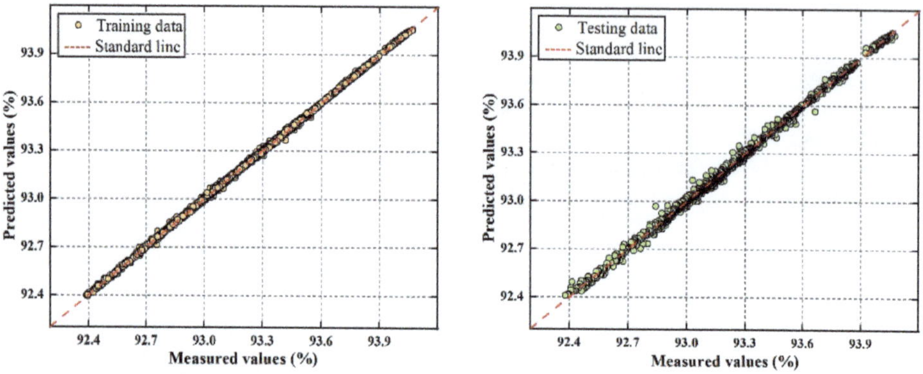

(**a**) $R^2 = 0.9998$ (**b**) $R^2 = 0.9982$

Figure 12. Comparison of the actual measured and predicted boiler efficiency values: (**a**) training dataset and (**b**) testing dataset.

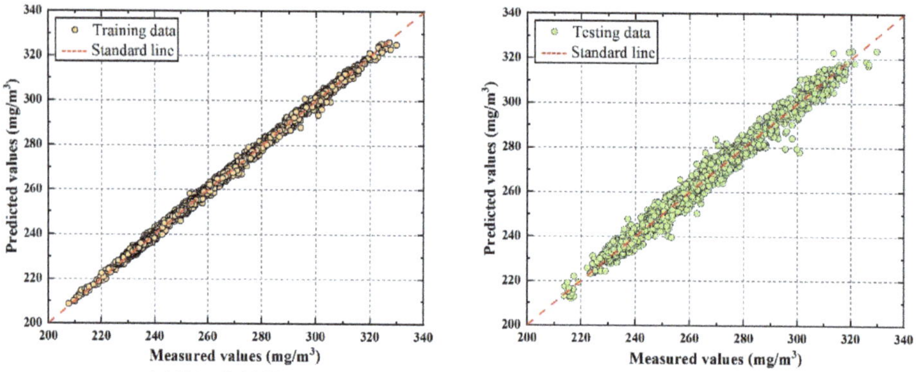

(**a**) $R^2 = 0.9982$ (**b**) $R^2 = 0.9831$

Figure 13. Comparison of the actual measured and predicted NO_x values: (**a**) training dataset and (**b**) testing dataset.

5.4. Construction of the Optimization Case Library

The multiobjective combustion optimization problem is described as Equation (20).

$$\begin{cases} Const = f_w(x_1, x_2, x_3) \\ \max \eta = f_\eta(f_w(x_1, x_2, x_3), x_4, x_5, \cdots, x_{30}) \\ \min NO_x = f_{NO_x}(f_w(x_1, x_2, x_3), x_4, x_5, \cdots, x_{30}) \\ s.t.\ x_{i\min} \leq x_i \leq x_{i\max} \\ 4 \leq i \leq 30 \end{cases} \quad (20)$$

where Const represents three working condition parameters; f_w represents the K-means++ algorithm or the online retrieval strategy; x_4~x_{30} are the operable parameters in Table 2, which are also the parameters for the multiobjective optimization solution in this paper; and f_η and f_{NOx} represent data-driven models for boiler efficiency and NO_x concentration, respectively. In the multiobjective optimization calculation, the upper limit and lower limit of oxygen content are set as 1.5% and 6%, respectively; the optimization range of damper opening is set as [0%, 100%].

Combustion optimization is conducted for different goals: (1) Mode 1: the boiler combustion efficiency is optimal; (2) Mode 2: the NO_x emission concentration is optimal; (3) Mode 3: the boiler combustion efficiency and NO_x emission concentration are balanced. Figures 14 and 15 show the comparison results of Mode 1 and Mode 2 using the query optimization strategy, intelligent optimization strategy, and hybrid optimization strategy proposed in this paper, respectively. Compared with the original operation condition, the three strategies all exhibit more effective optimization, especially the optimization method proposed in this paper, which performs well under all working conditions. Meanwhile, since the best optimization cases for Mode 1 and Mode 2 are primarily drawn from the Pareto front solution set, many operational conditions' best optimization cases achieve high efficiency or low pollution compared to the original cases while ensuring their optimal objectives. Notably, a very small number of operational conditions' best optimization cases are derived from historical observation data.

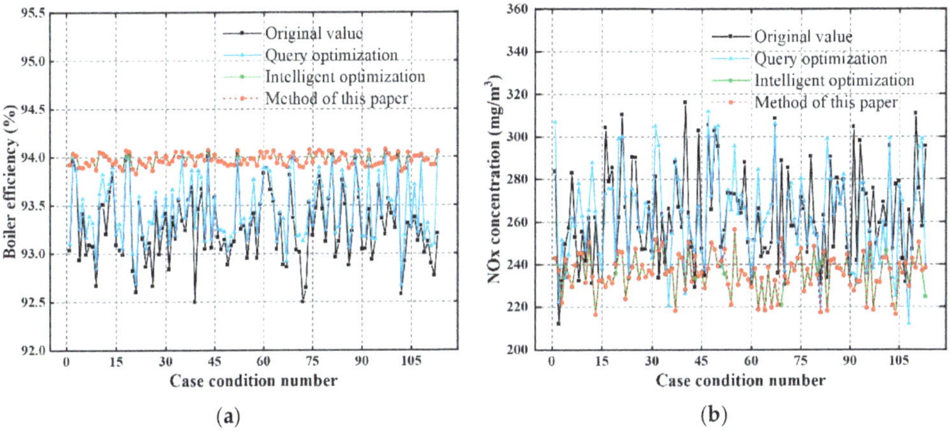

Figure 14. Optimization results of different methods in Mode 1. (**a**) The optimization results of boiler efficiency. (**b**) The optimization results of NOx concentration.

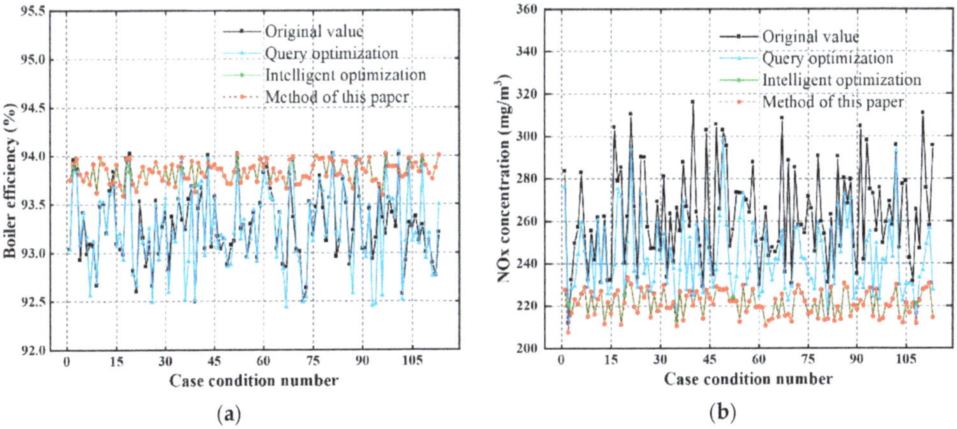

Figure 15. Optimization results of different methods in Mode 2. (**a**) The optimization results of boiler efficiency. (**b**) The optimization results of NOx concentration.

Figure 16a,b display the optimal solutions for all operational conditions in Modes 1 to 3, successfully demonstrating that Mode 3 is the most suitable combustion mode for achieving high efficiency and low pollution. Figure 16c illustrates the effectiveness of the inflection point attribution strategy through a working condition example, providing an adaptive computational approach for solving Mode 3 for online combustion optimization. Based on the above calculations and strategies, the best optimization schemes of Modes 1~3 are selected for all working conditions, and an initial optimization case library is constructed, providing a basic guarantee for boiler parameter adjustments.

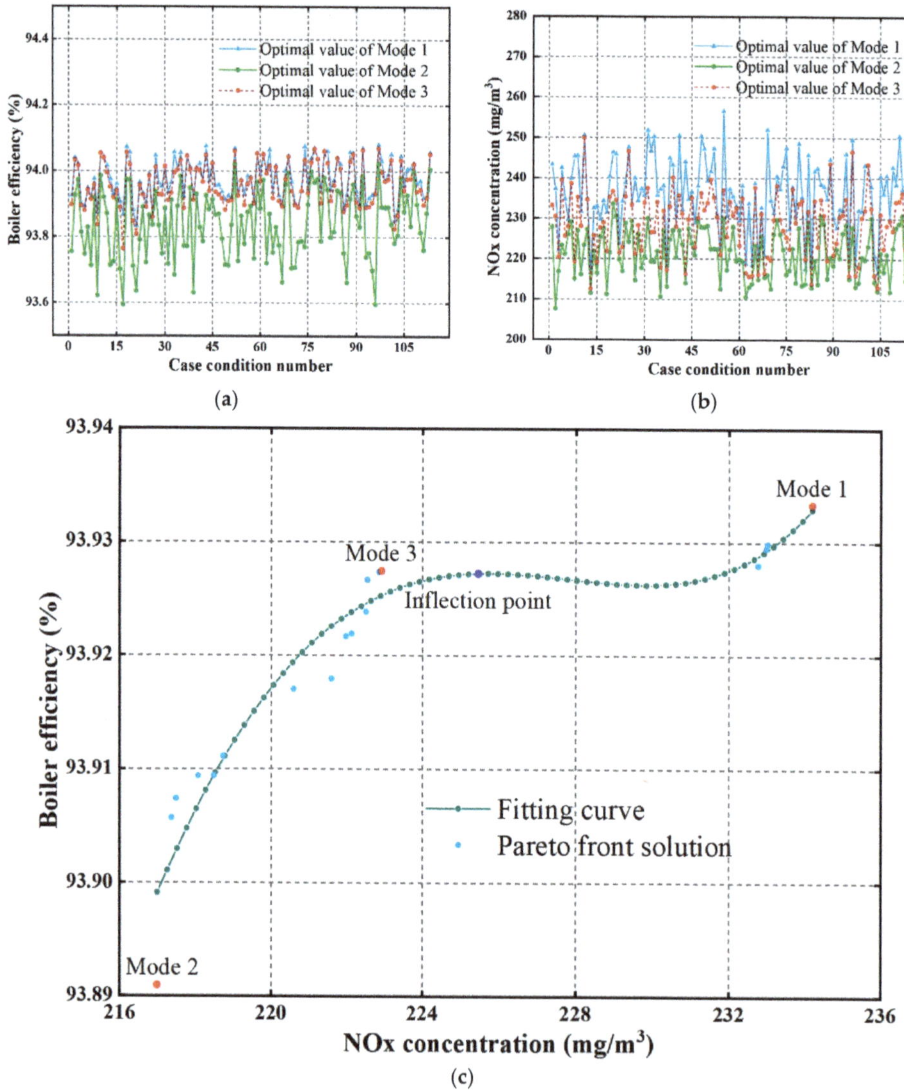

Figure 16. Comparison of the three optimization modes in this paper. (**a**) The optimization results of the three modes on the boiler efficiency under different case conditions. (**b**) The optimization results of the three modes on the NOx concentration under different case conditions. (**c**) The positions of the optimization solutions of the three modes in the frontier solution set.

5.5. Effective Analysis of the Proposed Strategy

5.5.1. Analysis of the Online Retrieval Strategy

In this section, we analyze the impact of the online retrieval process on the case library. Figure 17 shows the changes in the cluster center and data distribution in the normalized space with the working condition case library when a new set of high-quality data samples is observed by the system. As shown in Figure 17a, when the Euclidean distance between a new sample and the nearest cluster center is less than the CD, the new sample belongs to the working condition and is stored in the case library. As shown in Figure 17b, when the Euclidean distance between a new sample and the nearest cluster center is greater than the CD, there is no matching working condition in the case library. In this case, a new working condition is created in the case library with the new sample as the cluster center, and this condition is populated with the optimal solutions for the three optimization modes based on this sample. Before the update period T is reached, the case base is updated using this method during the real-time retrieval process of the case base. Figure 18a shows the change in distribution in the cluster center before adaptive updating.

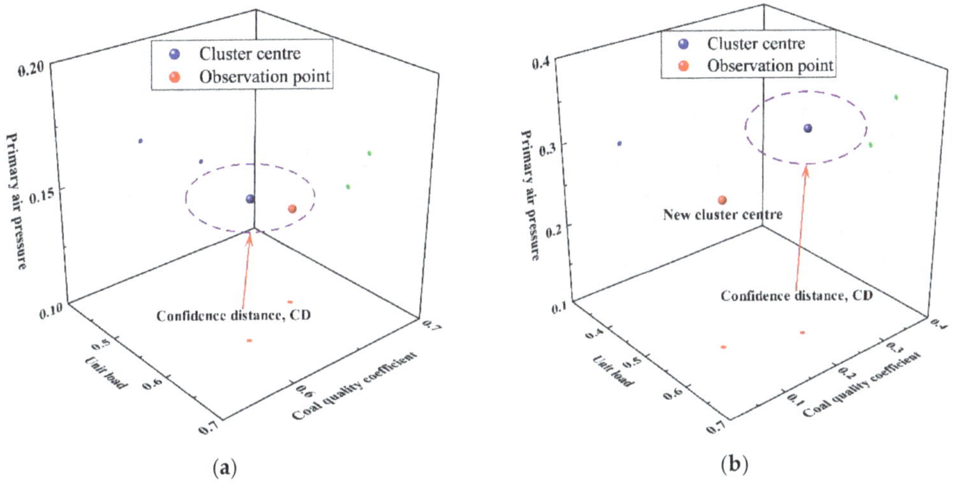

Figure 17. Effect diagram of the online retrieval strategy. (**a**) $D_{min} \leq CD$, (**b**) $D_{min} > CD$. The red dots are the projections of the observation point and the clustering center on the plane of unit load and coal quality coefficient. The blue dots are the projections on the plane of primary air pressure and coal quality coefficient, and the green dots are the projections on the plane of unit load and primary air pressure.

5.5.2. Analysis of the Reclustering Strategy

In this section, we verify the effectiveness of the proposed reclustering strategy. Figure 18a,b show the changes in the cluster center in the case library before and after the adaptive reclustering strategy, respectively. Due to the increasing richness of data samples in the case library during the online retrieval process, after reclustering, the data samples are divided into 313 working conditions. The maximum intracluster unit load range, the maximum intracluster coal quality coefficient range, and the maximum intracluster primary air pressure range are 13.2 MW, 0.049 MW·h/t, and 0.206 kPa, respectively. The results show that the proposed reclustering strategy can satisfy the online adaptive computation and the range of 2~3% working condition parameters.

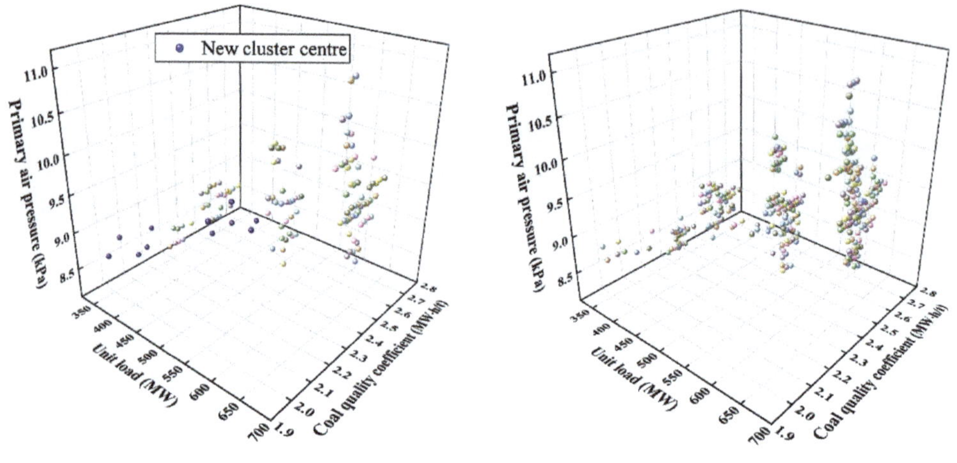

Figure 18. Effect diagram of the reclustering strategy. (**a**) Case library before adaptive reclustering strategy. (**b**) Case library after adaptive reclustering strategy. The dots of various colors represent cases in different regions and having different boundary values.

5.6. Online Combustion Optimization Experiment

In this section, we conduct online testing using future data for the same utility boiler, including multiple steady-state and transient processes. In the experiment, when a new data point is observed, the system immediately extracts its condition parameters $[x_1, x_2, x_3]$ and uses the online retrieval strategy to query optimization solutions from the case library and adjust the boiler operating parameters accordingly. The software and hardware configurations used in this study are detailed in Table 7.

Table 7. Software and hardware configurations.

Description	Configuration
OS	Windows 11 × 64
CPU	Intel(R) Core(TM) i7-9750H CPU @2.60 GHz 2.59 GHz
Memory	16 GB DDR4 3200 MHz
Python version	3.7.12

Figure 19a,b show the optimization results of Mode 1 and Mode 2, respectively; Figure 20 shows the optimization results of Mode 3. After testing, the results show that the operation optimization of the unit using the traditional data-driven model and intelligent optimization algorithm takes approximately 3 to 5 min, which obviously does not meet the real-time requirements of online combustion optimization. Since the proposed optimization method can basically obtain the optimal solution within 1 s (0.95~6.99 ms), it has an important calculation speed advantage in online combustion optimization. Finally, as shown in Figures 19 and 20, the proposed optimization method can improve boiler efficiency, reduce NO_x concentration, and effectively achieve the online boiler combustion optimization task.

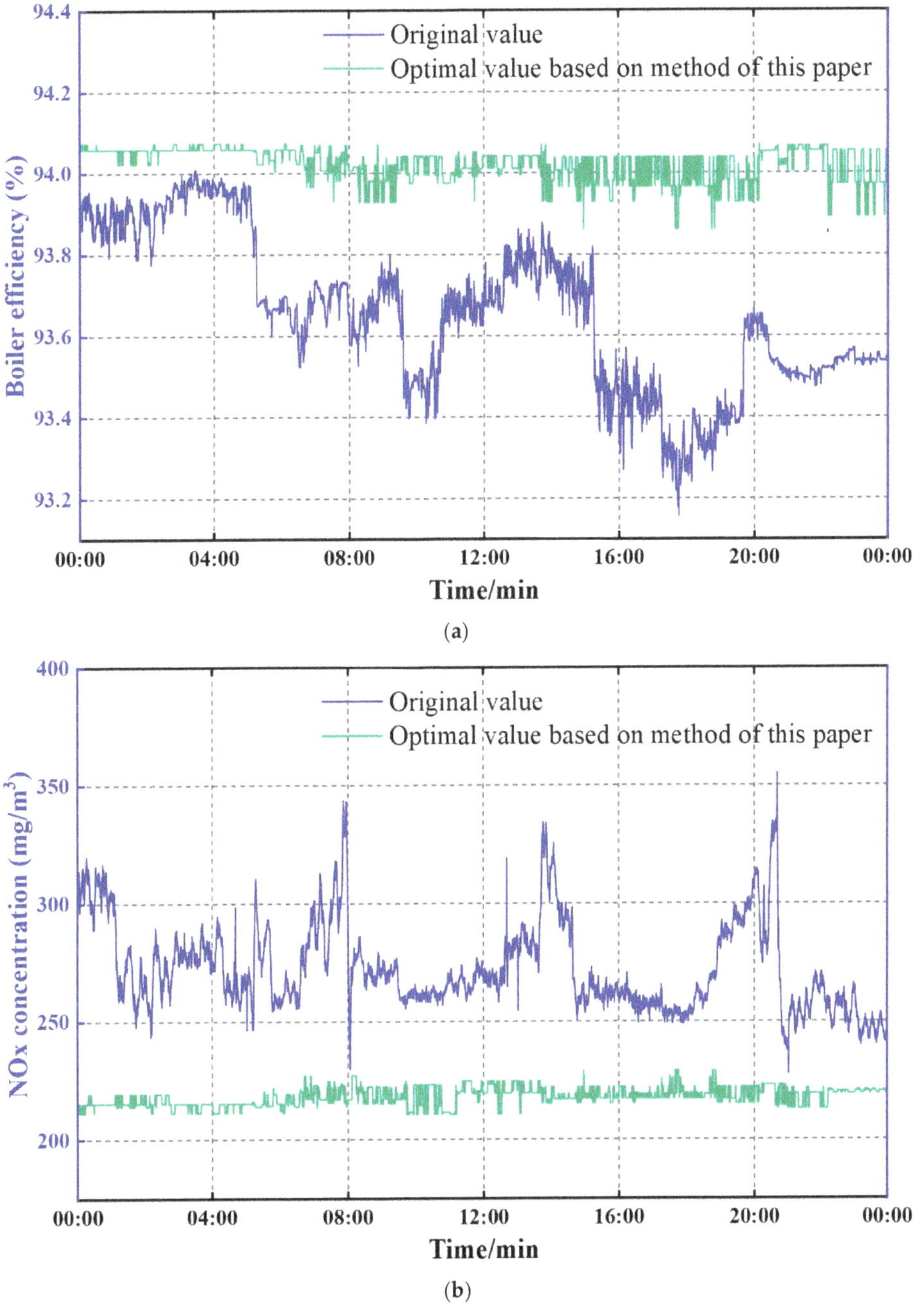

Figure 19. The effect diagram of single-objective online combustion optimization. (**a**) Mode 1, (**b**) Mode 2.

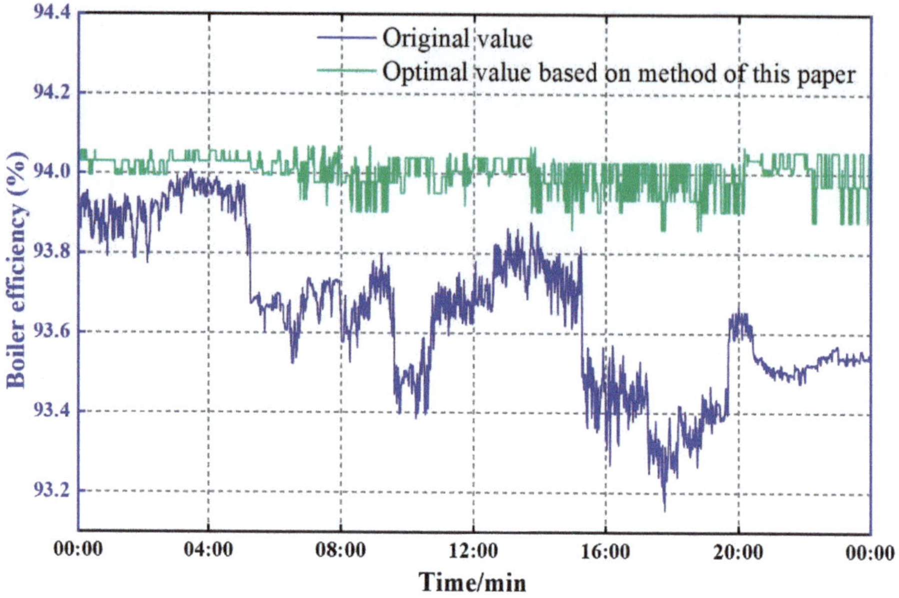

(**a**)

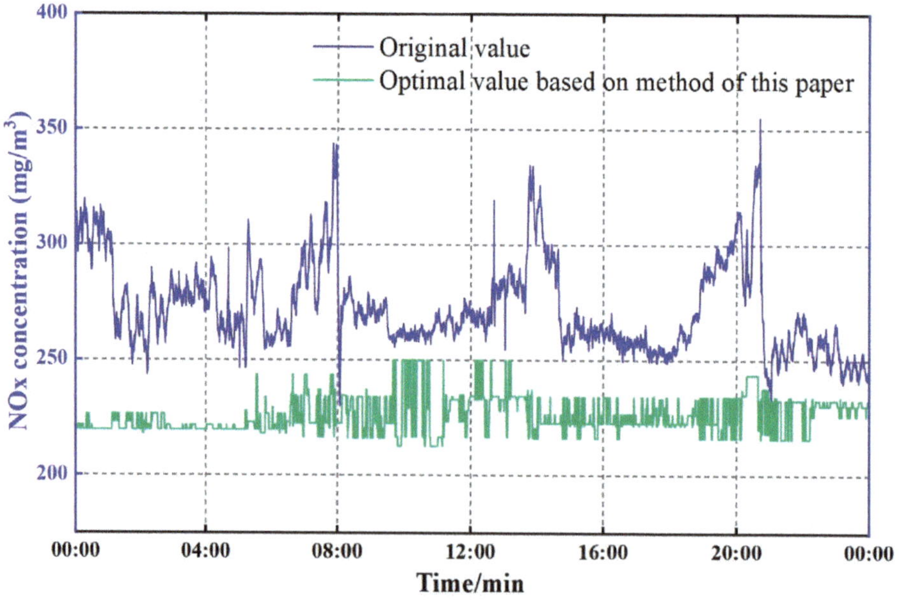

(**b**)

Figure 20. The effect diagram of multiobjective online combustion optimization. (**a**) Optimization of boiler efficiency. (**b**) Optimization of Nox concentration.

6. Conclusions

To address the time-consuming limitation of traditional combustion optimization methods, this study proposes a novel online combustion optimization framework integrating data filtering, case-based reasoning, data-driven modeling, and multiobjective optimization. The proposed method has been validated on a 660 MW utility boiler, demonstrating its ability to enhance combustion performance while ensuring a swift response to changing operating conditions. The key conclusions of this study are as follows:

(1) The robust data filter using Kernel-PCA, JSI, and VMD extracts 6392 high-quality steady-state samples from 72,000 original data points. The filter successfully retains steady-state data, removes anomalies, and identifies five steady-state periods within a 24 h dataset, demonstrating its efficiency in data pre-processing for combustion optimization.
(2) Based on the extracted samples, combustion system models were developed using XGBoost. The boiler efficiency model achieved R^2 values of 0.9998 and 0.9982 for the training and test sets, respectively, while the NOx emissions model attained R^2 values of 0.9982 and 0.9831, effectively capturing the relationships between operational parameters and combustion performance.
(3) K-means clustering was used to construct the working condition case library, while NSGA-II, combined with combustion system models and inflection point attribution strategy, optimized each condition to develop an optimization case library. This approach lowered NOx emissions from 329.90 mg/m^3 to 213.84 mg/m^3, a 35.2% reduction, while increasing boiler efficiency from 92.39% to 94.07%, enhancing both environmental and operational performance.
(4) In terms of speed, an online retrieval strategy based on CD quickly identifies the optimal case for real-time operation, enabling optimization in just 0.95–6.99 milliseconds, outperforming traditional methods requiring 3–5 min. A reclustering strategy dynamically updates the case library, expanding the number of working conditions from 113 to 313, ensuring that the optimization method remains effective under varying operating conditions.

Future research will focus on developing a dynamic combustion optimization approach based on a dynamic working condition case library. The proposed research direction involves employing deep learning techniques to extract latent features from time-series data of non-adjustable variables, serving as the basis for dynamic case library segmentation, aiming to further enhance the adaptability and effectiveness of intelligent combustion optimization.

Author Contributions: Conceptualization, C.Y. and L.S.; methodology, C.Y. and S.C.; software, C.Y. and S.C.; validation, H.Y. and Y.Z.; formal analysis, C.Y.; writing—original draft preparation, C.Y.; writing—review and editing, C.Y. and Q.W.; visualization, G.L.; supervision, L.S. All authors have read and agreed to the published version of the manuscript.

Funding: Funding was received from the Foundation of the Jiangsu Higher Education Institutions of China (Grant No. 23KJD470005), the Open Foundation of Hubei Key Laboratory of Industrial Fume and Dust Pollution Control, Jianghan University (Grant No. HBIK2023-02), and the National Natural Science Foundation of China (Grant No. 52006090).

Data Availability Statement: The original contributions presented in this study are included in the article. Further inquiries can be directed to the corresponding author.

Acknowledgments: The authors sincerely acknowledge the support provided by the Hubei Key Laboratory of Industrial Fume and Dust Pollution Control, Jianghan University. Special thanks are also extended to the editors and reviewers who made valuable comments to improve this paper.

Conflicts of Interest: The authors declare no conflicts of interest.

References

Wang, Y.N.; Zhao, C.H. A deep graph convolutional network model of NOx emission prediction for coal-fired boiler. *Can. J. Chem. Eng.* **2024**, *102*, 669–684. [CrossRef]

2. Baqain, M.; Neshumayev, D.; Konist, A. NO_x and N_2O emissions from Ca-rich fuel conversion in oxyfuel circulating fluidized bed combustion. *Therm. Sci. Eng. Prog.* **2023**, *42*, 101938. [CrossRef]
3. Khanali, M.; Akram, A.; Behzadi, J.; Rad, F.M.; Saber, Z.; Chau, K.W.; Pelesaraei, A.N. Multi-objective optimization of energy use and environmental emissions for walnut production using imperialist competitive algorithm. *Appl. Energy* **2021**, *284*, 116342. [CrossRef]
4. Liu, Y.; Zhou, J.X.; Fan, W. A novel robust dynamic method for NOx emissions prediction in a thermal power plant. *Can. J. Chem. Eng.* **2023**, *101*, 2391–2402. [CrossRef]
5. Chen, S.; Yu, C.; Zhu, Y.K.; Fan, W.; Yu, H.Q.; Zhang, T.H. NOx formation model for utility boilers using robust two-step steady-state detection and multimodal residual convolutional auto-encoder. *J. Taiwan Inst. Chem. Eng.* **2024**, *155*, 105252. [CrossRef]
6. Zhu, Y.K.; Yu, C.; Jin, W.; Shi, L.; Chen, B.; Xu, P. Mechanism-enhanced data-driven method for the joint optimization of boiler combustion and selective catalytic reduction systems considering gas temperature deviations. *Energy* **2024**, *291*, 130432. [CrossRef]
7. Chang, J.; Wang, W.; Zhou, Z.J.; Chen, H.G.; Niu, Y.G. CFD modeling of hydrodynamics, combustion and NOx emission in a tangentially fired pulverized-coal boiler at low load operating conditions. *Adv. Powder Technol.* **2021**, *32*, 290–303. [CrossRef]
8. Sun, D.P.; Fang, Q.Y.; Wang, H.J.; Zhou, H.C. A compact optimization strategy for combustion in a 125 MW tangentially anthracite-fired boiler by an artificial neural network. *Asia-Pac. J. Chem. Eng.* **2008**, *3*, 432–439. [CrossRef]
9. Aliramezani, M.; Koch, C.R.; Shahbakhti, M. Modeling, diagnostics, optimization, and control of internal combustion engines via modern machine learning techniques: A review and future directions. *Prog. Energy Combust. Sci.* **2002**, *88*, 100967. [CrossRef]
10. Yang, T.T.; Liu, J.Z.; Zeng, D.L.; Xie, X. Application of data mining in boiler combustion optimization. In Proceedings of the 2010 the 2nd International Conference on Computer and Automation Engineering (ICCAE), Singapore, 26–28 February 2010, Volume 2, pp. 225–228.
11. Zhao, W.J.; Liu, C. The optimizing for boiler combustion based on fuzzy association rules. In Proceedings of the 2011 International Conference of Soft Computing and Pattern Recognition (SoCPaR), Dalian, China, 14–16 October 2011; pp. 306–311.
12. Kusiak, A.; Song, Z. Combustion efficiency optimization and virtual testing: A data-mining approach. *IEEE Trans. Ind. Inform.* **2006**, *2*, 176–184. [CrossRef]
13. Kusiak, A.; Song, Z. Clustering-Based Performance Optimization of the Boiler-Turbine System. *IEEE Trans. Energy Convers.* **2008**, *23*, 651–658. [CrossRef]
14. Kusiak, A.; Song, Z. Constraint-Based Control of Boiler Efficiency: A Data-Mining Approach. *IEEE Trans. Ind. Inform.* **2007**, *3*, 73–83.
15. Zhou, H.; Cen, K.F.; Fan, J.R. Multi-objective optimization of the coal combustion performance with artificial neural networks and genetic algorithms. *Int. J. Energy Res.* **2005**, *29*, 499–510. [CrossRef]
16. Wang, C.L.; Liu, Y.; Zheng, S.; Jiang, A.P. Optimizing combustion of coal fired boilers for reducing NOx emission using Gaussian Process. *Energy* **2018**, *153*, 149–158. [CrossRef]
17. Song, J.G.; Romero, C.E.; Yao, Z.; He, B.S. Improved artificial bee colony-based optimization of boiler combustion considering NOx emissions, heat rate and fly ash recycling for on-line applications. *Fuel* **2016**, *172*, 20–28. [CrossRef]
18. Ma, Y.P.; Wang, H.Q.; Zhang, X.X.; Hou, L.K.; Song, J.C. Three-objective optimization of boiler combustion process based on multi-objective teaching–learning based optimization algorithm and ameliorated extreme learning machine. *Mach. Learn. Appl* **2021**, *5*, 100082. [CrossRef]
19. Rahat, A.A.M.; Wang, C.L.; Everson, R.M.; Fieldsend, J.E. Data-driven multi-objective optimisation of coal-fired boiler combustion systems. *Appl. Energy* **2018**, *229*, 446–458. [CrossRef]
20. Feng, K.; He, D.F.; Xu, A.J.; Wang, H.B. End Temperature Prediction of Molten Steel in RH Based on Case-based Reasoning with Optimized Case Base. *J. Iron Steel Res. Int.* **2015**, *22*, 68–74. [CrossRef]
21. Xia, J.; Chen, G.; Tan, P.; Zhang, C. An online case-based reasoning system for coal blends combustion optimization of thermal power plant. *Int. J. Electr. Power Energy Syst.* **2014**, *62*, 299–311. [CrossRef]
22. Zheng, W.; Wang, C.; Yang, Y.J.; Zhang, Y.F. Multi-objective combustion optimization based on data-driven hybrid strategy. *Energy* **2020**, *191*, 116478. [CrossRef]
23. Yu, C.; Xiong, W.; Ma, H.; Zhou, J.X.; Si, F.Q.; Jiang, X.M.; Fang, X.W. Numerical investigation of combustion optimization in a tangential firing boiler considering steam tube overheating. *Appl. Therm. Eng.* **2019**, *154*, 87–101. [CrossRef]
24. Fan, W.; Si, F.Q.; Ren, S.J.; Yu, C.; Cui, Y.F.; Wang, P. Integration of continuous restricted Boltzmann machine and SVR in NOx emissions prediction of a tangential firing boiler. *Chemom. Intell. Lab. Syst.* **2019**, *195*, 103870. [CrossRef]
25. Dragomiretskiy, K.; Zosso, D. Variational mode decomposition. *IEEE Trans. Signal Process.* **2014**, *62*, 531–544. [CrossRef]
26. Zhu, Y.K.; Yu, C.; Fan, W.; Yu, H.Q.; Jin, W.; Chen, S.; Liu, X. A novel NOx emission prediction model for multimodal operational utility boilers considering local features and prior knowledge. *Energy* **2023**, *280*, 128128. [CrossRef]
27. Fischer, J.; Doolan, C. An improved eigenvalue background noise reduction method for acoustic beamforming. *Mech. Syst. Signal Process.* **2020**, *140*, 106702. [CrossRef]

28. Arthur, D.; Vassilvitskii, S. K-means++ the advantages of careful seeding. In Proceedings of the Eighteenth Annual ACM-SIAM Symposium on Discrete Algorithms, New Orleans, Louisiana, 7–9 January 2007; pp. 1027–1035.
29. Chen, T.Q.; Guestrin, C. Xgboost: A scalable tree boosting system. In Proceedings of the 22nd ACM SIGKDD International Conference on Knowledge Discovery and Data Mining, San Francisco, CA, USA, 13–17 August 2016; pp. 785–794.
30. Deb, K.; Pratap, A.; Agarwal, S.; Meyarivan, T. A fast and elitist multiobjective genetic algorithm: NSGA-II. *IEEE Trans. Evol. Comput.* **2002**, *6*, 182–197. [CrossRef]

processes

MDPI

Article

Thermodynamic Model-Based Synthesis of Heat-Integrated Work Exchanger Networks

Aida Amini-Rankouhi, Abdurrafay Siddiqui and Yinlun Huang *

Department of Chemical Engineering and Materials Science, Wayne State University, Detroit, MI 48202, USA; aida.amini@wayne.edu (A.A.-R.); abdurrafay.siddiqui@wayne.edu (A.S.)
* Correspondence: yhuang@wayne.edu; Tel.: +1-313-577-3771

Abstract: Heat integration has been widely and successfully practiced for recovering thermal energy in process plants for decades. It is usually implemented through synthesizing heat exchanger networks (HENs). It is recognized that mechanical energy, another form of energy that involves pressure-driven transport of compressible fluids, can be recovered through synthesizing work exchanger networks (WENs). One type of WEN employs piston-type work exchangers, which demonstrates techno-economic attractiveness. A thermodynamic-model-based energy recovery targeting method was developed to predict the maximum amount of mechanical energy feasibly recoverable by piston-type work exchangers prior to WEN configuration generation. In this work, a heat-integrated WEN synthesis methodology embedded by the thermodynamic model is introduced, by which the maximum mechanical energy, together with thermal energy, can be cost-effectively recovered. The methodology is systematic and general, and its efficacy is demonstrated through two case studies that highlight how the proposed methodology leads to designs simpler than those reported by other researchers while also having a lower total annualized cost (TAC).

Keywords: work exchanger network synthesis; thermodynamic modeling; work integration; mechanical energy recovery; heat integration

1. Introduction

Continuous improvement of process sustainability is a significant endeavor for the industry, largely due to natural resource depletion, environmental pressures, and global competition. A key indicator of process sustainability is energy efficiency. Over the past few decades, the process industry has demonstrated great success in thermal energy recovery, frequently through the incorporation of cost-effective heat exchanger networks (HENs). In contrast, mechanical energy recovery, while having a large scope in industrial settings with about a 30% mechanical energy loss in production [1], has not drawn sufficient attention.

From a thermodynamic point of view, heat flow, for which temperature is the state variable, can be holistically controlled to ensure thermal energy efficiency. Workflow, on the other hand, for which pressure is the state variable, is yet to be fully characterized so that opportunities for recovering mechanical energy can be thoroughly identified. In process plants, many process streams need to be pressurized to different levels, which require work for compression, while many others should be depressurized, also to different levels, which produce work through expansion. This provides opportunities for work exchange among process streams for mechanical energy recovery. One of the technical approaches is to design work exchanger networks (WENs) for process plants. The concept of WEN synthesis was first introduced by Huang and Fan [2]. In their work, work exchangers (WEs) were piston-type devices that can be called direct WEs, and a detailed thermodynamic analysis of work integration was provided.

WEN synthesis did not catch sufficient attention initially until around 2010. Deng et al. conducted a basic thermodynamic analysis on gas–gas work exchange using direct WEs [3].

Chen and Feng described a graphical method for the design of a WEN for ammonia synthesis [4]. Liu et al. proposed a method using the pressure–work diagram that was initially introduced by Huang and Fan [5].

Razib et al. proposed the use of two-stream single-shaft-turbine-compressor (SSTC) units, a type of indirect work exchange unit, to design a WEN [6]. A superstructure-based modeling method was introduced, and a network was designed using a mixed-integer nonlinear programming (MINLP) technique. Onishi et al. described a new MINLP optimization model for WEN synthesis with hypothetic heat integration for optimal pressure recovery of gaseous processes that also uses SSTC units [7]. Huang and Karimi included a HEN in a WEN system and minimized total annualized cost (TAC) [8]. Meanwhile, Du et al. presented a transshipment model for WEN synthesis, where SSTC units were used under isothermal conditions [9]. The above studies have made significant strides in the advancement of WEN synthesis; however, many of them use SSTC units or indirect WEs. It is noted from Amini-Rankouhi and Huang's study that indirect WEs may greatly limit mechanical energy recovery [10].

In a study by Deng et al., both direct and indirect WEs were analyzed alongside heat integration [11]. A pinch analysis technique was used for the WEN design. However, the efficiency of mechanical energy recovery could not be guaranteed. Gundersen and associates discussed the importance of work integration alongside heat integration and analyzed the effect of compressor and expander placement on overall energy recovery in HENs [12–15]. They used these ideas to introduce the work–heat exchange network (WHEN) synthesis problem. This type of synthesis problem was also studied by Hamedi et al. [16], Huang et al. [17], Pavão et al. [18,19], Santos et al. [20,21], and Zhuang et al. [22], who used different optimization or meta-heuristic techniques in solution derivation. Very recently, Ibric et al. described an MINLP-based solution method for integrating a HEN into a WEN that was composed of compressors and expanders only [23]. These proposed methods, while effective at mechanical energy recovery, only use compressors and expanders. They do not make use of WEs for integration and, as such, do not fall under the scope of this study. The piston-type work exchanger is a cylindrical device for direct pressure exchange between a high-pressure stream on one side of the piston and a low-pressure stream on the other side of the piston. It is structurally and operationally different from the single-shaft-turbine-compressor (SSTC), which has a single shaft, with part of the turbine's output driving the compressor. Thus, the thermodynamic modeling and analysis and the model-based synthesis methodology proposed in this work cannot be directly applied to the SSTC-based work exchanger network design. On the other hand, the introduction of piston-type work exchangers is intended to reduce the number of compressors, which are generally more expensive.

As is clear from the extensive work conducted in the area, researchers have made substantial progress on the subject of WENs. However, there is still much room for advancement, especially for the design of WENs using piston-type direct WEs. Amini-Rankouhi and Huang introduced a thermodynamic analysis method for the evaluation of the maximum amount of feasibly recoverable mechanical energy by this type of WEN prior to flowsheet generation [24]. It shows that the preset mechanical energy recovery target is theoretically achievable and the energy efficiency of such a WEN is higher than in previous studies. The robustness of the proposed thermodynamic method allowed it to be used by the authors in this work as well. A further study by this group also reported a low capital cost feature of this type of WEN, as compared with the WENs that use other types of units [10]. Additionally, a comprehensive study on the design of direct WEs was conducted, and a detailed dynamic analysis of such a unit type in operation was performed [25].

WENs are highly complex because of the effects of pressurization and depressurization on the temperature of process streams. This, coupled with the fact that stream temperatures entering and leaving a WEN are usually determined by the operational requirements of the upstream and downstream process units, necessitates the incorporation of heat integration in the WEN design. In this work, we introduce a thermodynamic-model-based synthesis methodology, building upon the previous work introduced by Amini-Rankouhi and Huang [10,24], to develop a heat-integrated work exchanger network (HIWEN).

This paper is organized as follows: We first define the HIWEN synthesis task and summarize the thermodynamic-model-based method for the prediction of the maximum mechanical energy recovery prior to WEN synthesis. We then describe a synthesis strategy for HIWEN development. After that, we introduce a systematic synthesis methodology for stepwise HIWEN flowsheet development. Two case studies are then illustrated to detail the application of the synthesis methodology. The methodological efficacy is finally discussed, and future work is presented.

2. Synthesis Task and Prediction of Maximum Mechanical Energy Recovery

The WEN synthesis task can be described as follows: Given a set of high-pressure and low-pressure process streams, whose source and target pressures and temperatures, flowrates, heat capacities, and phases are specified, design a work exchanger network (WEN) using piston-type direct work exchangers (referred to simply as WEs in this paper), which maximizes energy recovery while also presenting the lowest possible cost for that scenario. The minimum driving pressure differential ($\Delta P_{\min}$) in each WE is pre-defined, and compressors and expanders are available to provide "external" power for pressurization and depressurization of process streams as needed.

To address such a problem statement, the maximum amount of mechanical energy recoverable by a WEN should be determined prior to network design. To do this, Amini-Rankouhi and Huang developed a rigorous and general thermodynamic modeling and analysis method to accurately identify the maximum amount of recoverable mechanical energy of any process plant [24]. The following is a summarized three-step procedure that briefly describes the method; for detailed formula derivations, please read Amini-Rankouhi and Huang [24].

(1) Identify the pressure intervals of the low-pressure streams (denoted as L streams) for pressurization by the high-pressure streams (denoted as H streams). Since WEN synthesis involves a series of stream matchings through which a certain amount of mechanical energy is transferred from each pair of H streams and L streams, it is necessary to identify the pressure intervals for the set of H streams and L streams. The formulations for this task are listed in boxes No. 1 and 2 in Figure 1, where matrix $\overline{\boldsymbol{P}}$ contains all of the identified pressure intervals between all of the pairs of H and L streams. Alongside the derivation of matrix $\overline{\boldsymbol{P}}$, two vectors should also be determined based on the given stream data: $\boldsymbol{W}_H$ (the mechanical energy that can be provided by H streams) and $\boldsymbol{W}_L$ (the mechanical energy required by L streams).

(2) Evaluate the mechanical energy transfer from H streams to L streams. By using the pressure interval information contained in matrix $\overline{\boldsymbol{P}}$, the mechanical energy transferrable from each pair of H and L streams can be calculated using the formulas listed in boxes No. 3 and 4 of Figure 1. Matrix $\boldsymbol{W}_\beta$ should be created to store all the calculated energy transfer data, and vector $\boldsymbol{W}_\gamma$ should be created to contain information about whether each individual H stream can provide sufficient mechanical energy to the L streams if they are matched.

(3) Determine the minimum amount of external energy required and the maximum amount of recoverable mechanical energy. This task is accomplished in two steps. The first step is to estimate the minimum amount of external energy required. The formulas listed in box No. 5 are for the calculation of $\boldsymbol{W}_C$ and $\boldsymbol{W}_E$, which are the compression and expansion power needed from external sources, respectively. With these derived, the maximum amount of recoverable mechanical energy, W_R^{tot}, can be determined using either of the two formulas listed in box No. 6 of Figure 1.

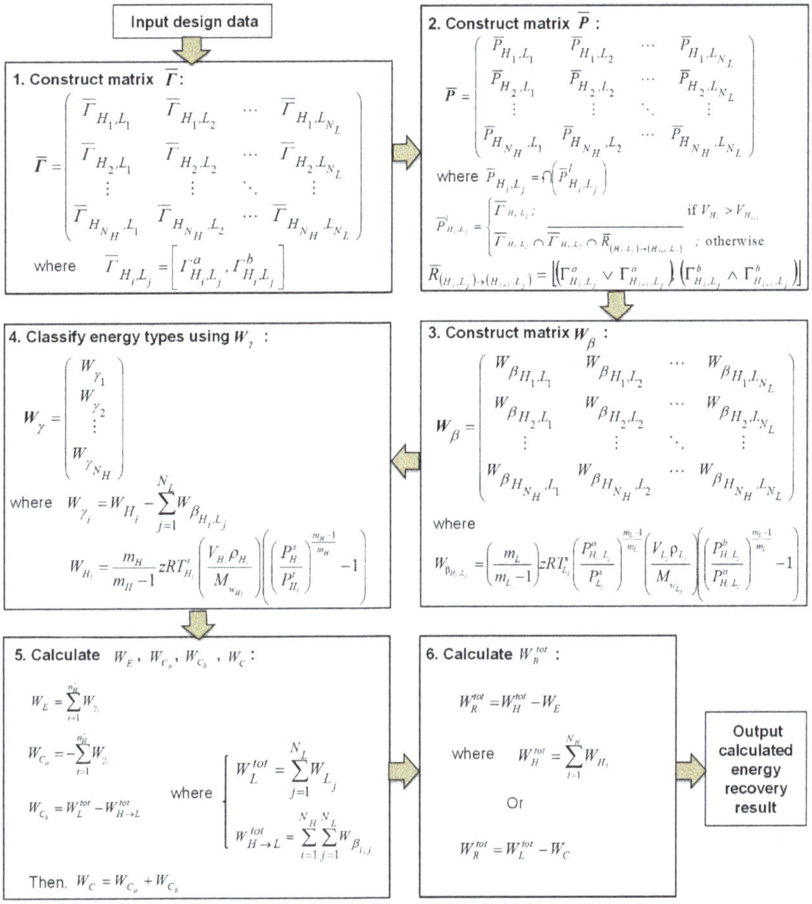

Figure 1. Systematic procedure for the evaluation of maximum recoverable mechanical energy.

3. Synthesis Strategy

As discussed above, meeting the temperature specifications for upstream and downstream processes requires the incorporation of heat integration into the design. This causes the WEN design to be coupled with an additional task: the integration of a HEN into a WEN, thereby rendering a heat-integrated WEN (HIWEN). To do this, two strategies for heat integration are proposed: either the integration is synthesized after the WEN synthesis or during the synthesis. Of the two strategies, synthesizing heat integration during WEN synthesis is not preferable. Such an integration strategy makes the whole synthesis not only very complex but, more importantly, also not efficient for thermal energy recovery. Alongside this, the dynamic characteristics and operational features of piston-type WEs explained by Amini-Rankouhi and Huang may bring additional operational challenges if a heat transfer unit is placed between two WEs [10]. Thus, it is more appropriate to synthesize a WEN first and use the resulting temperature profiles to define a heat integration problem. The simultaneous optimization of both networks is a direction that will be considered in the future. Thus, based on the identified hot and cold process streams of the synthesized WEN, a HEN can be developed. The resulting HEN can then be integrated into the WEN to generate a HIWEN that should result in the maximum recovery of mechanical and thermal energies at the lowest total annualized cost.

4. Thermodynamic-Model-Guided WEN Synthesis

For a synthesis problem involving N_H high-pressure (H) streams and N_L low-pressure (L) streams, the maximum amount of recoverable energy by a WEN can be determined by the methodology developed by Amini-Rankouhi and Huang prior to network design [24]. The main tasks of the synthesis can be accomplished using the thermodynamic models to determine: (1) the placement of the minimum number of WEs between a set of H streams (N_H) and a set of L streams (N_L) such that the maximum amount of mechanical energy is recovered, and (2) the placement of the minimum number of compressors and expanders such that the least amount of external energy is consumed.

4.1. Placement of Work Exchangers

Determining the optimal placement of WEs requires the selection of stream pairs within the proper pressure intervals for recovering the maximum amount of mechanical energy. To select stream pairs, two matrices, $\boldsymbol{W_W}$ and $\overline{\boldsymbol{P}}_{\boldsymbol{M}}$, are used, referring to the workload and pressure interval of each WE match, respectively. Figure 2 shows how matrices $\boldsymbol{W_W}$ and $\overline{\boldsymbol{P}}_{\boldsymbol{M}}$ are derived.

Detailed steps for matrix construction are described below.

Step 1. Determination of pairs of H streams and L streams for matching and the amount of mechanical energy to recover in each WE. This can be realized by constructing matrix $\boldsymbol{W_W}$ below.

$$\boldsymbol{W_W} = \begin{pmatrix} W_{W_{H_1,L_1}} & W_{W_{H_1,L_2}} & \cdots & W_{W_{H_1,L_{N_L}}} \\ W_{W_{H_2,L_1}} & W_{W_{H_2,L_2}} & \cdots & W_{W_{H_2,L_{N_L}}} \\ \vdots & \vdots & \ddots & \vdots \\ W_{W_{H_{N_H},L_1}} & W_{W_{H_{N_H},L_2}} & \cdots & W_{W_{H_{N_H},L_{N_L}}} \end{pmatrix} \tag{1}$$

where $W_{W_{H_i,L_j}}$ is the workload of the WE that is used to match the H_i stream with the L_j stream.

To determine the element values of the matrix, the information contained in the following matrices and vectors is used: $\boldsymbol{W_H}$ ($N_H \times 1$), $\boldsymbol{W_L}$ ($N_L \times 1$), $\overline{\boldsymbol{P}}$ ($N_H \times N_L$), $\boldsymbol{W_\beta}$ ($N_H \times N_L$), and $\boldsymbol{W_\gamma}$ ($N_H \times 1$). Each of these matrices and vectors were derived when predicting the maximum recoverable mechanical energy, and as a result, their derivations are not detailed here. For more detailed explanations, refer to Figure 1 or Amini-Rankouhi and Huang's previous work [24].

The procedure for deriving matrix $\boldsymbol{W_W}$ is shown in Figure 2. Note that matrix $\boldsymbol{W_\beta}$ contains the amount of mechanical energy that can be feasibly transferred from each H stream to each L stream. However, the amount of energy that can be provided by an H stream can be either higher, equal, or lower than that of an L stream based on the data in vector $\boldsymbol{W_\gamma}$. Therefore, to generate matrix $\boldsymbol{W_W}$, matrix $\boldsymbol{W_\beta}$ and vector $\boldsymbol{W_L}$ need to be used to find the most appropriate locations for placing WEs. Any positive term in vector $\boldsymbol{W_\gamma}$ indicates that the relevant H stream can provide sufficient mechanical energy to the L streams assigned to it. As shown in Figure 2, for the situation where W_{γ_i} has a positive value of the i-th H stream or an absolute value less than $W_{\beta_{H_i,L_j}}$, a WE should be placed with a workload equal to $W_{\beta_{H_i,L_j}}$. If W_{γ_i} has a negative value and its absolute value is higher than $W_{\beta_{H_i,L_j}}$, a WE should also be placed, but the workload should be $W_{\beta_{H_i,L_j}} - |W_{\gamma_i}|$ unless $W_{\beta_{H_i,L_j}}$ is the largest number in the j-th column of $\boldsymbol{W_\beta}$.

Step 2. Determination of the pressure intervals of the stream pairs where work exchangers are placed. This can be realized by constructing matrix $\overline{\boldsymbol{P}}_{\boldsymbol{M}}$ below.

$$\overline{P}_M = \begin{pmatrix} \overline{P}_{M_{H_1,L_1}} & \overline{P}_{M_{H_1,L_2}} & \cdots & \overline{P}_{M_{H_1,L_{N_L}}} \\ \overline{P}_{M_{H_2,L_1}} & \overline{P}_{M_{H_2,L_2}} & \cdots & \overline{P}_{M_{H_2,L_{N_L}}} \\ \vdots & \vdots & \ddots & \vdots \\ \overline{P}_{M_{H_{N_H},L_1}} & \overline{P}_{M_{H_{N_H},L_2}} & \cdots & \overline{P}_{M_{H_{N_H},L_{N_L}}} \end{pmatrix} \tag{2}$$

where

$$\overline{P}_{M_{H_i,L_j}} = \left[P^a_{M_{H_i,L_j}},\ P^b_{M_{H_i,L_j}} \right] \tag{3}$$

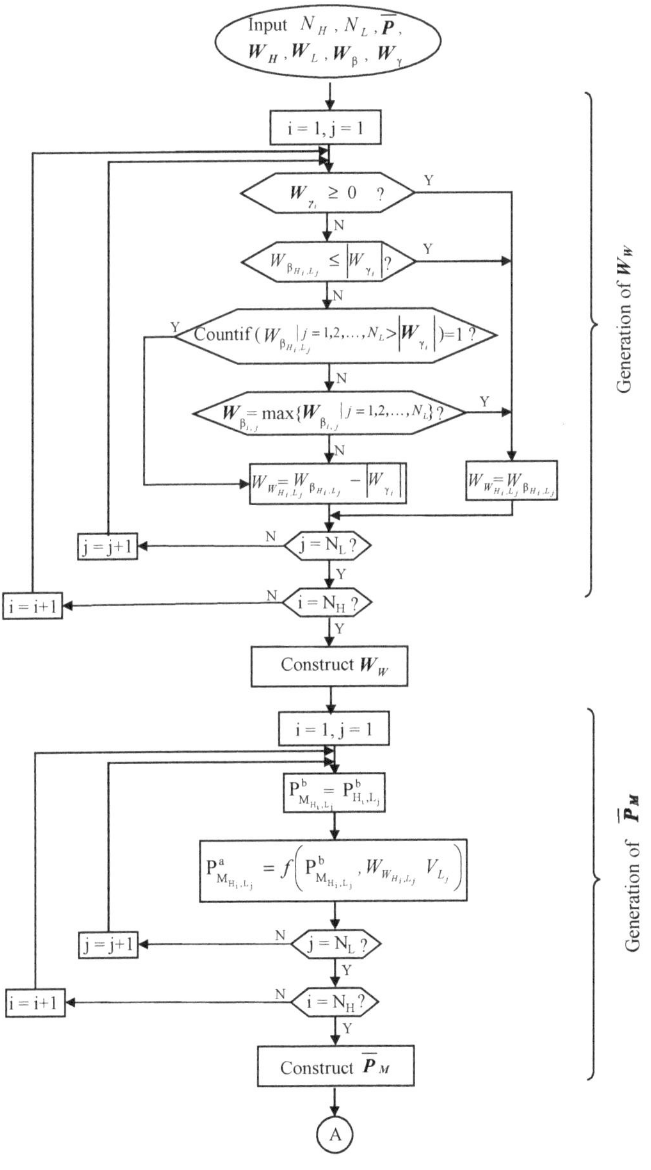

Figure 2. Generation of W_W and $\overline{P}_M$.

Note that $\overline{P}_{M_{H_i,L_j}}$ gives the pressure interval of low-pressure stream L_j such that within this interval it can receive mechanical energy from high-pressure stream H_i. The high-pressure stream is also within its own given pressure range of $\left[P^t_{H_i}, P^s_{H_i}\right]$. Once matrix $\overline{\boldsymbol{P}}_{\boldsymbol{M}}$ is completed, each element in it, as long as it is not a null interval (i.e., [0, 0]), should be indicative of one WE to be placed in the interval of $\left[P^t_{H_i}, P^s_{H_i}\right]$ for the H_i stream and in the interval of $\left[P^a_{M_{H_i,L_j}}, P^b_{M_{H_i,L_j}}\right]$ for the L_j stream. The energy exchange between the two streams is equal to $W_{W_{H_i,L_j}}$, which was calculated in Step 1.

Note that the amount of mechanical energy that needs to be removed from the H_i stream in the pressure range of $\left[P^t_{H_i}, P^s_{H_i}\right]$ may be transferrable to more than one low-pressure stream. In this case, stream H_i can be split into two or more branches.

The interval values of element $\overline{P}_{M_{H_i,L_j}}$ are determined using the following equations. The upper bound is set to be the upper bound of the interval $\overline{P}_{H_i,L_j}$, i.e.,

$$P^b_{M_{H_i,L_j}} = P^b_{H_i,L_j} \tag{4}$$

and the lower bound is to be calculated depending on the operating condition below:

$$P^a_{M_{H_i,L_j}} = \begin{cases} P^b_{M_{H_i,L_j}} exp\left(-\dfrac{W_{W_{H_i,L_j}}}{zRT^s_{L_j}\left(\frac{V_{L_j}\rho_{L_j}}{Mw_{L_j}}\right)}\right); & \text{Isothermal condition} \\ \left(\left(P^b_{M_{H,L}}\right)^{\left(\frac{k_L}{k_L-1}\right)} - \dfrac{W_{W_{H_i,L_j}}\left(P^s_L\right)^{\frac{k_L-1}{k_L}}}{\left(\frac{k_L}{k_L-1}\right)zRT^s_L\left(\frac{V_L\rho_L}{Mw_L}\right)}\right)^{\left(\frac{k_L}{k_L-1}\right)}; & \text{Isentropic condition (adiabatic)} \\ \left(\left(P^b_{M_{H,L}}\right)^{\left(\frac{m_L}{m_L-1}\right)} - \dfrac{W_{W_{H_i,L_j}}\left(P^s_L\right)^{\frac{m_L-1}{m_L}}}{\left(\frac{m_L}{m_L-1}\right)zRT^s_L\left(\frac{V_L\rho_L}{Mw_L}\right)}\right)^{\left(\frac{m_L}{m_L-1}\right)}; & \text{Polytropic condition (adiabatic)} \end{cases} \tag{5}$$

In the above formula, parameter k is a ratio of the heat capacity at constant pressure and volume (i.e., $k = c_P/c_V$), and parameter m is related to parameter k (i.e., $m = k\eta_p/[1 - k(1 - \eta_p)]$), where η_p is the polytropic efficiency. If η_p reaches 100% (i.e., no friction), then parameters k and m are equal. Hence, $(k-1)/k = (m-1)/m = R/c_P$. For more detailed information about formula derivation, see Walas and Liu et al. [5,26].

For cases where a HEN will be located before the WEN, the formulation shown in Equation (5) will be slightly changed. This will be discussed later.

4.2. Placement of Compressors and Expanders

For any L stream, if its outlet pressure when leaving a WEN after receiving mechanical energy from one or more H streams still has not reached its target pressure, then one or more compressors will be needed. The location(s) and power(s) of such compressor(s) can be determined by constructing two vectors: $\boldsymbol{W}_{\boldsymbol{C_a}}$ and $\boldsymbol{W}_{\boldsymbol{C_b}}$. Figure 3 shows a procedure for generating these vectors.

Vector $\boldsymbol{W}_{\boldsymbol{C_a}}$ is of the following structure:

$$\boldsymbol{W}_{\boldsymbol{C_a}} = \begin{pmatrix} W_{C_{a_1}} \\ W_{C_{a_2}} \\ \vdots \\ W_{C_{a_{N_L}}} \end{pmatrix} \tag{6}$$

where

$$W_{C_{a_j}} = \sum_{i=1}^{N_H} \left(W_{\beta_{H_i,L_j}} - W_{W_{H_i,L_j}} \right) \tag{7}$$

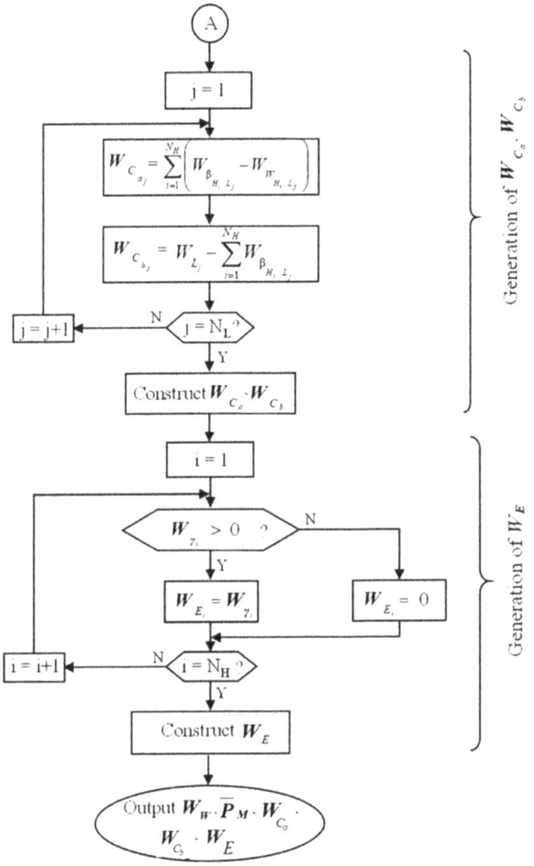

Figure 3. Generation of W_{C_a}, W_{C_b}, and $\mathbf{W}_E$.

Note that $W_{\beta_{H_i,L_j}}$ is the amount of mechanical energy that can be feasibly transferred from the H_i stream to the L_j stream. However, this does not mean the energy needs to be fully transferred; it is possible that a certain amount of energy from the H_i stream may be transferred to some other low-pressure stream(s); $W_{W_{H_i,L_j}}$ is the amount of mechanical energy that the L_j stream has received from the H_i stream through a WE that matches this pair of streams. Thus, $W_{C_{a_j}}$ gives the power needed for a compressor to be placed on the L_j stream if its value is not zero.

It is possible that the L_j stream, even after having a compressor placed to raise its pressure, may still have not received all the necessary amount of compression in order to be pressurized to its target pressure ($P^t_{L_j}$). For this reason, vector $\mathbf{W}_{C_b}$ is introduced to identify the location(s) and the power(s) for additional compressor(s) for all L streams. Figure 3 shows how to generate this vector. A more detailed explanation is provided below:

$$\boldsymbol{W_{C_b}} = \begin{pmatrix} W_{C_{b_1}} \\ W_{C_{b_2}} \\ \vdots \\ W_{C_{b_{N_L}}} \end{pmatrix} \tag{8}$$

where

$$W_{C_{b_j}} = W_{L_j} - \sum_{i=1}^{N_H} W_{\beta_{H_i, L_j}} \tag{9}$$

Note that W_{L_j} is the total amount of mechanical energy needed in order to raise the low-pressure stream's source pressure ($P^s_{L_j}$) to its target pressure ($P^t_{L_j}$). If it is still greater than the total amount of mechanical energy that can be feasibly transferred from all the high-pressure streams (i.e., $\sum_{i=1}^{N_H} W_{\beta_{H_i, L_j}}$), then additional compressor(s) need to be placed.

It is possible that a high-pressure stream (H_i), after a feasible transfer of mechanical energy to one or more L streams, has an outlet pressure that is higher than its target pressure ($P^t_{H_i}$). If this is the case, then an expander is needed. Thus, vector $\boldsymbol{W_E}$ is introduced, which has the following structure:

$$\boldsymbol{W_E} = \begin{pmatrix} W_{E_1} \\ W_{E_2} \\ \vdots \\ W_{E_{N_H}} \end{pmatrix} \tag{10}$$

where

$$W_{E_i} = \begin{cases} 0; & W_{\gamma_i} \le 0 \\ W_{\gamma_i}; & W_{\gamma_i} > 0 \end{cases} \tag{11}$$

Note that a positive value of W_{γ_i} means the amount of mechanical energy left in the H_i stream is unable to be feasibly transferred to any L streams. Under this situation, an expander is needed so that the H_i stream can reduce its pressure to its target pressure ($P^t_{H_i}$).

5. Heat Integration with a WEN Design

The WEN synthesis methodology described in the last section is applicable to any design problem where stream target temperatures are not defined. However, in industrial applications, there are requirements for the temperatures of the streams leaving the WENs that are set by the downstream systems for their operations. In particular, Deng et al. suggested that more attention should be given to a WEN where streams are in the gas phase [3]. As such, processes with adiabatic conditions where the outlet temperatures of streams, after each stage of pressurization or depressurization, are different from the desired temperatures are of particular focus. These systems will require heating or cooling, and thus heat integration will have to be applied alongside work integration.

Heat integration is realized through synthesizing a cost-effective HEN. The HEN can be placed either before or after the WEN, as shown in Figure 4, where vectors $\boldsymbol{P^s}$ and $\boldsymbol{P^t}$ are the source and target pressures of all high-pressure and low-pressure streams, respectively, and vectors $\boldsymbol{T^s}$ and $\boldsymbol{T^t}$ are their source and target temperatures, respectively.

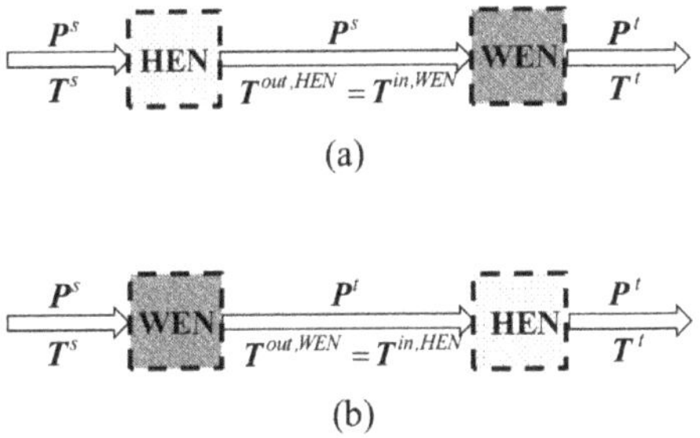

Figure 4. Options for the placement of a HEN: (**a**) HEN located before WEN and (**b**) WEN located before HEN.

Note that the pressures of the streams entering and leaving the HEN are the same, as it is commonly assumed that the pressure drop of a stream through a heat transfer unit is negligible. However, the stream temperatures in the middle of the two networks, in Case (a) and Case (b), are different from either $\boldsymbol{T}^s$ or $\boldsymbol{T}^t$. These stream temperatures are calculated as follows for each case:

Case (a)—the HEN is positioned before the WEN. As shown in Figure 4a, the inlet temperatures of each stream in the HEN are known, i.e., $T_{H_i}^{in,HEN} = T_{H_i}^s$ and $T_{L_j}^{in,HEN} = T_{L_j}^s$. In addition, the outlet temperatures and pressures of the WEN are also known, i.e., $T_{H_i}^{out,WEN} = T_{H_i}^t$, $T_{L_j}^{out,WEN} = T_{L_j}^t$, $P_{H_i}^{out,WEN} = P_{H_i}^t$, and $P_{L_j}^{out,WEN} = P_{L_j}^t$. Between the HEN and the WEN, $P_{H_i}^{out,HEN} = P_{H_i}^{in,WEN} = P_{H_i}^s$ and $P_{L_j}^{out,HEN} = P_{L_j}^{in,WEN} = P_{L_j}^s$, as it is assumed that there is no pressure drop in any heat transfer unit. The values of $T_{H_i}^{in,WEN}$ (i.e., $T_{H_i}^{out,HEN}$) and $T_{L_j}^{in,WEN}$ (i.e., $T_{L_j}^{out,HEN}$) are unknown and can be calculated based on the given stream target temperatures (i.e., T^t). These temperatures can be calculated based on each stream's target temperatures (T^t) and source and target pressures (P^s and P^t, respectively), which are also known.

As an example, the i-th stream's temperature when entering the WEN ($T_i^{in,WEN}$) can be calculated using the equation below, assuming it is under the polytropic (adiabatic frictional) condition.

$$T_i^{in,WEN} = \frac{T_i^t}{\left(\frac{P_i^t}{P_i^s}\right)^{\frac{R}{C_P}}} \tag{12}$$

Case (b)—the HEN is positioned after the WEN. As shown in Figure 4b, the only unknown parameters are the stream temperatures between the WEN and the HEN, which are $T_{H_i}^{out,WEN}$ (i.e., $T_{H_i}^{in,HEN}$) and $T_{L_j}^{out,WEN}$ (i.e., $T_{L_j}^{in,HEN}$). The outlet temperatures of the streams when leaving the WEN (i.e., $T^{out,WEN}$) can be calculated based on the given stream source temperatures (i.e., T^s). As an example, the i-th stream's temperature when leaving the WEN ($T_i^{out,WEN}$) can be calculated using the equation below, again assuming a polytropic (adiabatic frictional) condition.

$$T_i^{out,WEN} = T_i^s \left(\frac{P_H^t}{P_H^s}\right)^{\frac{R}{C_P}} \tag{13}$$

The above two formulas are applicable to both high-pressure and low-pressure streams.

After determining the streams' temperatures in the middle of the two networks in either Case (a) or Case (b), the HEN design problems are fully defined, as is the WEN design problem. Figure 5 shows a procedure for selecting the location of the HEN based on an estimation of energy recovery and TAC.

Note that in the estimation process, the prediction of the maximum recoverable mechanical energy by the WEN should be performed using the methodology developed by Amini-Rankouhi and Huang [24]. Thermal energy recovery alongside external heating and cooling can be developed by various matured methods, such as mixed-integer nonlinear programming (MILP)-based methods [27] or Pinch Analysis-based methods [28].

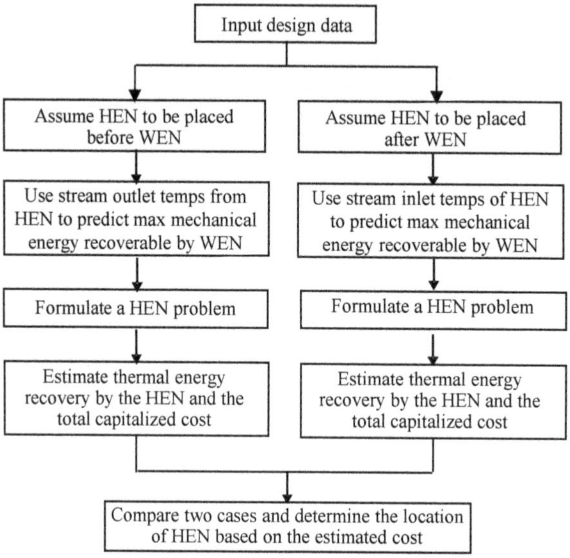

Figure 5. Flowchart for the determination of a HEN location.

6. WEN Design Modification after Heat Integration

The synthesized WEN, when connected to the derived HEN, generates a heat-integrated WEN (i.e., HIWEN) that can recover both mechanical and thermal energies at the lowest possible cost. Note that there could be further opportunities to recover more mechanical energy in the integrated network because the temperatures and flowrates of some high-pressure and low-pressure streams may be adjustable within some small ranges. Thus, some compressor(s) and expander(s) in the WEN design may be replaceable by additional WE(s). Therefore, the derived HIWEN needs to be evaluated for an improvement opportunity. The following procedure can be followed to identify such an opportunity for further improvement of the energy recovery in the system:

Step 1. Identify the compressor on the L stream that consumes the highest external power in the flowsheet.

Step 2. Identify an expander on an H stream that can provide the mechanical energy required by the identified compressor. If no expander is left, then identify the next compressor and repeat Step 2; otherwise, go to Step 4.

Step 3. Check the matching feasibility using the thermodynamic models.

(a) If feasible, then determine the maximum amount of mechanical energy recoverable by a WE, which is to be used to replace the compressor and the expander. The relevant streams' inlet and outlet pressures and temperatures should be calculated. This may result in a smaller compressor or expander and a heater or cooler to meet the process specifications. Whether this replacement is acceptable or not depends on the cost estimation. If acceptable, then modify the process.

(b) If infeasible, then identify another compressor on an L stream that consumes the next highest external power in the flowsheet and go to Step 2.

Step 4. Stop the process modification. The resultant HIWEN is the final design.

7. Case Studies

The synthesis methodology introduced in this work has been successfully used to solve a number of HIWEN design problems. In this section, two of them, selected from the open literature, are illustrated to show the robustness of the introduced methodology. For the work transfer units (piston-type WEs, compressors, and expanders), isentropic (adiabatic) conditions, reversibility, and 100% efficiency are assumed. Stream heat capacities are assumed to be constant, and pressure drops in heat transfer units (heat exchangers, heaters, and coolers) are assumed negligible.

7.1. Case 1—HIWEN Synthesis with Detailed Steps

The design problem reported by Razib et al. is selected for study [6]. In their work, single-shaft-turbine-compressor (SSTC) units are used as work exchangers. In this study, piston-type direct work exchangers are employed. This problem was analyzed in previous studies for predicting the maximum recoverable mechanical energy, and thus the prediction result in that study is adopted [24]. Table 1 lists the stream data for this synthesis problem, which has three high-pressure streams (H_1 to H_3) and two low-pressure streams (L_1 and L_2).

Table 1. Process stream data for Case 1.

Stream No.	Supply Pressure (P^s, kPa)	Target Pressure (P^t, kPa)	Flowrate (kg/s)	Source Temperature (T^s, K)	Target Temperature (T^t, K)	Heat Capacity (C_P, kJ/kg·K)
H_1	850	100	3	600	430	1.432
H_2	960	160	5	580	300	0.982
H_3	800	300	2	960	300	1.046
L_1	100	510	3	300	700	1.432
L_2	100	850	3	300	600	1.432

To implement the methodology introduced in this work, the WEN and the HEN design problems should be defined by following the steps listed in Figure 5. For the HEN, $\Delta T_{\min}$ was defined as 10 K.

Option 1—HEN placed before the WEN. Taking streams H_1 and L_1 as an example, $T_{H_1}^{in,WEN}$ and $T_{L_1}^{in,WEN}$ are evaluated as follows:

$$T_{H_1}^{in,WEN} = \frac{T_{H_1}^t}{\left(\frac{P_{H_1}^t}{P_{H_1}^s}\right)^{\frac{R}{C_P}}} = \frac{430}{\left(\frac{100}{850}\right)^{\frac{0.35}{1.43}}} = 722.96\,\text{K} \tag{14}$$

and

$$T_{L_1}^{in,WEN} = \frac{T_{L_1}^t}{\left(\frac{P_{L_1}^t}{P_{L_1}^s}\right)^{\frac{R}{C_P}}} = \frac{700}{\left(\frac{510}{100}\right)^{\frac{0.64}{1.43}}} = 337.21\,\text{K} \tag{15}$$

where C_P is the heat capacity and R is the specific gas constant given by Razib et al. [6].

Figure 6a lists all the pressure and temperature data of the process streams for the WEN and HEN synthesis problems. As shown, for the HEN problem, there are three hot streams (H_2, H_3, and L_2) and two cold streams (H_1 and L_1).

At the same time, a WEN synthesis problem is completely defined, as shown in Figure 6a. The maximum recoverable mechanical energy of this problem can be predicted using the methodology by Amini-Rankouhi and Huang [24]. The three high-pressure streams (H_1 to H_3) listed in Table 1 and Figure 6a can provide a total of 2,676 kW of mechanical energy, while the two low-pressure streams (L_1 and L_2) also in Table 1 and Figure 6a require a total of 3,132.02 kW. However, the transferable mechanical energy from the high-pressure streams to the low-pressure streams is calculated as 1,700.04 kW. This accounts for 63.5% of the total mechanical energy of the three high-pressure streams and 54.3% of the total mechanical energy required by the two low-pressure streams.

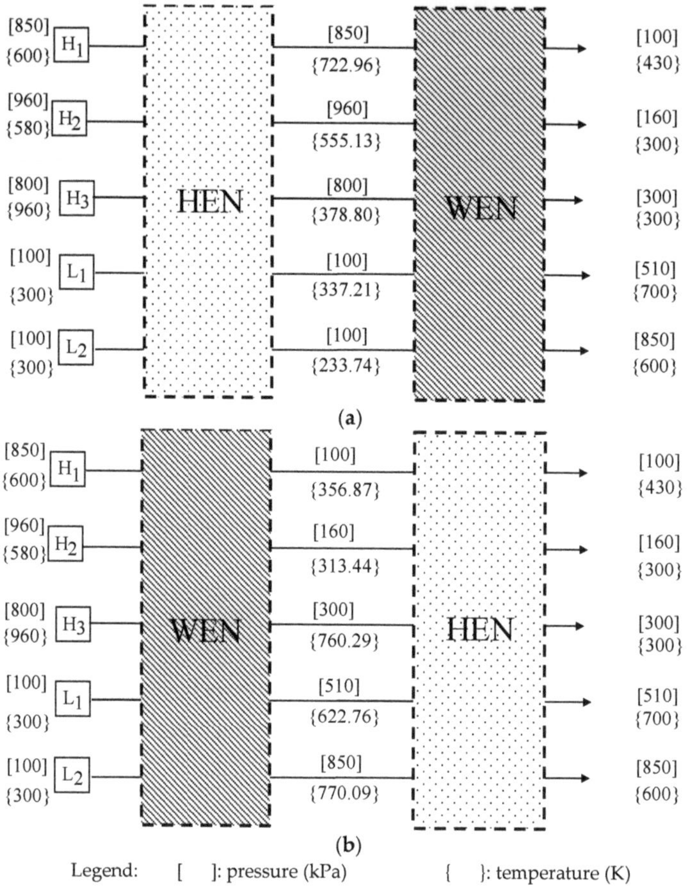

Figure 6. (**a**) HEN located before the WEN design and (**b**) HEN located after the WEN design for Case 1.

For the HEN design problem, the hot streams (H_2, H_3, and L_1) need to remove a total of 1,623 kW in different temperature ranges, while the cold streams (H_1 and L_2) require a total of 688 kW. By the Pinch Analysis method, the pinch point is located at 960 K for the hot streams and 950 K for the cold streams. This is a simple synthesis problem as all the streams are below the pinch temperature. The maximum amount of thermal energy recoverable is found to be 688 kW, and the minimum number of heat transfer units is 5. Note that a few design alternatives for the HEN design use the Pinch Analysis method. The one with the lowest TAC is chosen. It is also of note that because all the streams are below the pinch point, each derived HEN can recover the maximum amount of thermal energy while also using the minimum number of heat transfer units.

Option 2—HEN placed after the WEN. Taking streams H_1 and L_1 as an example, $T_{H_1}^{out,WEN}$ and $T_{L_1}^{out,WEN}$ are evaluated as follows:

$$T_{H_i}^{out,WEN} = T_{H_i}^{s}\left(\frac{P_{H_i}^{t}}{P_{H_i}^{s}}\right)^{\frac{R}{C_P}} = 600\left(\frac{100}{850}\right)^{\frac{0.35}{1.43}} = 355.62\ \mathrm{K} \tag{16}$$

and

$$T_{L_j}^{out,WEN} = T_{L_j}^{s}\left(\frac{P_{L_j}^{t}}{P_{L_j}^{s}}\right)^{\frac{R}{C_P}} = 300\left(\frac{510}{100}\right)^{\frac{0.64}{1.43}} = 622.01 \text{ K} \tag{17}$$

Figure 6b lists all the pressure and temperature data of the process streams for the WEN and HEN synthesis problems. As shown, for the HEN problem, there are also three hot streams (H_2, H_3, and L_2) and two cold streams (H_1 and L_1). Note that the temperature ranges of the streams in Figure 6b are very different from those in Figure 6a.

The WEN synthesis problem is also completely defined, as shown in Figure 6b. The maximum recoverable mechanical energy of this problem design can also be predicted using the methodology by Amini-Rankouhi and Huang [24]. The three high-pressure streams (H_1 to H_3) listed in Table 1 and Figure 6b can provide 2,771 kW of mechanical energy, and the two low-pressure streams (L_1 and L_2) require 3,406 kW. However, the transferable mechanical energy from the high-pressure streams to the low-pressure streams is calculated as 1,777.97 kW. This accounts for 64.15% of the total mechanical energy of the high-pressure streams and 52.2% of the total mechanical energy required by the low-pressure streams.

For the HEN design problem, the hot streams (H_2, H_3, and L_2) need to remove a total of 1,759.84 kW in different temperature ranges, while the cold streams (H_1 and L_1) require a total of 646.25 kW. By the Pinch Analysis method, the pinch point is identified as 770.09 K for the hot streams and 760.09 K for the cold streams. This makes the design simple, as all the streams are below the pinch point.

The results of both options are summarized in Table 2. If comparing the operating costs of the two designs, the HEN should be placed before the WEN. These two networks can be designed separately.

Table 2. Energy recovery and operating cost comparison of the HIWENs with a HEN placed before or after a WEN for Case 1.

Parameter	Placement of the HEN	
	Before WEN	After WEN
Mech. energy recovery by WEs (kW)	1,700.04	1,777.97
External compression power (kW)	1,431.98	1,628.13
External expansion power (kW)	976.04	993.13
Thermal energy recovery by HEs (kW)	688.07	645.95
External cooling (kW)	934.54	1,113.14
Operating cost (k$/year)	1,382	1,572
C_{elec} = 0.12 $/kWh; C_{steam} = 0.035 $/kWh; C_{CW} = 0.001 kWh; Operational time = 8000 h/yr		

WEN flowsheet development. The process stream data are provided in Table 1 and Figure 6a. The following matrices are calculated using the prediction methodology by Amini-Rankouhi and Huang, which is shown in Figure 1 [24]. For this design problem involving three high-pressure streams (H_1 to H_3) and two low-pressure streams (L_1 and L_2), matrix $\overline{\boldsymbol{P}}$ contains the identified pressure intervals between all of the pairs of high-pressure and low-pressure streams, which are as follows:

$$\overline{\boldsymbol{P}} = \begin{pmatrix} [170,\ 230] & [170,\ 230] \\ [230,\ 510] & [230,\ 850] \\ [0,\ 0] & [0,\ 0] \end{pmatrix} \tag{18}$$

Vector $\mathbf{W}_H$ contains the mechanical energy that can be provided by each of the three high-pressure streams; it is obtained as follows:

$$\mathbf{W}_H = \begin{pmatrix} 1,258.55 \\ 1,252.67 \\ 164.86 \end{pmatrix} \tag{19}$$

Vector $\mathbf{W}_L$ lists the required mechanical energy of each of the two low-pressure streams; it is calculated as follows:

$$\mathbf{W}_L = \begin{pmatrix} 1,558.56 \\ 1,573.46 \end{pmatrix} \tag{20}$$

Matrix $\mathbf{W}_\beta$ contains information about the transferrable mechanical energy between each pair of high-pressure and low-pressure streams. It is derived as follows:

$$\mathbf{W}_\beta = \begin{pmatrix} 266.69 & 180.68 \\ 902.83 & 1,128.36 \\ 0 & 0 \end{pmatrix} \tag{21}$$

Vector $\mathbf{W}_\gamma$ includes the data that show whether each individual high-pressure stream can provide mechanical energy to the low-pressure streams. It is calculated as follows:

$$\mathbf{W}_\gamma = \begin{pmatrix} 811.18 \\ -778.52 \\ 164.86 \end{pmatrix} \tag{22}$$

Determination of the placement of work exchangers (WEs). WEs are to be placed by identifying pairs of streams for matching. Matrix $\mathbf{W}_W$ shown below is designed to contain all the stream matching information.

$$\mathbf{W}_W = \begin{pmatrix} W_{W_{H_1,L_1}} & W_{W_{H_1,L_2}} \\ W_{W_{H_2,L_1}} & W_{W_{H_2,L_2}} \\ W_{W_{H_3,L_1}} & W_{W_{H_3,L_2}} \end{pmatrix} \tag{23}$$

The six elements in the matrix are derived as follows:

Since W_{γ_1} is a positive value (811.18 kW), the two elements related to high-pressure stream H_1 can be readily obtained as follows:

$$W_{W_{H_1,L_1}} = W_{\beta_{H_1,L_1}} = 266.69\,\text{kW} \tag{24}$$

$$W_{W_{H_1,L_2}} = W_{\beta_{H_1,L_2}} = 180.68\,\text{kW} \tag{25}$$

Corresponding to stream H_2, W_{γ_2} has a negative value (−778.52 kW), but its absolute value is smaller than $W_{\beta_{H_2,L_1}}$ (902.83 kW). Thus, we have the following:

$$W_{W_{H_2,L_1}} = W_{\beta_{H_2,L_1}} = 902.83\,\text{kW} \tag{26}$$

As for $W_{W_{H_2,L_2}}$, since W_{γ_2} has a negative value (−778.52 kW) and its absolute value is less than $W_{\beta_{H_2,L_2}}$ (1128.36 kW), we have the following:

$$W_{W_{H_2,L_2}} = W_{\beta_{H_2,L2}} - |W_{\gamma_2}| = 1128.36 - 778.52 = 349.84\text{ kW} \tag{27}$$

Since W_{γ_3} is a positive value (164.86 kW), we have the following:

$$W_{W_{H_3,L_1}} = W_{\beta_{H_3,L_1}} = 0\,\text{kW} \tag{28}$$

and

$$W_{W_{H_3,L_2}} = W_{\beta_{H_3,L_2}} = 0\,\text{kW} \tag{29}$$

Thus, matrix $\boldsymbol{W}_W$ can be obtained as follows:

$$\boldsymbol{W}_W = \begin{pmatrix} 266.69 & 180.68 \\ 902.83 & 349.86 \\ 0 & 0 \end{pmatrix} \tag{30}$$

As shown, four WEs need to be placed, and the workload of each exchanger is listed in the matrix.

Determination of the pressure intervals of the WEs placed. From the operational stability point of view, if two or more WEs are used to transfer mechanical energy from a high-pressure stream, it is better to split the high-pressure stream so that the WEs are operated in parallel. This strategy is used here to split both streams H_1 and H_2 that are to be matched with streams L_1 and L_2. The pressure intervals of the low-pressure streams are contained in matrix $\overline{\boldsymbol{P}}_M$ as follows:

$$\overline{\boldsymbol{P}}_M = \begin{pmatrix} \overline{P}_{M_{H_1,L_1}} & \overline{P}_{M_{H_1,L_2}} \\ \overline{P}_{M_{H_2,L_1}} & \overline{P}_{M_{H_2,L_2}} \\ \overline{P}_{M_{H_3,L_1}} & \overline{P}_{M_{H_3,L_2}} \end{pmatrix} \tag{31}$$

The derivations of $\overline{P}_{M_{H_1,L_1}}$ and $\overline{P}_{M_{H_2,L_2}}$ are shown below.

Calculation of $\overline{P}_{M_{H_1,L_1}}$. This interval is derived from $\overline{P}_{H_1,L_1}$. According to Equation (4), the upper bounds of both intervals are the same, i.e.,

$$\overline{P}^b_{M_{H_1,\,L_1}} = \overline{P}^b_{H_1,\,L_1} = 230\,\text{kPa} \tag{32}$$

The lower bound is calculated as follows:

$$\overline{P}^a_{M_{H_1,\,L_1}} = \overline{P}^b_{M_{H_1,\,L_1}} \left(1 - \frac{W_{W_{H_1,L_1}} \left(P^t_{L_1}\right)^{\frac{R}{C_{P_{L_1}}}}}{F_{L_1} C_{P_{L_1}} T^t_{L_1} \left(P^b_{M_{H_1,\,L_1}}\right)^{\frac{R}{C_{P_{L_1}}}}} \right)^{\left(\frac{C_{P_{L_1}}}{R}\right)} = 170\,\text{kPa} \tag{33}$$

Thus,

$$\overline{P}_{M_{H_1,L_1}} = \left[\overline{P}^b_{M_{H_1,L_1}}, \overline{P}^a_{M_{H_1,L_1}}\right] = [170,\ 230] \tag{34}$$

Calculation of $\overline{P}_{M_{H_2,L_2}}$. This interval is derived from $\overline{P}_{H_2,L_2}$. According to Equation (4), the upper bounds of both intervals are the same, i.e.,

$$\overline{P}^b_{M_{H_2,\,L_2}} = \overline{P}^b_{H_2,\,L_2} = 850\,\text{kPa} \tag{35}$$

The lower bound is calculated in the same way as Equation (33).

$$P^a_{M_{H_2,\,L_2}} = P^b_{M_{H_2,\,L_2}} \left(1 - \frac{W_{W_{H_2,L_2}} \left(P^t_{L_2}\right)^{\frac{R}{C_{P_{L_2}}}}}{F_{L_2} C_{P_{L_2}} T^t_{L_2} \left(P^b_{M_{H_2,\,L_2}}\right)^{\frac{R}{C_{P_{L_2}}}}} \right)^{\left(\frac{C_{P_{L_2}}}{R}\right)} = 610.40\,\text{kPa} \tag{36}$$

Therefore,

$$\overline{P}_{M_{H_1,L_1}} = [610.40,\ 850] \tag{37}$$

The other four elements in matrix $\overline{\boldsymbol{P}}_M$ can be calculated using the same approach. The complete matrix is as follows:

$$\overline{\boldsymbol{P}}_M = \begin{pmatrix} [170, 230] & [170, 230] \\ [230, 510] & [610.40, 850] \\ [0, 0] & [0, 0] \end{pmatrix} \tag{38}$$

Placement of the compressors. The amount of mechanical energy transferred from high-pressure streams H_1 and H_2 is insufficient for the low-pressure streams L_1 and L_2 to be pressurized to their target pressures. Thus, compressors are needed, which require the determination of the number of compressors, their powers, and placement locations. This task can be accomplished below.

Using Equations (6) and (7) and the data in Equations (23) and (32), vector $\boldsymbol{W}_{C_a}$ is obtained as follows:

$$\boldsymbol{W}_{C_a} = \begin{pmatrix} W_{C_{a_1}} \\ W_{C_{a_2}} \end{pmatrix} = \begin{pmatrix} (266.69 - 266.69) + (902.83 - 902.83) \\ (180.18 - 180.68) + (1,128.36 - 349.86) \end{pmatrix} = \begin{pmatrix} 0 \\ 778.52 \end{pmatrix} \tag{39}$$

As shown above, one compressor, named C_1, should be placed to pressurize stream L_2 in the pressure range of [230, 610.40], and the power is 778.52 kW. This pressure interval is found by comparing the pressure intervals from the second columns and the second rows of matrices $\overline{\boldsymbol{P}}$ in Equation (18) and $\overline{\boldsymbol{P}}_M$ in Equation (38).

Matrix $\boldsymbol{W}_{C_b}$ is derived below using Equations (8) and (9) and the data in Equations (20) and (21).

$$\boldsymbol{W}_{C_b} = \begin{pmatrix} W_{C_{b_1}} \\ W_{C_{b_2}} \end{pmatrix} = \begin{pmatrix} 1558.57 - (266.69 + 902.83) \\ 1573.46 - (180.68 + 1,128.36) \end{pmatrix} = \begin{pmatrix} 389.04 \\ 264.42 \end{pmatrix} \tag{40}$$

After comparing the pressure intervals in matrix $\overline{\boldsymbol{P}}$ in Equation (18) with the source and target pressures of each low-pressure stream from Table 1, it is determined that two compressors, named C_2 and C_3, should be placed. C_2 is used to pressurize L_1 in the pressure interval of [100, 170] with a workload of 389.04 kW, and C_3 is used for L_2 with a workload of 264.42 kW in the pressure interval of [100, 170].

Placement of the expanders. Note that the mechanical energy possessed by high-pressure stream H_3 cannot be used to pressurize any low-pressure stream at this stage. Vector $\boldsymbol{W}_E$ is derived below using Equations (10) and (11) and the data in Equation (22).

$$\boldsymbol{W}_E = \begin{pmatrix} W_{E_1} \\ W_{E_2} \\ W_{E_3} \end{pmatrix} = \begin{pmatrix} 811.18 \\ 0 \\ 164.86 \end{pmatrix} \tag{41}$$

As shown, two expanders, named E_1 and E_2, are needed. E_1 is for H_1 with a workload of 811.18 kW in the pressure interval of [100, 850], and E_2 is for H_3 with a workload of 164.86 kW in the pressure interval of [300, 800].

Summary of the placement of the work transfer units and the process flowsheet. Table 3 summarizes the information derived above, which includes four work exchangers (W_1 to W_4), three compressors (C_1 to C_3), two expanders (E_1 and E_2), and the pressure intervals of the streams and their workloads. Figure 7a gives the derived flowsheet of the WEN.

Table 3. Placement of work transfer units in the WEN for Case 1.

Work Transfer Units and Symbols		Stream(s)	Pressure Interval of the *H* Stream	Pressure Interval of the *L* Stream	Workload (kW)
Work Exchanger	W_1	H_1/L_1	[850, 100]	[170, 230]	266.69
	W_2	H_1/L_2	[850, 100]	[170, 230]	180.68
	W_3	H_2/L_1	[960, 160]	[230, 510]	902.93
	W_4	H_2/L_2	[960, 160]	[610, 850]	1,128.36
Compressor	C_1	L_2		[230, 610]	778.52
	C_2	L_1		[170, 230]	389.04
	C_3	L_2		[170, 230]	264.42
Expander	E_1	H_1	[100, 850]		811.18
	E_2	H_3	[300, 800]		164.86

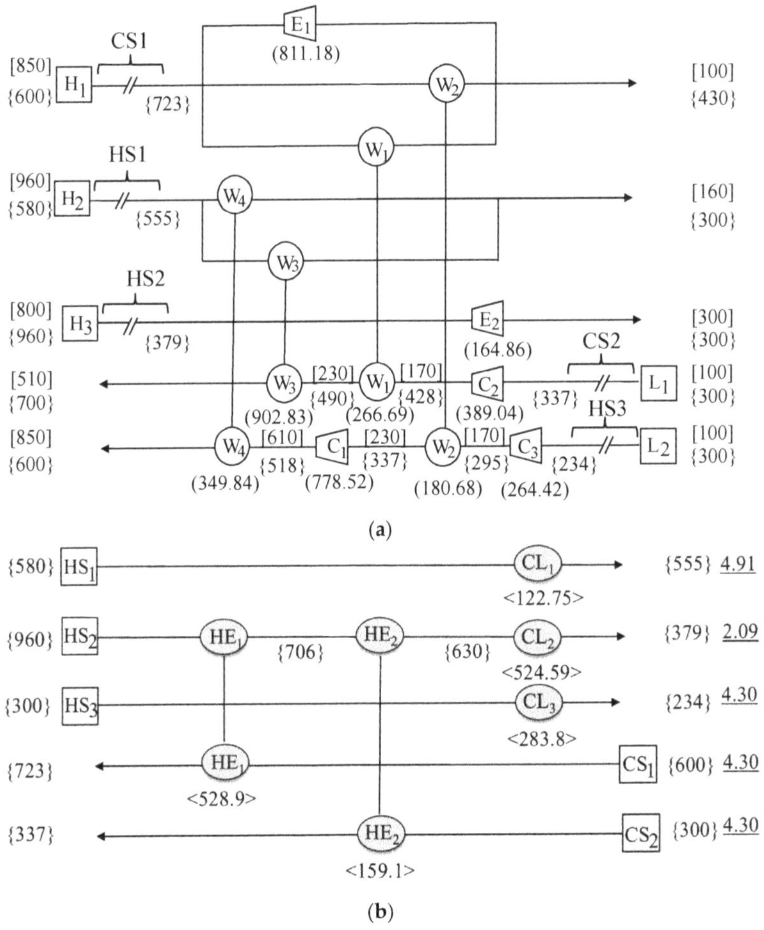

Figure 7. *Cont.*

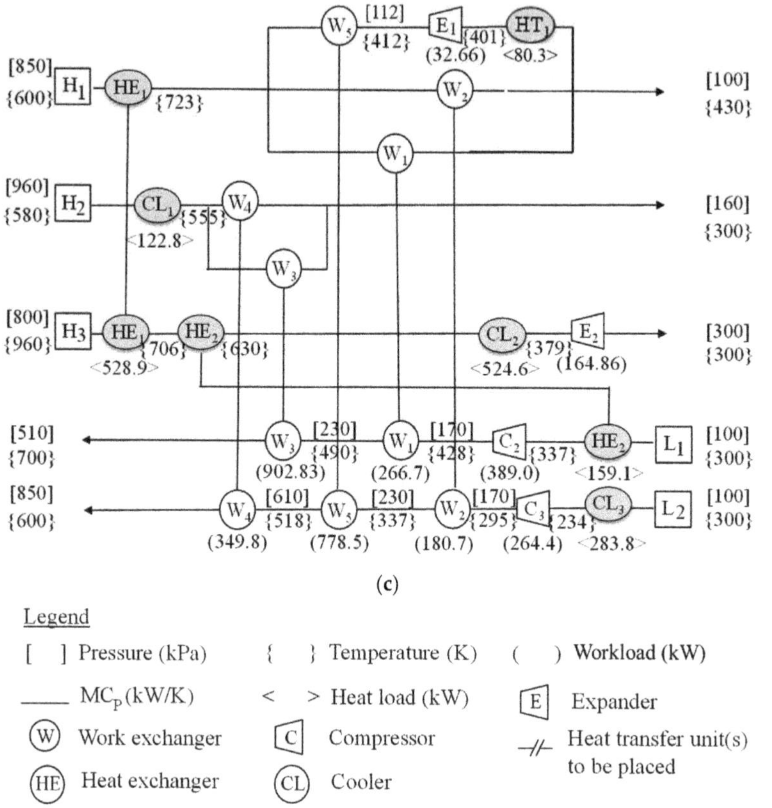

Figure 7. Flowsheet of the heat-integrated work exchange network for Case 1: (**a**) the WEN, (**b**) the HEN, and (**c**) the HIWEN.

HEN design. It has been decided that the HEN is to be placed before the WEN. Figure 6a has listed the inlet and outlet temperatures of the three hot and two cold streams. These streams are marked in Figure 7a as well. They are hot streams named HS_1, HS_2, and HS_3, which are the first part of high-pressure streams H_2 and H_3 and low-pressure stream L_2, respectively, and cold streams named CS_1 and CS_2, which are the first part of high-pressure stream H_1 and low-pressure stream L_1, respectively. This is a simple synthesis problem, as all of the streams are below the pinch point. By using the Pinch Analysis method, an optimal flowsheet is obtained, which is shown in Figure 7b, where two heat exchangers (HE_1 and HE_2) and three coolers (CL_1 to CL_3) are employed to recover the maximum amount of thermal energy of 646.25 kW.

Design improvement. The HEN in Figure 7b should be integrated into the WEN in Figure 7a to generate a HIWEN. The design should be evaluated using the four-step method in Section 5 for possible design improvement. By executing Step 1, compressor C_1 on stream L_2 is identified. As shown in vector $\mathbf{W}_{C_a}$, the workload of the compressor is 778.52 kW in the pressure interval of [230, 610]. In Step 2, expander E_1 on a branch of stream H_1 is selected, which has a workload of 811.18 kW in the pressure interval of [850, 100]. In Step 3, it is confirmed that the mechanical energy of H_1 is 778.52 kW and can be used to pressurize L_2 by adding exchanger W_5. In this way, compressor C_1 is no longer needed, and expander E_1 can be much smaller as the workload is reduced from 811.18 kW to 32.66 kW. With this design modification, the total mechanical energy recovery is improved by 45.8% (from 1,700 kW to 2,479 kW), and the external compression and expansion utility requirements are decreased, respectively, by 54% (from 1,432 kW to 653 kW) and 83% (976.04 kW to 164.86 kW). The TAC is also decreased from $2,964,000 to $2,097,000. In particular, the operating cost decreased by more than 52% from $1,382,000 to $657,000. A complete HIWEN flowsheet is shown in Figure 7c.

This HIWEN design was compared to the design by Razib et al. [6] and can be seen in Table 4. It is of note that the TAC reported in this study for the design made by Razib et al. is different from their report. This is because the method and assumptions for calculating the operating cost were changed to those used by the authors of this study in the interest of consistency. However, the capital cost calculations were not changed. Assumptions for capital cost can be found in Razib et al. [6]. As is shown in Table 4, the design by Razib et al. has a slightly lower capital cost, but the operating cost for the design proposed in this study is significantly less. In fact, the TAC decreased by more than 42% when compared to that of the optimal design proposed by Razib et al., showing clearly the efficacy of this design and the significant improvements it can make on proposed designs from the literature.

Table 4. Performance comparison of HIWENs to different methods for Case 1.

Design	This Work	Razib et al. [6]
Mech. energy exchange (kW)	2,479	1,573
Thermal energy exchange (kW)	688	-
Compression utility (kW)	653	875
Expansion utility (kW)	198	1,415
Heating utility (kW)	80	688
Cooling utility (kW)	935	1,619
No. of WEs	5 piston-type (3 Vessel) WEs	4 SSTC compressors plus 3 SSTC turbines
No. of HEs	2	-
No. of HT and CL	4	11
No. of compressors	2	2
No. of expanders/valves	2	2
CAPEX (k$/yr)	1,440	1,253
OPEX (k$/yr)	657	2,404
TAC (k$/yr)	2,097	3,657

7.2. Case 2—HIWEN Synthesis with Comprehensive Cost Comparison

The design problem studied by Onishi et al. [7] and Huang and Karimi [8] is selected for this case study, which makes a comprehensive cost comparison possible due to the data available in those studies. Both studies used single-shaft-turbine-compressor (SSTC) units, and the designs were obtained using mathematical programming techniques to achieve the lowest possible TAC. The data for the synthesis problem are summarized in Table 5.

Table 5. Process stream data for Case 2.

Stream No.	Source Pressure (P^s, kPa)	Target Pressure (P^t, kPa)	Flowrate (kg/s)	Source Temperature (T^s, K)	Target Temperature (T^t, K)	Heat Capacity (C_P, kJ/kg·K)
H_1	900	100	15	350	350	2.454
H_2	850	150	15	350	350	0.982
H_3	700	200	15	400	400	1.432
L_1	100	700	18	390	390	1.432
L_2	100	900	15	420	420	2.454

HIWEN design. The same synthesis methodology used in Case 1 is used here. The network development steps are exactly followed. Thus, the calculation details are omitted here. Figure 8 shows two design options: (a) a HEN before the WEN and (b) a HEN after the WEN.

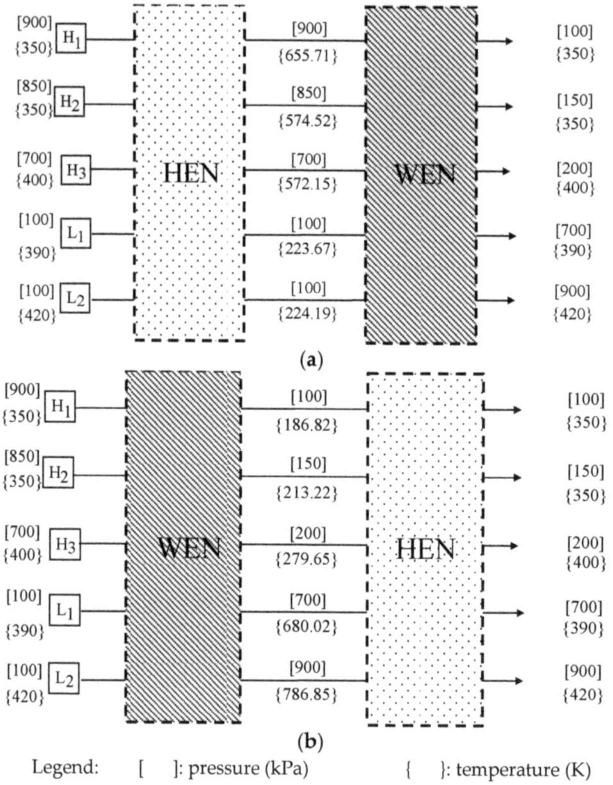

Figure 8. (**a**) HEN located before the WEN design and (**b**) HEN located after the WEN design for Case 2.

A more detailed study of their energy recovery capabilities and annual operating costs is listed in Table 6.

Table 6. Energy recovery and operation cost comparison based on the HEN location for Case 2.

Statistic	Placement of the HEN	
	Before WEN	After WEN
Mech. Energy Recovery by WEs (kW)	8,846.94	6,006.59
External Compressor Energy (kW)	2,648.29	14,972.56
External Expander Energy (kW)	9,410.97	4,599.90
Thermal Energy Recovery by HEs (kW)	2,908.20	10,606.54
External Heater Energy (kW)	15,349.71	-
External Cooler Energy (kW)	8,587.03	10,372.76
OPEX (k$/year)	6,909	14,457
C_{elec} = 0.12 $/kWh; C_{steam} = 0.035 $/kWh; C_{CW} = 0.001 kWh; Operational time = 8000 h/yr		

It is shown that the HIWEN design where a HEN is placed before the WEN would cost 6,909 k\$/yr, which is 48.4% of the cost for the HIWEN with a HEN placed after the WEN (14,457 k\$/yr). Note that the total number of work transfer units (5) and heat transfer units (7) for the designs are estimated to be the same, and the annualized capital costs are also very close. Thus, a HIWEN is synthesized with a HEN placed before the WEN. Figure 9a,b depict, respectively, the flowsheets of the WEN and the HEN. A final design of the HIWEN is obtained by combining the WEN and the HEN, which is shown in Figure 9c.

Cost comparison with known designs. The TAC, including the annualized capital cost (CAPEX) and the operating cost (OPEX), is used to compare the design in Figure 9c with the designs by Onishi et al. [7] and Huang and Karimi [8]. To ensure a fair comparison, the same assumptions are used in cost calculation, such as formulation, equipment cost coefficient, fixed cost, and utility cost. Their studies used SSTC units for mechanical energy recovery. The capital cost of external expanders ($CAPEX_E$) is computed as follows:

$$CAPEX_E = FC_E^U + FC_E(F) \tag{42}$$

where FC_E^U is the external expander fixed cost and is assumed to be 200 k\$/year; FC_E is the cost coefficient and is equal to 1 k\$/year; and F is the flowrate of stream flows through the expander. The capital cost of external compressors ($CAPEX_C$) is computed using the following formula:

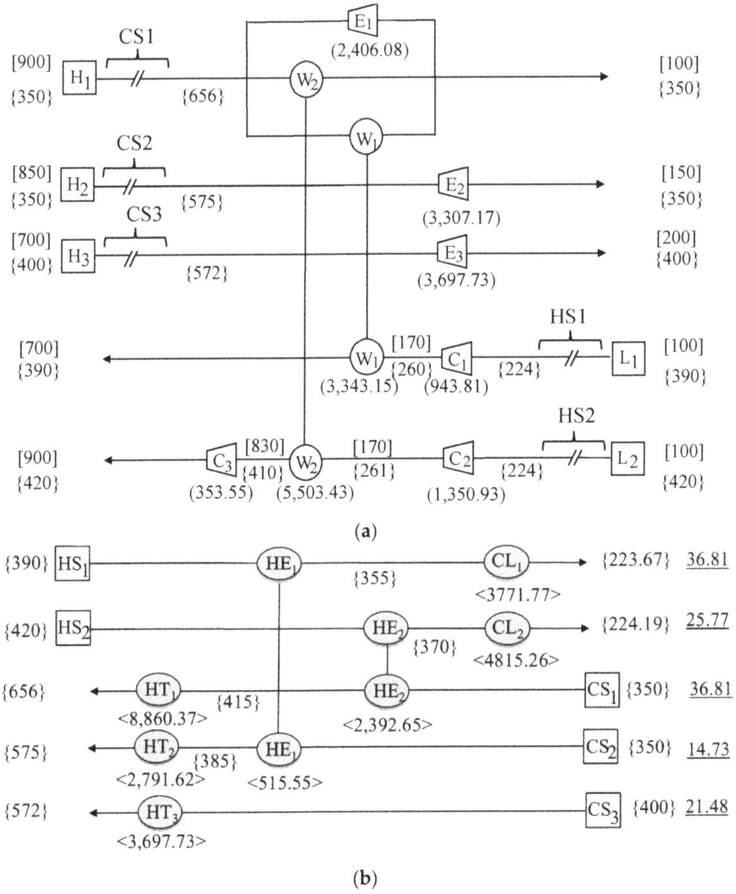

Figure 9. *Cont.*

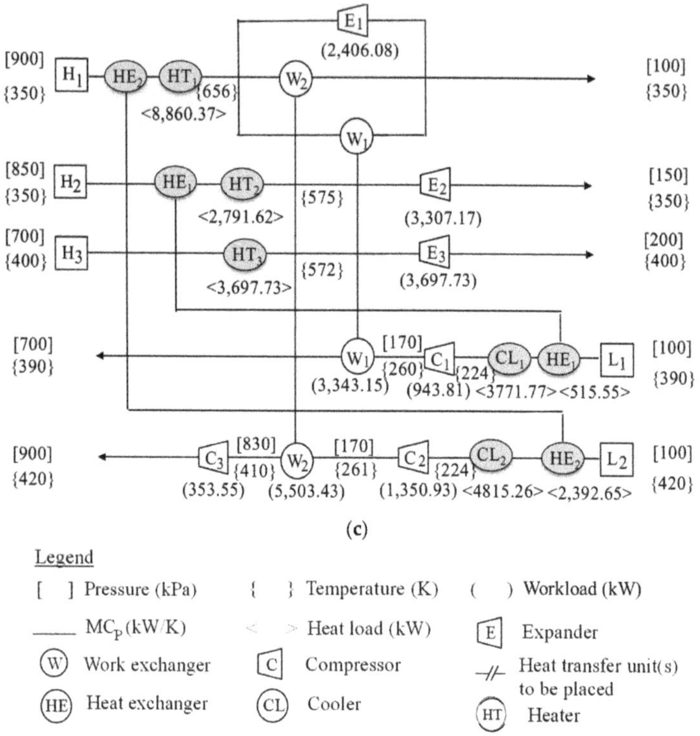

Figure 9. Flowsheet of the heat-integrated work exchange network for Case 2: (**a**) the WEN, (**b**) the HEN, and (**c**) the HIWEN.

$$CAPEX_C = FC_C^U + FC_C(F) \tag{43}$$

where FC_C^U is the external compressor fixed cost and is assumed to be 250 k\$/year; FC_C is the cost coefficient and is equal to 1 k\$/year; and F is the flowrate of stream flows through the compressor. The capital costs of heat exchangers, heaters, and coolers ($CAPEX_{HE}$) are computed as follows:

$$CAPEX_{HE} = FC_H + C(A)^{\delta} \tag{44}$$

where FC_H is the heat exchanger fixed cost and is assumed to be 3 k\$/year; C is the cost coefficient and is equal to 0.03 k\$/year; A is the heat exchanger area; and δ is the exponent for the area cost of the HE and is equal to 1.

The capital cost of the work exchanger ($CAPEX_{WE}$) is computed as follows:

$$CAPEX_{WE} = \alpha S^{\beta} \tag{45}$$

where S is the volume of one vessel and cost parameters assuming stainless steel, the pressure tolerance of 1034 MPa for a vessel and 5 MPa for valves are 995.78 and 0.36 for α and β, respectively. Amini-Rankouhi and Huang provide a chart for the cost estimation of a WE based on the volume of the displacement vessel made of stainless steel, working in different conditions with a maximum pressure tolerance of the vessels and valves [10].

The operating cost is estimated using the following formula:

$$OPEX = C_{elec}\left(\sum_{j=1}^{N_L}\left(W_{C_{1_j}} + W_{C_{2_j}}\right)\right) + C_{steam}\left(\sum_{i=1}^{N_{HT}} Q_m\right) + C_{CW}\left(\sum_{i=1}^{N_{CL}} Q_n\right) \tag{46}$$

where C_{elec} is the cost of electricity (\$/kWh); C_{steam} is the cost of steam used as a heating utility (\$/kWh); C_{CW} is the cost of cooling water used as a cooling utility; Q_m and Q_n represent the heat duty of the heating and cooling utilities, respectively; N_{HT} is the total number of heaters; and N_{CL} is the total number of coolers.

The utility cost is listed in Table 6. The cost estimation of all of the units for Case 2 shown in Figure 9c is summarized in Table 7, which gives a TAC of \$9,666,994.

Table 7. CAPEX and OPEX of units in Case 2.

Process Units		Size Factor	CAPEX (\$/Year)	OPEX (\$/Year)
Heat exchanger	HE_1	A = 9570 m^2	290,118	-
	HE_2	A = 952 m^2	31,554	-
Heater	HT_1	A = 734 m^2	118,226	2,480,904
	HT_2	A = 182 m^2	76,816	781,654
	HT_3	A = 245 m^2	25,021	1,035,364
Cooler	CL_1	A = 3841 m^2	8469	38,522
	CL_2	A = 2461 m^2	10,351	30,174
Compressor	C_1	F = 25.77 kW/k	275,776	906,058
	C_2	F = 36.81 kW/k	286,810	339,408
	C_3	F = 36.81 kW/k	286,810	1,296,893
Expander	E_1	F = 7.87 kW/k	207,871	-
	E_2	F = 14.73 kW/k	214,730	-
	E_3	F = 21.48 kW/k	221,480	-
Work exchanger	W_1	S = 20 L (10 vessels)	351,993	-
	W_2	S = 20 L (10 vessels)	351,993	-
	Total		2,758,018	6,908,976

Table 8 provides a detailed design comparison of this HIWEN system with those by Onishi et al. [7] and Huang and Karimi [8], whose TACs are, respectively, 5.67% and 8.64% higher.

Table 8. Performance comparison of HIWENs to different methods for Case 2.

Design	This Work	Onishi et al. [7]	Huang and Karimi [8]
Mech. energy exchange (kW)	8,847	10,474	11,579
Thermal energy exchange (kW)	2,908	8,794	15,920
Compression utility (kW)	2,648	8,840	7,734
Expansion utility (kW)	9,411	-	-
Heating utility (kW)	15,349	1,680	5,276
Cooling utility (kW)	8,587	10,520	13,010
No. of WEs	2 piston-type WEs	3 SSTC compressors plus 3 SSTC turbines	3 SSTC compressors plus 3 SSTC turbines
No. of HEs	2	8	6
No. of HT and CL	5	5	9
No. of compressors	3	2	1
No. of expanders/valves	3	1	-
CAPEX (k\$/yr)	2,758	-	1,180
OPEX (k\$/yr)	6,909	-	9,006
TAC (k\$/yr)	9,667	10,502	10,187

For the capital cost estimation of the piston-type WEs in our design, the cost formula by Amini-Rankouhi and Huang is used [10].

In general, the design in Figure 9c is less complicated and easier to implement than the known designs. Note that due to pressure and temperature correlations, the assumptions made regarding the heat integration made an impact on the total amount of energy that can be provided by the high-pressure streams for pressurizing low-pressure streams. In this case, the total mechanical energy recovery is not necessarily the best parameter for comparison. The capital cost for our design is higher than the design by Huang and Karimi; this can be explained as the result of more compressors and expanders, in addition to the cost of work exchangers [8]. The operating cost of our design is significantly lower than the design by Huang and Karimi because of lower utility consumption [8].

8. Concluding Remarks

In this paper, a thermodynamic-model-based synthesis methodology for designing cost-effective heat-integrated work exchanger networks (HIWENs) that can recover significant amounts of mechanical and thermal energies was introduced. Using this methodology, the maximum amount of mechanical energy recoverable by piston-type work exchangers estimated by the methodology by Amini-Rankouhi and Huang can be fully implemented into the network design, and heat integration can ensure that all the process stream specifications regarding temperature can be achieved [24]. The synthesis methodology is systematic and general. The two case studies have demonstrated methodological efficacy. Each of these studies created simple HIWEN designs, which lowered the TAC compared to previous works. The first study decreased the TAC by 42% due to the addition of a HEN and a decrease in compression utility. The second case study, while having only a slightly lower TAC than previous studies, significantly improved the operating costs, lowering them very significantly.

Future work will be in two directions. First, the current method for WEN design modification after heat integration is basically an exhaustive search approach for design modification. In fact, the authors of this study are fully aware that even after the implementation of the design modification to produce the final HIWEN, the approach used in this study does not guarantee the global optimality of the solution. The simultaneous synthesis of a hybrid network containing work transfer units (work exchangers, compressors, and expanders) and heat transfer units (heat exchangers, heaters, and coolers) would produce better results for the TAC than the current approach, especially when the minimum TAC is targeted. However, such a derived hybrid network may be more difficult to control, especially as the controls of the intermediate and final target pressures and temperatures of each stream have close interaction. The current approach is more straightforward, and the resultant networks are relatively easier to control. Thus, the authors of this study intend to develop a simultaneous synthesis approach that minimizes the TAC. Second, the current methodology will be applied to a number of real-world problems, such as ammonia manufacturing, hydrogen manufacturing, and LNG production in the gas processing industry.

Author Contributions: Conceptualization: Y.H.; Data curation: A.A.-R. and A.S.; Formal analysis: Y.H., A.A.-R., and A.S.; Methodology: Y.H. and A.A.-R.; Project administration and supervision: Y.H.; Resources: Y.H., A.A.-R., and A.S.; Software: Y.H., A.A.-R., and A.S.; Validation: Y.H., A.A.-R., and A.S.; Visualization: Y.H., A.A.-R., and A.S.; Writing—original draft: A.A.-R. and A.S.; Writing—review and editing: Y.H. and A.S. All authors have read and agreed to the published version of the manuscript.

Funding: This work is supported in part by the U.S. National Science Foundation (No. 2031385 and 2348993).

Data Availability Statement: Available upon request.

Conflicts of Interest: The authors declare no conflicts of interest.

Nomenclature

N_H	the number of high-pressure streams
N_L	the number of low-pressure streams
$\overline{\boldsymbol{P}}$	the matrix containing all of the identified pressure intervals between all of the pairs of high-pressure and low-pressure streams
$P^s_{H_i}$	the source pressure of the high-pressure stream
$P^t_{H_i}$	the target pressure of the high-pressure stream
$P^s_{L_j}$	the source pressure of the low-pressure stream
$P^t_{L_j}$	the target pressure of the low-pressure stream
$\overline{\boldsymbol{P}}_{\boldsymbol{M}}$	the matrix containing the pressure intervals of the work exchangers
$P^a_{M_{H_i,L_j}}$	the lower bound of the pressure exchange interval for the low-pressure stream
$P^b_{M_{H_i,L_j}}$	the upper bound of the pressure exchange interval for the low-pressure stream
$\boldsymbol{P}^s$	the vector of the streams' source pressures
$\boldsymbol{P}^t$	the vector of the streams' target pressures
$T^{in,WEN}$	the vector of the streams' inlet temperatures of a work exchanger network
$T^{out,WEN}$	the vector of the streams' outlet temperatures of a work exchanger network
$\boldsymbol{T}^s$	the vector of the streams' source temperatures
$\boldsymbol{T}^t$	the vector of the streams' target temperatures
$\boldsymbol{W}_C$	the vector containing the compression needed from external sources
$\boldsymbol{W}_E$	the vector containing the expansion needed from external sources
$\boldsymbol{W}_H$	the vector of the mechanical energy that can be provided by high-pressure streams
$\boldsymbol{W}_L$	the vector of the mechanical energy required by low-pressure streams
W^{tot}_R	the maximum amount of recoverable mechanical energy
$\boldsymbol{W}_W$	the matrix containing the workloads of the work exchangers
$\boldsymbol{W}_\beta$	the matrix storing all of the calculated energy transfer data
$\boldsymbol{W}_\gamma$	the vector containing information about whether each individual high-pressure stream can provide sufficient mechanical energy to low-pressure streams
$\Delta P_{\min}$	the minimum driving pressure differential
$\Delta T_{\min}$	the minimum driving temperature differential

References

1. Energetics Inc. *Manufacturing Energy and Carbon Footprint. Prepared for the U.S. Department of Energy, Advanced Manufacturing Office*; Energetics Inc.: Washington, DC, USA, 2014.
2. Huang, Y.L.; Fan, L.T. Analysis of a work exchanger network. *Ind. Eng. Chem. Res.* **1996**, *35*, 3528–3538. [CrossRef]
3. Deng, J.Q.; Shi, J.Q.; Zhang, Z.X.; Feng, X.A. Thermodynamic analysis on work transfer process of two gas streams. *Ind. Eng. Chem. Res.* **2010**, *49*, 12496–12502. [CrossRef]
4. Chen, H.; Feng, X. Graphical approach for targeting work exchange networks. *World Acad. Sci. Enginerring Technol.* **2012**, *6*, 977–981.
5. Liu, G.L.; Zhou, H.; Shen, R.J.; Feng, X. A graphical method for integrating work exchange network. *Appl. Energy* **2014**, *114*, 588–599. [CrossRef]
6. Razib, M.S.; Hasan, M.M.F.; Karimi, I.A. Preliminary synthesis of work exchange networks. *Comput. Chem. Eng.* **2012**, *37*, 262–277. [CrossRef]
7. Onishi, V.C.; Ravagnani, M.; Caballero, J.A. Simultaneous synthesis of work exchange networks with heat integration. *Chem. Eng. Sci.* **2014**, *112*, 87–107. [CrossRef]
8. Huang, K.; Karimi, I. Work-heat exchanger network synthesis (WHENS). *Energy* **2016**, *113*, 1006–1017. [CrossRef]
9. Du, J.; Zhuang, Y.; Liu, L.L.; Li, J.L.; Fan, J.; Meng, Q.W. Synthesis of indirect work exchanger network based on transhipment model. *Comput. Aided Chem. Eng.* **2015**, *37*, 1139–1144.
10. Amini-Rankouhi, A.; Huang, Y. Mechanical energy recovery through work exchanger network integration: Challenges and opportunities. *Comput. Aided Chem. Eng.* **2018**, *44*, 409–414.
11. Deng, J.Q.; Cao, Z.; Zhang, D.B.; Feng, X. Integration of energy recovery network including recycling residual pressure energy with pinch technology. *Chin. J. Chem. Eng.* **2017**, *25*, 453–462. [CrossRef]
12. Fu, C.; Gundersen, T. Heat, and work integration: Fundamental insights and applications to carbon dioxide capture processes. *Energy Convers. Manag.* **2016**, *121*, 36–48. [CrossRef]
13. Yu, H.; Gendersen, T. Review of work exchange networks (WENs) and work and heat exchange networks (WHENs). *Chem. Eng. Trans.* **2017**, *61*, 1345–1356.
14. Yu, H.; Fu, C.; Vikse, M.; He, C.; Gundersen, T. Identifying optimal thermodynamic paths in work and heat exchange network synthesis. *AIChE J.* **2018**, *65*, 549–561. [CrossRef]

15. Yu, H.; Vikse, M.; Anantharaman, R.; Gundersen, T. Model reformulations for work and heat exchange network (WHEN) synthesis problems. *Comput. Chem. Eng.* **2019**, *125*, 89–97. [CrossRef]
16. Hamedi, H.; Karimi, I.; Gundersen, T. Simulation-based approach for integrating work within heat exchange networks for sub-ambient processes. *Energy Convers. Manag.* **2020**, *203*, 112276. [CrossRef]
17. Huang, Y.; Zhuang, Y.; Zhuang, L.; Du, J. Synthesis of Work-integrated Heat Exchanger Networks Coupled with Organic Rankine Cycles. *Chem. Eng. Trans.* **2023**, *103*, 139–144.
18. Pavão, L.; Costa, C.; Ravagnani, M. A new framework for work and heat exchange network synthesis and optimization. *Energy Convers. Manag.* **2019**, *183*, 617–632. [CrossRef]
19. Pavão, L.; Mirandaa, C.B.; Caballerob, J.A.; Ravagnani, M.; Costa, C. Multiperiod work and heat integration. *Energy Convers. Manag.* **2021**, *227*, 113587. [CrossRef]
20. Santos, L.; Costa, C.; Caballero, J.; Ravagnani, M. Synthesis and optimization of work and heat exchange networks using an MINLP model with a reduced number of decision variables. *Appl. Energy* **2020**, *262*, 114441. [CrossRef]
21. Santos, L.; Costa, C.; Caballero, J.; Ravagnani, M. MINLP model for work and heat exchange networks synthesis considering unclassified streams. *Comput. Aided Chem. Eng.* **2022**, *51*, 793–798.
22. Zhuang, Y.; Zhang, L.; Liu, L.; Du, J.; Shen, S. An upgraded superstructure-based model for simultaneous synthesis of direct work and heat exchanger networks. *Chem. Eng. Res. Des.* **2020**, *159*, 377–394. [CrossRef]
23. Ibric, N.; Fu, C.; Gundersen, T. Simultaneous optimization of work and heat exchange networks. *Energies* **2024**, *17*, 1753 [CrossRef]
24. Amini-Rankouhi, A.; Huang, Y. Prediction of maximum recoverable mechanical energy via work integration: A Thermodynamic Modeling and Analysis Approach. *AIChE J.* **2017**, *63*, 4814–4826. [CrossRef]
25. Amini-Rankouhi, A.; Huang, Y. Modeling and simulation of a piston-type work exchanger for mechanical energy recovery. *Ind. Eng. Chem. Res.* **2019**, *58*, 18292–18300. [CrossRef]
26. Walas, S.M. Chemical Process Equipment—Selection and Design. In *Butterman-Heinemam Series*; Gulf Professional Publishing: Houston, TX, USA, 1990.
27. Yee, T.; Grossman, I. Simultaneous optimization models for heat integration—II. Heat exchanger network synthesis. *Comput Chem. Eng.* **1990**, *14*, 1165–1184. [CrossRef]
28. Linnhoff, B. Pinch analysis—A state-of-the-art overview. *Chem. Eng. Res. Des.* **1993**, *71*, 503–522.

processes

Article

Fault Intelligent Diagnosis for Distribution Box in Hot Rolling Based on Depthwise Separable Convolution and Bi-LSTM

Yonglin Guo [1], Di Zhou [1,*], Huimin Chen [1], Xiaoli Yue [1] and Yuyu Cheng [2]

[1] School of Mechanical Engineering, Donghua University, Shanghai 200051, China; 2231083@mail.dhu.edu.cn (Y.G.); ch_huimin@dhu.edu.cn (H.C.); xlyue@dhu.edu.cn (X.Y.)
[2] School of Information Science and Technology, Donghua University, Shanghai 200051, China; 210910401@mail.dhu.edu.cn
* Correspondence: zhoudi@dhu.edu.cn

Abstract: The finishing mill is a critical link in the hot rolling process, influencing the final product's quality, and even economic efficiency. The distribution box of the finishing mill plays a vital role in power transmission and distribution. However, harsh operating conditions can frequently lead to distribution box damage and even failure. To diagnose faults in the distribution box promptly, a fault diagnosis network model is constructed in this paper. This model combines depthwise separable convolution and Bi-LSTM. Depthwise separable convolution and Bi-LSTM can extract both spatial and temporal features from signals. This structure enables comprehensive feature extraction and fully utilizes signal information. To verify the diagnostic capability of the model, five types of data are collected and used: the pitting of tooth flank, flat-headed sleeve tooth crack, gear surface crack, gear tooth surface spalling, and normal conditions. The model achieves an accuracy of 97.46% and incorporates a lightweight design, which enhances computational efficiency. Furthermore, the model maintains approximately 90% accuracy under three noise conditions. Based on these results, the proposed model can effectively diagnose faults in the distribution box, and reduce downtime in engineering.

Keywords: fault diagnosis; finishing mill; distribution box; depthwise separable convolution; Bi-LSTM

1. Introduction

As an important component of the hot rolling process, the finishing mill performs a key operation in the control of the final quality of products [1]. Considering the harsh external environmental conditions and the complexity of internal mechanisms, finishing mills inevitably experience performance degradation and failure [2]. In a finishing mill, the distribution box functions like the power transmission and distribution in the rolling system. The variable speed and load conditions can also result in a higher probability of failure compared to other mechanisms in the finishing mill [3]. The inability to diagnose faults in a timely manner may increase downtime, and this downtime will cause potential economic losses. To mitigate the economic impact of downtime, it is essential to develop effective fault diagnosis methods for distribution boxes. The difficulties in fault diagnosis for distribution boxes stem from two primary issues. First, the hot rolling process is characterized by lengthy procedures and interconnected structures, introducing numerous uncertainties and nonlinearities [4]. Second, the complex working environment of the distribution box makes fault pulses easily masked by ambient noise. Consequently, traditional physical models and machine learning (ML) methods may fall short of addressing these challenges effectively. Specifically, the manual feature extraction process and the limited depth of models restrict their ability to accurately classify faults in noisy environments.

With the enhancement of data acquisition and processing capabilities, big data approaches have emerged as new methods for fault diagnosis [5]. In recent years, methods based on machine learning have demonstrated promising initial results in the fault diagnosis of distribution boxes in finishing mills. Considering the environmental conditions in hot rolling operations, Yuan [6] proposes a novel approach combining multiwavelet sliding window neighboring coefficient denoising and optimal blind deconvolution techniques. Zhao [7] combined improved multivariate variational mode decomposition with multivariate composite multiscale weighted permutation entropy to extract bearing fault features. Shin [8] proposed a new diagnosis method for roll eccentricity under roll speed changes. Liu [9] constructed the architecture of a remote fault diagnosis system based on empirical mode decomposition and support vector machines for heavy mills. Zhang [10] employed a KPCA-based approach to nonlinear activation issues in a hot rolling automation system. Chen [11] suggested a customized maximal-overlap multiwavelet denoising technique for fault identification. Although the above methods have achieved promising results in the study of fault diagnosis in a rolling mill, machine learning and signal process methods still face some difficulties. In machine learning, feature extraction and fault classification are usually performed separately. In particular, the manual feature extraction process limits the performance accuracy of fault classification results. Additionally, only a limited number of layers are employed in the model, which makes it difficult to extract features in a noisy environment.

Considering the shortcomings of traditional ML, recent optimizations and innovations in neural networks have shown promising results. Hinton [12] introduced the concept of deep learning (DL). Since its introduction, DL has emerged as a beneficial technology widely applied in computer vision, natural language processing, and other fields, and has demonstrated remarkable performance. Compared to traditional ML, DL can deal with large-scale data and extract deeper features. By increasing the number of neurons in the neural network and the depth of the network, the accuracy of classification can be improved. Additionally, in the fault diagnosis of a rolling mill, DL has gradually become a research focus [13]. To enhance the feature extraction capabilities, Zhang [14] integrated the attention mechanism into both CNN and LSTM. Yu [15] fused signals from multiple sensors as inputs into the DBN, which enhanced the performance under limited datasets. Considering the coupling relationship of the signals of the finishing mill, Hou [16] built a graph transformer model for fault diagnosis. Shi [17] used time–frequency images as inputs for a dual attention-guided feature enhancement network, improving the model's focus on both temporal and spectral features. Zhao [18] presented a multisource domain adversarial graph convolutional networks framework to overcome variable working conditions. That model can achieve the fault diagnosis of a rolling mill under complex conditions. From these studies, we see that DL can mine deeper features in the input data, reduce the impact of manual feature extraction, and better address the challenges that traditional ML faces to achieve desirable feature extraction and fault recognition.

Benefiting from multilayer convolution and pooling operations, a CNN can more effectively mine deeper failure features from data [19]. Chen [20] employed group normalization in CNN to normalize the feature maps of the network, which reduces the impact of data distribution discrepancy. Dash [21] combined the bond graph technique with CNN to enhance fault diagnosis, even with a minimal amount of labeled data. Considering the calculation cost, Zhao [22] proposed an efficient and lightweight model for fault diagnosis. Zhang [23] investigated the use of spatial dropout regularization to control gradient explosion and improve network stability. Zhou [24] combined inverted residuals with a lightweight network to reduce interference from unclear and small datasets. Liu [25] created a method with depthwise separable convolutions to simultaneously extract different features from vibration signals. Wang [26] developed a lightweight CNN for fault diagnosis of bearings, which can satisfy the need for fewer parameters and storage space. In addition, considering the temporal nature of fault signals, several time series-based models were investigated for fault diagnosis. Lei [27] presented an end-to-end long short-term memory model to learn features directly from multivariate time series data. Zou [28] linked multiscale weighted entropy morphological filtering and Bi-LSTM to overcome the low degree of fault discrimination and high computational complexity. Cao [29] suggested a method employing deep bidirectional long short-term memory to address time-varying and non-stationary operating conditions. Based on the above research, CNNs excel at extracting local features from signals, while LSTMs are adept at capturing temporal information. Consequently, many scholars combined CNNs and LSTMs to enhance the feature extraction capability of the model. Huang [30] constructed a CNN-LSTM model to extract feature information and time delay information, demonstrating both the accuracy and noise immunity of the model. Zhi [31] integrated joint wavelet regional correlation threshold denoising with a CNN-LSTM model, finding that the model effectively mines hidden features after denoising. Wang [32] employed a CNN and an LSTM network to address feature nonlinearity and complex conditions in the motor drive control system. Considering the varying scales of fault features, Chen [33] proposed an MRJDCNN-LSTM model that effectively reduces the loss of essential features. Qiao [34] constructed a model that employs two different convolutions and two LSTMs, enhancing diagnostic ability under variable load and noise conditions. Liu [35] established a Siamese CNN-Bi-LSTM model to address imbalanced sample classification and varying working conditions. Several related fields also modeled the CNN-LSTM class benefit after its superior feature extraction capabilities. Ullah [36] constructed a CNN and Bi-LSTM model for real-time anomaly detection in complex surveillance scenarios. Sun [37] proposed a novel intrusion detection model based on CNN-LSTM with an attention mechanism, improving both convergence speed and prediction accuracy. Xia [38] proposed an ensemble framework that fuses convolutional bidirectional long short-term memory with multiple time windows for accurate remaining useful life prediction. Huang [39] used a transfer depthwise separable convolutional recurrent network to estimate the remaining useful life of the bearing with incomplete data. The widespread application and effective results of CNN-LSTM models demonstrate their enhanced performance in feature extraction.

Considering that engineering fault data are often limited, irreproducible, and prone to noise interference, we designed a fault diagnosis model that integrates a depthwise separable convolution block with a Bi-LSTM block. This structure allows the model to extract both spatial and temporal features separately from the input signals, maximizing the information captured from the signals. By leveraging spatial features through convolution and temporal features through Bi-LSTM, the model takes full advantage of the inherent information in the signals. The Bi-LSTM improves the ability to capture time-related dependencies more effectively than a standard LSTM. Meanwhile, depthwise separable convolution reduces the number of parameters and computational complexity, ensuring the model remains lightweight without reducing accuracy.

To address the issue of timely fault diagnosis and minimize downtime for distribution boxes, this paper introduces a novel intelligent fault diagnosis model for the distribution box. The proposed model incorporates a depthwise separable convolution block and a Bi-LSTM block, designed to separately capture spatial and temporal features from fault signals. This model is capable of accurately diagnosing four types of faults: the pitting of tooth flanks, flat-headed sleeve tooth cracks, gear surface cracks, and gear tooth surface spalling. Compared to existing diagnostic models, the proposed model demonstrates superior accuracy, enhanced noise resistance, and a lightweight design.

This paper is organized as follows: Section 2 describes the distribution box data detection condition and diagnostic process within a hot rolling process. Section 3 introduces fundamental algorithm theories and the construction of the proposed model. The performance of the proposed model and comparison results are shown in Section 4. Conclusions are presented in Section 5.

2. Fault Diagnosis for Distribution Box

2.1. Distribution Box in Rolling Mill

The hot rolling process is regarded as the key stage for ensuring the quality and performance of the final product. As depicted in Figure 1, this process generally comprises four major stages: ironmaking, the continuous casting of steel, hot rolling, and cold rolling. Among these stages, the finishing mill plays a significant role in shaping steel processing. Additionally, relevant information about the hot rolling process is provided in Table 1.

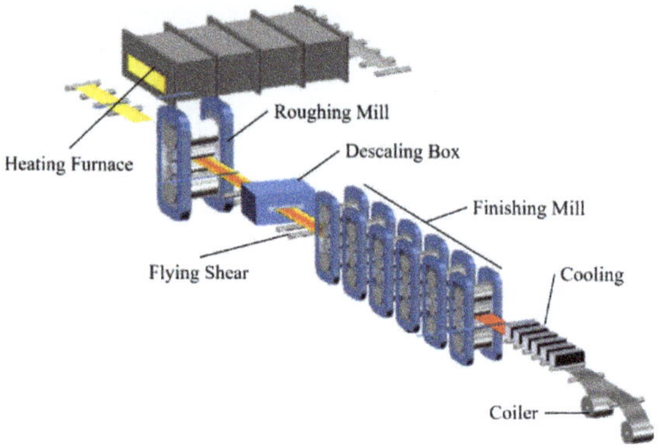

Figure 1. Hot rolling line.

Table 1. Specifications of the hot rolling.

Specifications	Parameters
Production line name	Hot rolling 1580
Capacity	3 million tons per year
Thickness range	1.2 mm to 12.5 mm
Maximum width	1580 mm
Production speed	5 m/s to 15 m/s
On year	Since 2000
Number of stands	7 finishing mills

Considering extensive and continuous production demands, the finishing mill is vital for maintaining the quality of strip steel. In the finishing process, there are seven finishing mills working and labeled F1–F7. The rolling mill F2 was chosen as the object of this study. The finishing mill column is a four-roll irreversible horizontal mill, a continuous arrangement, with the finishing mill-specific structure schematic diagram shown in Figure 2.

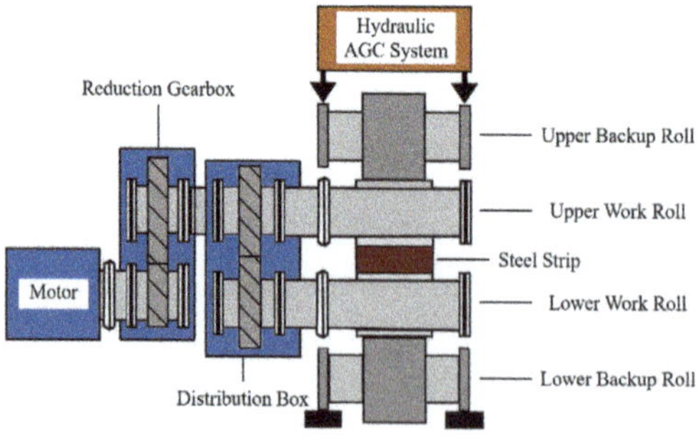

Figure 2. Finishing mill structure.

As depicted in Figure 2, a finishing mill consists primarily of seven components: a motor, a reduction gearbox, the main coupling, a distribution box, a hydraulic AGC system, upper and lower work rolls, and upper and lower support rolls. The motor drives the upper work roll rotation via the gearbox and the main coupling. The upper work roll is linked to the lower work roll through gears in the distribution box. The above mechanism enables the upper and lower work rolls to process steel plates. As a critical component in the transmission system, the distribution box is always working at variable speeds and load conditions. Therefore, it has a higher probability of failure compared to other equipment in the finishing mill.

2.2. Installation of Sensors

During hot rolling production, various sensors are used to monitor the real-time conditions of the hot rolling process to ensure safety and stability. The distribution box contains several rotating parts, such as gears, bearings, and couplings. These rotating parts generate periodic mechanical vibration pulses during operation. Acceleration vibration sensors can effectively capture the pulse features of vibration signals in the rotating component. Triaxial vibration acceleration sensors are typically installed on the shell of the rolling mill to collect operational data. The locations of the sensors are shown in Figure 3. To ensure the accuracy and completeness of the detected signals, four sensors are installed near the rotating components. The collected data are then stored in a database for subsequent studies.

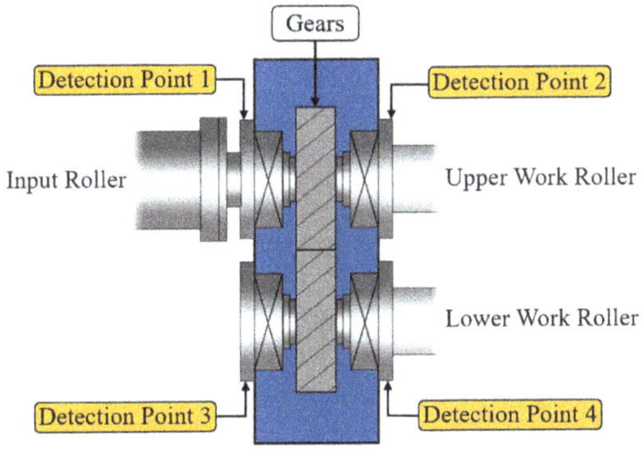

Figure 3. Sensor spatial layout.

2.3. Intelligent Fault Diagnosis for Distribution Box

With the development of intelligent technologies, digital intelligence technology has been applied to monitoring numerous types of industrial equipment. As an important part of machinery monitoring, fault diagnosis can detect faults in time, ensuring operating conditions and improving the reliability of the equipment. The diagnostic process of the distribution box is illustrated in Figure 4. When the distribution box is operational, sensors continuously collect signals, which are then stored in a database for further processing. The system applies a fault diagnosis model, depicted in the upper right section, to analyze the abnormal signals. Once a specific fault is identified, it can be quickly addressed to minimize downtime. Finally, the diagnostic results are fed back into the engineering system for further action.

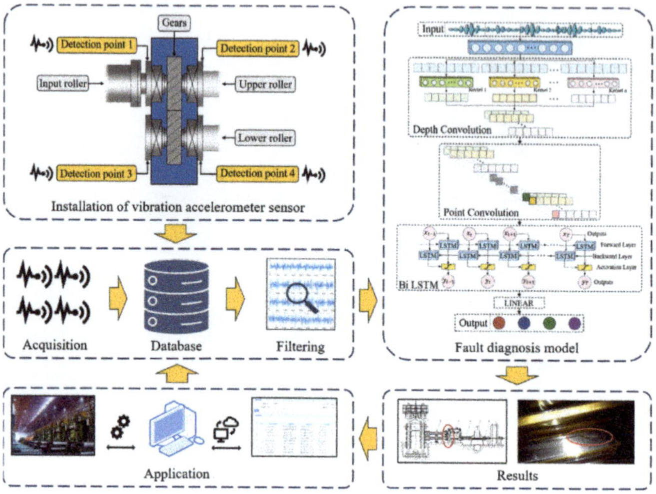

Figure 4. Flow of intelligent fault diagnosis for distribution box.

3. Framework of Proposed Model

3.1. Spatial Feature Extraction Based on Convolutional Neural Networks

CNNs offer robust spatial feature extraction through convolutional layers, where kernels slide over the input signals to capture local features [40]. A CNN model can be established from pooling layers, activation layers, linear layers, and so on [41].

The convolutional layer processes the input data using kernels to extract features and output the computed properties. The mathematical expressions of the convolution operation are shown in Equations (1) and (2):

$$x_j^l = \sum_{i \in M_j} x_i^{l-1} * \omega_{ij}^l + b_j^l \tag{1}$$

and

$$y_j = f(x_j^l) = f(\sum_{ieM_j} x_i^{l-1} * \omega_{ij}^l + b_j^l), \tag{2}$$

where x_j^l and x_i^{l-1} are the output of layer l and the input of layer l, respectively; M_j is the features set of layer $l-1$; ω_{ij}^l is the weight parameter of the convolutional kernel; b_i^l is the offset; and $f(\cdot)$ indicates the activation function.

The pooling layer reduces parameters and mitigates overfitting by downsampling data from the previous layer [42]. The mathematical expression for the maximum pooling is shown in Equation (3):

$$P_i^{l+1}(j) = \max_{(j-1)W+1 \le l \le jW} \left\{ q_i^l(t) \right\}, \tag{3}$$

where $P_i^{l+1}(j)$ represents the output value of layer $l+1$; W is the width of the pooling kernel; and $q_i^l(t)$ denotes the output of the t-th neuron in the i-th channel of the l layer.

Activation functions introduce nonlinearity, allowing the network to learn complex patterns [43]. ReLU, a rectification function, enhances generalization, and its activation function is described in the following equation:

$$\sigma_{\text{ReLU}} = \max(0, x) \tag{4}$$

Generally, the output values of ReLU are normalized via batch normalization (BN) [44]. The mathematical form of BN is shown in Equations (5) and (6):

$$\hat{x}_{ij} = \frac{x_{ij} - \mu_i}{\sqrt{\sigma_i^2 + \in}} \tag{5}$$

and

$$y_{ji} = \gamma_i \hat{x}_{ji} + \beta_i \tag{6}$$

where $\hat{x}_{ij}$ represents the normalized input feature; μ_i and σ_i^2 denote the mean and variance of the i-th feature, respectively; γ_i and β_i are learnable parameters that scale and translate the normalized features, enabling the model to represent richer features; and factor $\in$ is a small constant added to prevent division by zero.

3.2. Temporal Feature Extraction Based on Long Short-Term Memory Network

LSTM (long short-term memory), a type of RNN, excels at capturing long-term dependencies [45]. With more parameters to control retained and discarded information, LSTM effectively analyzes periodicity, trends, and patterns in time series [46].

An LSTM mainly consists of the input gate, the forget gate, and the output gate. In an LSTM, the forget gate manages memory by determining which information to retain or discard. The output of the forget gate is expressed as follows:

$$f_t = \sigma(W_f \cdot x_t + R_f \cdot h_{t-1} + b_f) \tag{7}$$

where x_t is the data value at time t; h_{t-1} is the output at time $t-1$; W_f and R_f are the weight matrices associated with f_t; b_f is the corresponding bias vector and indicates the dot product operation.

The information needs to be input and the candidate $\widetilde{C}_t$ for the state value of the unit is identified on the basis of Equations (8) and (9):

$$i_t = \sigma(W_i \cdot x_t + R_i \cdot h_{t-1} + b_i) \tag{8}$$

and

$$\widetilde{C}_t = \varphi(W_c \cdot x_t + R_c \cdot h_{t-1} + b_c) \tag{9}$$

where W_i and R_i are the weight matrices associated with it; b_i is the corresponding bias vector; W_i and Rc are the weight matrices associated with Ct; and b_c is the corresponding bias vector.

By integrating f_t, i_t, $\widetilde{C}_{t-1}$ and the state value C_{t-1} of the unit at time $t-1$, the state value C_t of the unit at time t can be obtained via Equation (10):

$$C_t = f_t \times C_{t-1} + i_t \times \widetilde{C}_t \tag{10}$$

where factor $\times$ is the element-wise multiplication of vectors.

Finally, adding state value C_t, the input x_t of the unit, and the output h_{t-1} of the previous unit can produce the final output h_t of the unit, as shown in Equations (11) and (12):

$$O_t = \sigma(W_o \cdot x_t + R_o \cdot h_{t-1} + b_o) \tag{11}$$

and

$$h_t = O_t * \frac{e^{C_t} - e^{C_t}}{e^{C_t} + e^{C_t}} \tag{12}$$

3.3. Fault Diagnosis Model Based on Spatiotemporal Feature Extraction

Considering the irreproducibility of fault signals in distribution boxes, a model combining depthwise separable convolution with Bi-LSTM (bidirectional long short-term memory) was established to utilize the fault signals fully. Compared to a convolution model, depthwise separable convolution can capture local features in each channel. It can help prevent subtle fault features from being overlooked. Additionally, incorporating Bi-LSTM can capture the temporal features of signals. A Bi-LSTM consists of two independent LSTM networks, which can be divided so that one processes the sequence forward, and the other processes it in reverse. Compared to LSTM, a Bi-LSTM can extract features more comprehensively. The framework of the proposed model is shown in Figure 5. The main components are the input layer, the 1D convolution layer, the depthwise separable convolution, the Bi-LSTM, the linear layer, and the output layer.

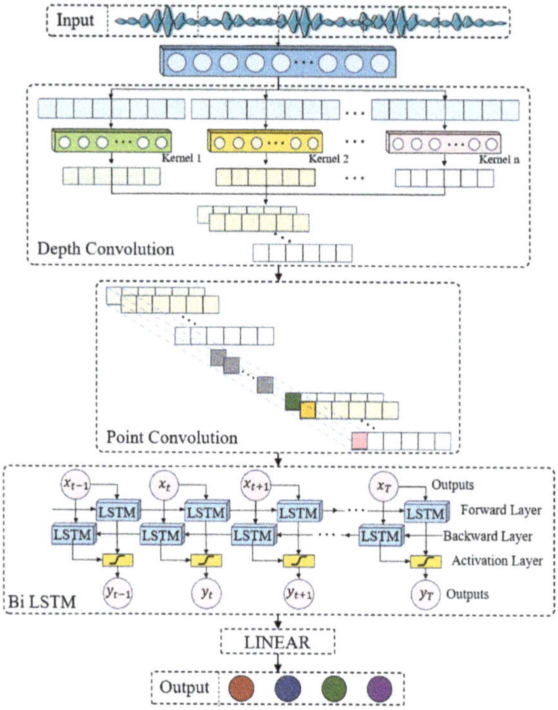

Figure 5. Structure of the proposed model.

As input data for the proposed model, the raw signals need to be divided into standardized samples. These divided signals are then passed through the 1D convolution layer to generate several feature maps, which are subsequently fed into the depthwise separable convolution.

Depthwise separable convolution consists of deep convolution and point convolution. During deep convolution, each feature map is processed by an independent kernel and can capture the local feature individually. The point convolution integrates information across channels to achieve spatial interaction.

After the depthwise separable convolution, the raw signals are transformed into feature maps with spatial features. To analyze the temporal information from these feature maps, a Bi-LSTM is constructed to extract temporal features. The Bi-LSTM contains the forward LSTM layer and the backward LSTM layer. The forward LSTM deals with data from the start to the end of a time series, while the backward LSTM processes data from the end to the beginning.

Finally, the final feature maps are fed into a linear layer. The linear layer transforms the input feature maps into a one-dimensional vector, which is then used for the final classification tasks.

4. Model Performance Analysis of Fault Diagnosis

4.1. Dataset Introduction

The distribution box we investigated had a working component rotational speed of approximately 600 rpm, and a response frequency of sensor at 1250 Hz. From December 2022 to April 2023, engineers monitored and assessed the vibration signals of four different faults. These signals were collected from sensors installed in Detection Point 3, as shown in Figure 3, ensuring that they accurately reflected real-world operating conditions. The four fault conditions included the pitting of tooth flanks, flat-headed sleeve tooth cracks, gear surface cracks, and gear tooth surface spalling. These are common faults in distribution boxes, and accurately diagnosing them is the primary objective of this study. Additionally, the signal of the normal conditions was recorded. In the latter part of the experiment, faults were replaced with their corresponding labels. The signals collected for the four faults and their normal conditions were organized and segmented. There were 322 samples for each class, with each sample containing 512 data points. The ratio of the training set to the test set for each fault was 7:3. The final segmentation results are shown in Table 2.

Table 2. Description of sample distribution.

Fault Location	Label	Sample Number	Training/Testing
Pitting of tooth flanks	A	322	225/97
Flat-headed sleeve tooth crack	B	322	225/97
Gear surface crack	C	322	225/97
Gear tooth surface spalling	D	322	225/97
Normal conditions	E	322	225/97

Considering that a static 7:3 split can be a simplistic evaluation setup, we implemented leave-P-out cross-validation for a more rigorous performance assessment. In our experiments, we conducted five iterations, each time removing a different subset of 50 samples from each class, leaving 272 samples for training and 82 for validation. The final accuracy of our proposed model, as well as the comparison models, are reported as the average accuracy obtained from cross-validation results.

The time domain signals are shown in Figure 6, illustrating the variations in signal amplitude over time.

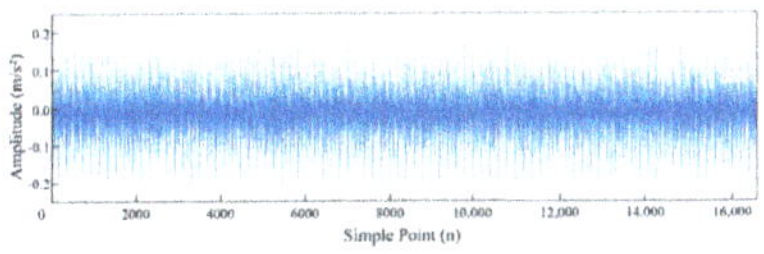

(**a**) pitting of tooth flanks

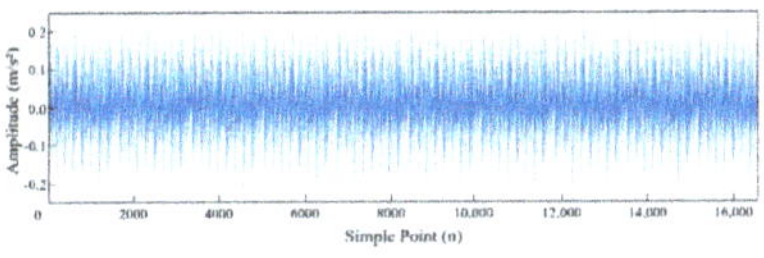

(**b**) flat-headed sleeve tooth crack

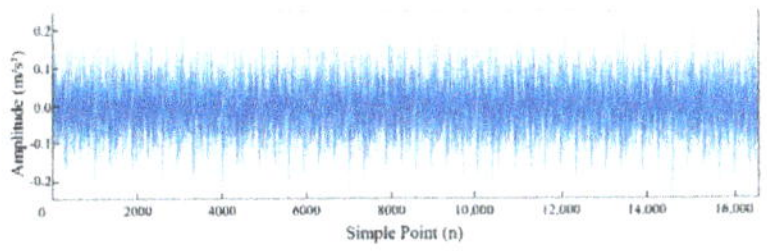

(**c**) gear surface crack

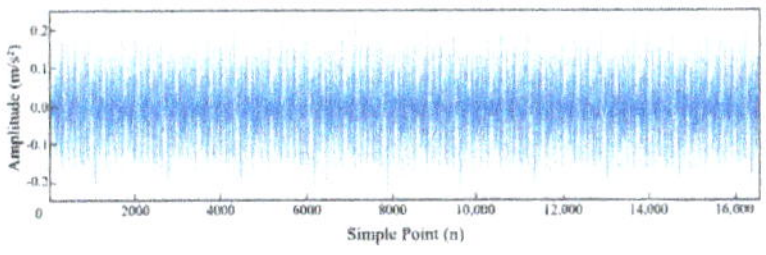

(**d**) gear tooth surface spalling

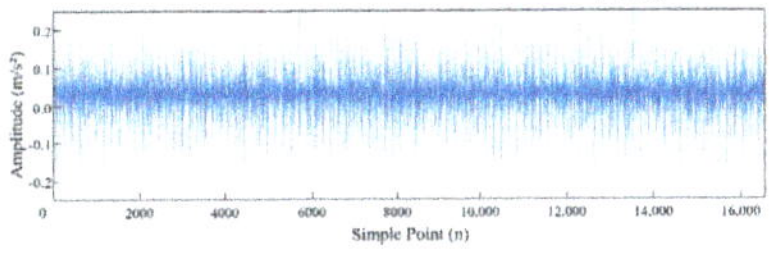

(**e**) normal conditions

Figure 6. Time domain waveform of original signals: (**a**) pitting of tooth flanks; (**b**) flat-headed sleeve tooth crack; (**c**) gear surface crack; (**d**) gear tooth surface spalling (**e**) normal conditions.

In addition, a spectrogram of the Short-Time Fourier transform (STFT) is shown in Figure 7. The horizontal axis represents time, the vertical axis represents frequency, and the color intensity represents the magnitude of the frequency components.

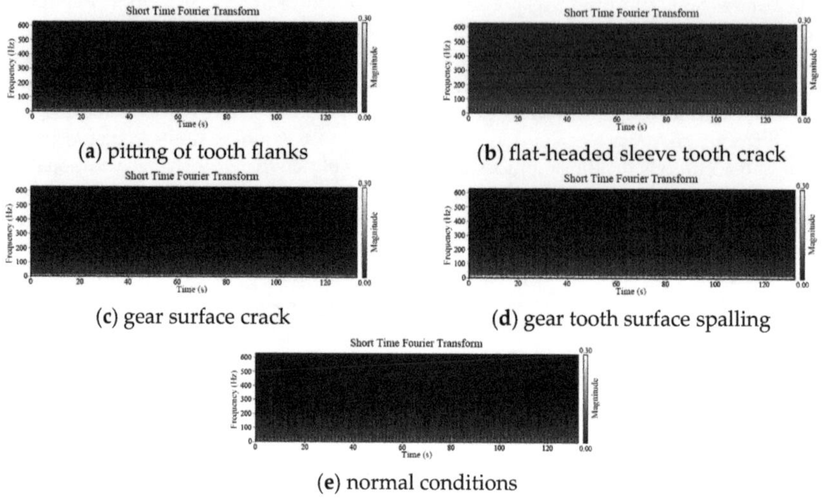

Figure 7. STFT analysis of raw signals: (**a**) pitting of tooth flanks; (**b**) flat-headed sleeve tooth crack; (**c**) gear surface crack; (**d**) gear tooth surface spalling; (**e**) normal conditions.

4.2. Model Parameters and Evaluation Metrics

The proposed model consists of 12 layers. The main computational components include an input layer, a 1D convolution layer, a depthwise separable convolution layer, pointwise convolution layers, a Bi-LSTM layer, an adaptive average pooling layer, and a linear layer. Additionally, we incorporated dropout layers, ReLU activation layers, batch normalization layers, and two permute layers to appropriately adjust the inputs for subsequent layers. The model framework parameters are shown in Table 3.

Table 3. Structural parameters of the proposed model.

Network Layer	Parameter	Output Size
Input		512×1
ConV	Kernel Size = 32; Stride = 4; Channel Size = 24	121×24
Deep_ConV	Kernel Size = 16; Stride = 1; Channel Size = 24; Padding = same	121×24
ReLU		
BatchNormal		
Point_ConV	Kernel Size = 1; Stride = 1; Channel Size = 24; Padding = same	121×24
Dropout	Dropout = 0.5	
Permute		
BiLSTM	Input size = 24; hidden size = 16; batch first = True; bidirectional = True	2×16
Permute		
AdaptiveAvgPool		
Linear	16×2; out features = 5	1×5

In classification problems, common metrics for evaluating model performance include accuracy, precision, recall, and F1 score. As a type of classification problem, the performance of the diagnosis model is also evaluated using the aforementioned metrics. The specific formulas are presented in Equations (13), (14), (15), and (16):

$$Accuracy = \frac{TP + TN}{TP + FP + TN + FN} \times 100\% \quad (13)$$

and

$$Precision = \frac{TP}{TP + FP} \times 100\% \quad (14)$$

and

$$Recall\text{-}rate = \frac{TP}{TP + FN} \times 100\% \quad (15)$$

and

$$\text{F1-}score = \frac{2 \times Recall \times Precision}{Recall + Precision} \quad (16)$$

where TP represents the number of true positive samples, which are samples correctly predicted as positive by the classifier; TN represents the number of true negative samples, which are samples correctly predicted as negative by the classifier; FN represents the number of false negative samples, which are samples incorrectly predicted as negative by the classifier; and FP represents the number of false positive samples, which are samples incorrectly predicted as positive by the classifier.

For this study, all models were run on a computer equipped with an Intel Core i7-9750H CPU, an Nvidia GeForce GTX 1660 Ti GPU, and 16 G of RAM, and the program was written with Python 3.7 and Pytorch. Additionally, all models were run under the same conditions: the learning rate was set to 0.01, the batch size was 32, and the number of epochs was 100. Graphs of the proposed model were plotted to illustrate the entire training history visually and effectively. The horizontal axis represents the number of epochs, while the vertical axis represents accuracy, as shown in Figure 8.

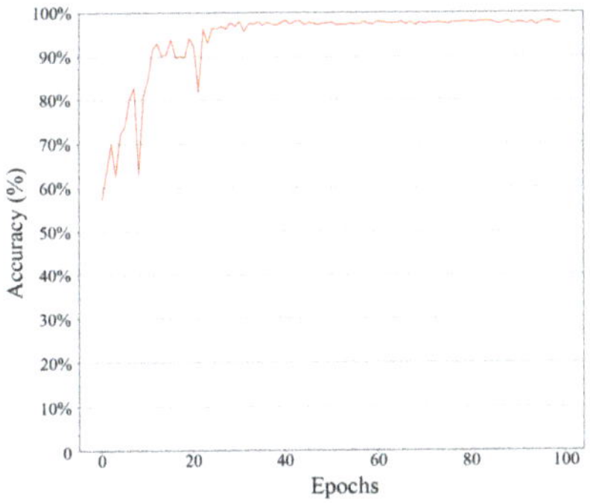

Figure 8. Accuracy curve of proposed model.

Figure 8 illustrates that the proposed model achieved reasonably acceptable accuracy in the fault diagnosis of the distribution box. The loss function is commonly used to evaluate and monitor the training process of a model. It indicates whether the model is overfitting during training. If the loss value stabilizes after a number of training iterations, it indicates that the model converges, reaching an optimal state. The loss function rate curves of the models are presented in Figure 9. The curves show that the values of loss on the training and test datasets changed during the training process. After 40 iterations, the function loss rate of the model dropped below 0.06, indicating excellent convergence and stability.

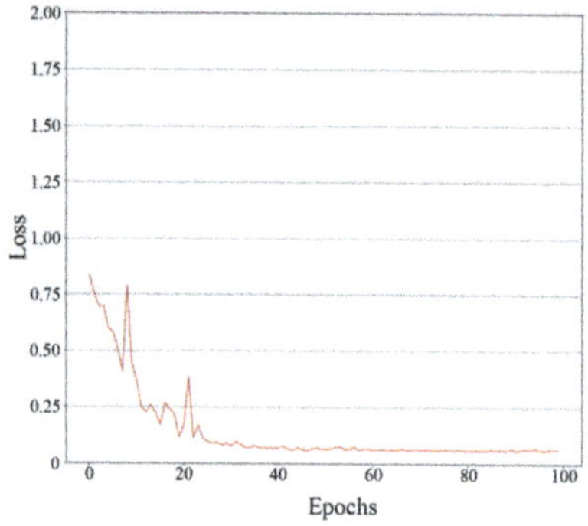

Figure 9. Loss curve of the proposed model.

Accuracy provides an overall measure of diagnostic performance. To illustrate the diagnostic results for each individual fault, we plotted a confusion matrix. The confusion matrix is a tool used to evaluate the performance of a classification model. The *X*-axis represents the predicted labels, while the *Y*-axis represents the true labels. Intuitively, elements on the diagonal indicate the number of samples correctly classified by the model, while elements off the diagonal indicate the number of samples incorrectly classified by the model. As shown in Figure 10, the proposed model correctly classified the vast majority of fault categories. All but 12 fault samples were recognized correctly.

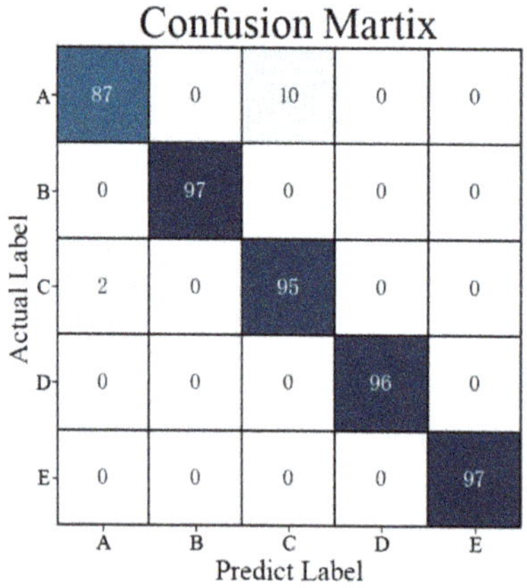

Figure 10. Confusion matrices of the proposed model.

To analyze the distribution of each type, t-SNE was used to visually examine the clustering patterns, classification boundaries, and anomalies in the data. As shown in Figure 11, the left image was generated from the original data, where data points were evenly distributed without distinct clustering or structural patterns. The figure on the right displays the t-SNE plot generated post-training. Purple represents the pitting of tooth flanks, blue represents flat-headed sleeve tooth crack, cyan represents gear tooth surface crack, yellow represents gear surface spalling, and green represents normal conditions.

As shown in the t-SNE visualization, the four types of faults are distinctly separated into four clusters across two dimensions. Among them, the samples of normal conditions, gear surface spalling, and the flat-headed sleeve tooth crack are well-differentiated. But the pitting of tooth flanks and gear tooth surface cracks are close. This is also reflected in the confusion matrix that misclassification occurred in the two classes.

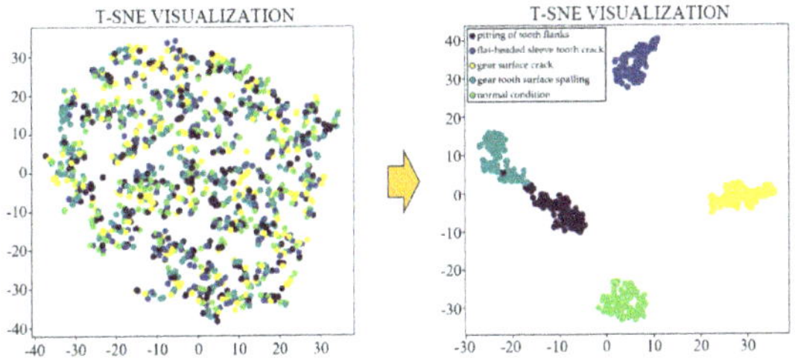

Figure 11. t-SNE visualization of the proposed model.

4.3. Model Comparison

To verify the capability of the proposed method, three other methods were used to diagnose the same fault dataset for comparison. The compared models included CNN-LSTM [29], multi-channels convolutional neural network (MCCNN), WDCNN [47], Bi-LSTM, and standard convolutional neural network (CNN). The same test set was used for evaluation, and the classification accuracies were calculated. Figure 12 depicts the training process of the six diagnostic models. The accuracies of the CNN, MCCNN, Bi-LSTM, and proposed models, obtained through cross-validation, were 86.78%, 90.08%, 87.81%, 76.86%, 90.08%, and 97.46%, respectively.

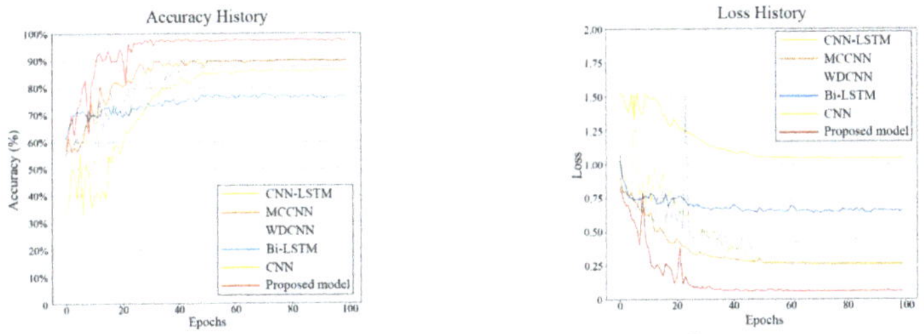

(**a**) accuracy curves (**b**) loss curves

Figure 12. Training history of different models: (**a**) accuracy curves; (**b**) loss curves.

For a comprehensive comparison, the metrics of each model, including precision, recall, F1 score, and accuracy, were collected and are listed in Table 4. The proposed model demonstrates competent performance compared to all other models, achieving over 97% in all metrics. In contrast, the results obtained for the other models are lower than those of the model proposed in this study.

Table 4. Performance comparison of different models.

Model	Precision	Recall	F1-Score	Accuracy
CNN-LSTM	86.99%	86.78%	86.87%	86.78%
MCCNN	90.13%	90.08%	9011%	90.08%
WDCNN	88.11%	87.81%	87.87%	87.81%
Bi-LSTM	76.73%	76.86%	76.57%	76.86%
CNN	90.62%	90.08%	90.10%	90.08%
Proposed Model	97.64%	97.46%	97.46%	97.46%

Considering computer hardware and optimization schemes, parameters and FLOPs were introduced to measure model complexity. In deep learning, a smaller value indicates greater computational efficiency. As shown in Table 5, the proposed model had fewer parameters and FLOPs, indicating better efficiency compared to other models.

Table 5. Comparison of model parameters and FLOPs.

Model	Parameters	FLOPs
CNN-LSTM	6.2×10^4	3.2×10^7
MCCNN	5.7×10^4	3.2×10^7
WDCNN	8.7×10^4	5.1×10^6
Bi-LSTM	1.2×10^6	2.1×10^6
CNN	2.5×10^5	4.7×10^6
Proposed Model	7.1×10^3	4.6×10^6

The confusion matrices of each model are plotted in Figure 13. By analyzing the confusion matrix of the comparison models, it is clear to see that the proposed model had better classification performance than other methods.

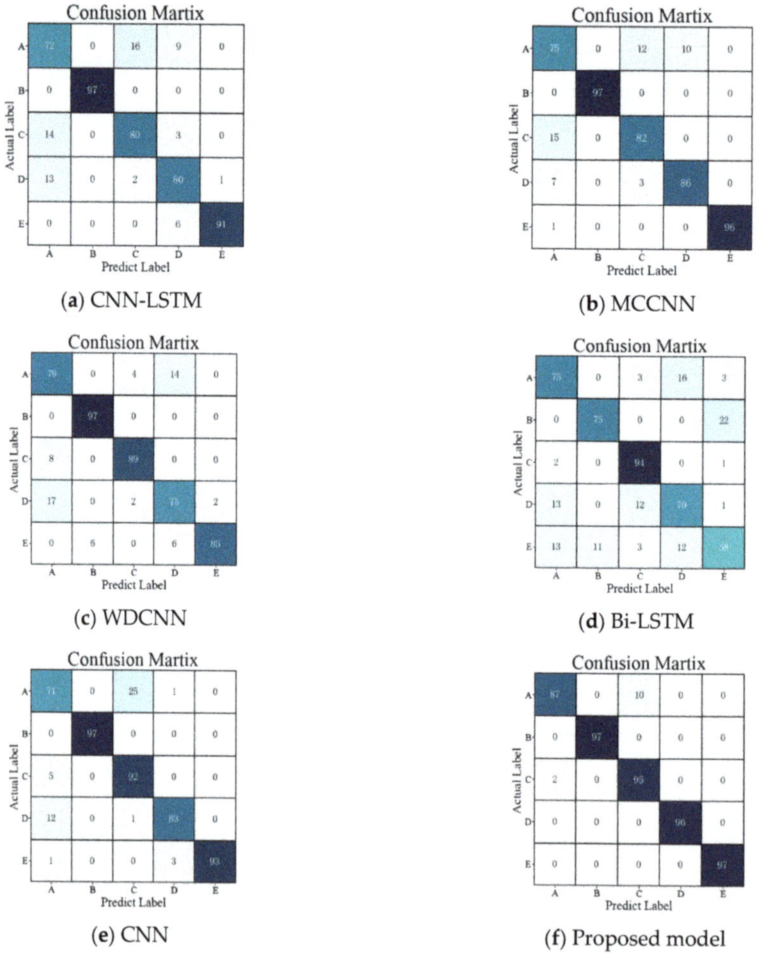

(**a**) CNN-LSTM (**b**) MCCNN (**c**) WDCNN (**d**) Bi-LSTM (**e**) CNN (**f**) Proposed model

Figure 13. Confusion matrix of the compared models: (**a**) CNN-LSTM; (**b**) MCCNN; (**c**) WDCNN; (**d**) Bi-LSTM; (**e**) CNN; (**f**) Proposed model.

4.4. Verification of Noise Robustness

In the working environment of a distribution box, there may be varying levels of white noise interference. White Gaussian noise can be effectively used to simulate noise interference. Three different signal-to-noise ratios (SNRs) of 4 dB, 6 dB, and 8 dB were added to validate the performance of the models. The accuracy of each model under these SNR conditions is shown in Table 6. The accuracy of the proposed model was approximately 90%, showing better resistance to noise compared to other models.

Table 6. Accuracy of each model under different SNR conditions.

Model	4DB	6DB	8DB
CNN-LSTM	61.36%	68.18%	58.06%
MCCNN	83.88%	72.52%	70.87%
WDCNN	88.22%	73.55%	78.10%
Bi-LSTM	69.42%	66.94%	70.04%
CNN	85.74%	44.21%	59.09%
Proposed model	88.22%	87.40%	91.12%

Considering the complex working environment of distribution boxes, noise always impacts the quality of extracted signals. The proposed model has the ability to resist the noise up to a certain extent. Different faults can be distinguished, and engineering accidents can be analyzed. Also, the proposed model can be utilized for intelligent fault diagnosis of the distribution box in the hot milling process.

5. Conclusions

Considering the high load operating conditions, distribution boxes are prone to gradual failure in the hot milling process. To diagnose faults in the distribution box in a timely manner, it is crucial to develop an effective diagnostic method for stable operation. This paper proposes an intelligent fault diagnosis model that integrates depthwise separable convolution with Bi-LSTM. The model achieves an accuracy of 97.46% in diagnosing four types of faults: the pitting of tooth flanks, flat-headed sleeve tooth cracks, gear surface cracks, and gear tooth surface spalling. Additionally, the model demonstrates robustness by maintaining approximately 90% accuracy under different SNR levels, indicating its ability to resist noise. Therefore, the proposed model is suitable for engineering applications. It effectively diagnoses these faults and facilitates timely repairs, which reduces downtime and enhances safety in the hot milling process.

Compared to the five existing diagnostic models, the proposed model offers significant advantages. It not only delivers superior diagnostic performance but also has fewer parameters and lower FLOPs, addressing the limitations associated with the computational load. Future research will focus on reducing the model's longer inference times compared to other approaches. We will work on optimizing the model to improve its computational efficiency without compromising accuracy and performance.

Author Contributions: Methodology, writing, Y.G.; software, formal analysis, D.Z.; supervision, validation, H.C.; supervision, investigation, X.Y.; conceptualization, methodology, Y.C. All authors have read and agreed to the published version of the manuscript.

Funding: This project is supported by the Shanghai Scientific Research project (No. 22511103604) and the Fundamental Research Fund for the Central Universities (2232023D-17). Also, this project is supported by the Foundation of Key Laboratory of Vibration and Control of Aero-Propulsion System, Ministry of Education (VCAME202104), which received support from Northeastern University.

Data Availability Statement: The data presented in this study are available on request from the corresponding author. The data are not publicly available due to privacy or ethical restrictions.

Conflicts of Interest: The authors declare no conflicts of interest.

References

1. Hu, B.; Chen, Z. System-Level Predictive Maintenance Optimization for No-Wait Production Machine–Robot Collaborative Environment under Economic Dependency and Hybrid Fault Mode. *Processes* **2024**, *12*, 1690. [CrossRef]
2. Yildiz, S.K.; Forbes, J.F.; Huang, B.; Zhang, Y.; Wang, F.; Vaculik, V.; Dudzic, M. Dynamic modelling and simulation of a hot strip finishing mill. *Appl. Math. Model.* **2009**, *33*, 3208–3225. [CrossRef]
3. Zhou, X.; Ben, X. Maintenance modelling for work rolls in hot finishing mill group with constraint of thermal character. *Int. J. Prod. Res.* **2023**, *62*, 1846–1861. [CrossRef]
4. He, H.-N.; Shao, J.; Wang, X.-C.; Yang, Q.; Liu, Y.; Xu, D.; Sun, Y.-Z. Research and application of approximate rectangular section control technology in hot strip mills. *J. Iron Steel Res. Int.* **2021**, *28*, 279–290. [CrossRef]
5. Hu, X.; Cao, Y.; Tang, T.; Sun, Y. Data-driven technology of fault diagnosis in railway point machines: Review and challenges. *Transp. Saf. Environ.* **2022**, *4*, 36. [CrossRef]
6. Yuan, J.; He, Z.; Zi, Y.; Liu, H. Gearbox fault diagnosis of rolling mills using multiwavelet sliding window neighboring coefficient denoising and optimal blind deconvolution. *Sci. China Technol. Sci.* **2009**, *52*, 2801–2809. [CrossRef]
7. Zhao, C.; Sun, J.; Lin, S.; Peng, Y. Rolling mill bearings fault diagnosis based on improved multivariate variational mode decomposition and multivariate composite multiscale weighted permutation entropy. *Measurement* **2022**, *195*, 111190. [CrossRef]
8. Lee, C.W.; Kang, H.K.; Park, C.J.; Shin, K.H. Fault diagnosis of roll shape under the speed change in hot rolling mill. *IFAC Proc. Vol.* **2005**, *38*, 45–50. [CrossRef]
9. Liu, J.F.; Chen, M.; Gu, J.Y.; Cheng, L. Remote fault diagnosis system based on EMD and SVM for heavy rolling-mills. *Adv. Mater. Res.* **2014**, *889–890*, 681–686. [CrossRef]
10. Zhang, F.; Zong, S.; Ling, Z. Fault diagnosis using kernel principal component analysis for hot strip mill. *J. Eng.* **2017**, *2017*, 527–535. [CrossRef]
11. Chen, J.; Wan, Z.; Pan, J.; Zi, Y.; Wang, Y.; Chen, B.; Sun, H.; Yuan, J.; He, Z. Customized maximal-overlap multiwavelet denoising with data-driven group threshold for condition monitoring of rolling mill drivetrain. *Mech. Syst. Signal Process.* **2016**, *68–69*, 44–67. [CrossRef]
12. Hinton, G.E.; Osindero, S.; Teh, Y.-W. A Fast Learning Algorithm for Deep Belief Nets. *Neural Comput.* **2006**, *18*, 1527–1554 [CrossRef] [PubMed]
13. Hu, X.; Tang, T.; Tan, L.; Zhang, H. Fault Detection for Point Machines: A Review, Challenges, and Perspectives. *Actuators* **2023**, *12*, 391. [CrossRef]
14. Zhang, K.; Zhang, X.; Peng, K. A novel parallel feature extraction-based multibatch process quality prediction method with application to a hot rolling mill process. *J. Process. Control* **2024**, *135*, 103166. [CrossRef]
15. Yu, Y.; Shi, P.; Tian, J.; Xu, X.; Hua, C. Rolling mill health states diagnosing method based on multi-sensor information fusion and improved DBNs under limited datasets. *ISA Trans.* **2023**, *134*, 529–547. [CrossRef]
16. Hou, D.; Zhang, B.; Chen, J.; Shi, P. Improved GNN based on Graph-Transformer: A new framework for rolling mill bearing fault diagnosis. *Trans. Inst. Meas. Control* **2024**. [CrossRef]
17. Shi, P.; Gao, H.; Yu, Y.; Xu, X.; Han, D. Intelligent fault diagnosis of rolling mills based on dual attention- guided deep learning method under imbalanced data conditions. *Measurement* **2022**, *204*, 111993. [CrossRef]
18. Zhao, S.; Bao, L.; Hou, C.; Bai, Y.; Yu, Y. Multi-source domain adversarial graph convolutional networks for rolling mill health states diagnosis under variable working conditions. *Struct. Health Monit.* **2024**. [CrossRef]
19. Hu, X.; Zhang, X.; Wang, Z.; Chen, Y.; Xia, J.; Du, Y.; Li, Y. Railway Switch Machine Fault Diagnosis Considering Sensor Abnormality Scenarios. In Proceedings of the 2023 IEEE 26th International Conference on Intelligent Transportation Systems (ITSC), Bilbao, Spain, 24–28 September 2023; Volume 9, pp. 4834–4839.
20. Chen, Z.; Mauricio, A.; Li, W.; Gryllias, K. A deep learning method for bearing fault diagnosis based on Cyclic Spectral Coherence and Convolutional Neural Networks. *Mech. Syst. Signal Process.* **2020**, *140*, 106683. [CrossRef]
21. Dash, B.M.; Bouamama, B.O.; Boukerdja, M.; Pekpe, K.M. Bond Graph-CNN based hybrid fault diagnosis with minimum labeled data. *Eng. Appl. Artif. Intell.* **2024**, *131*, 107734. [CrossRef]
22. Zhao, Z.; Jiao, Y. A Fault Diagnosis Method for Rotating Machinery Based on CNN With Mixed Information. *IEEE Trans. Ind. Inform.* **2023**, *19*, 9091–9101. [CrossRef]
23. Zhang, J.; Kong, X.; Li, X.; Hu, Z.; Cheng, L.; Yu, M. Fault diagnosis of bearings based on deep separable convolutional neural network and spatial dropout. *Chin. J. Aeronaut.* **2022**, *35*, 301–312. [CrossRef]
24. Zhou, S.; Liu, J.; Fan, X.; Fu, Q.; Goh, H.H. Thermal Fault Diagnosis of Electrical Equipment in Substations Using Lightweight Convolutional Neural Network. *IEEE Trans. Instrum. Meas.* **2023**, *72*, 5005709. [CrossRef]
25. Ling, L.; Wu, Q.; Huang, K.; Wang, Y.; Wang, C. A Lightweight Bearing Fault Diagnosis Method Based on Multi-Channel Depthwise Separable Convolutional Neural Network. *Electronics* **2022**, *11*, 4110. [CrossRef]
26. Wang, Y.; Yan, J.; Sun, Q.; Jiang, Q.; Zhou, Y. Bearing Intelligent Fault Diagnosis in the Industrial Internet of Things Context: A Lightweight Convolutional Neural Network. *IEEE Access* **2020**, *8*, 87329–87340. [CrossRef]
27. Lei, J.; Liu, C.; Jiang, D. Fault diagnosis of wind turbine based on Long Short-term memory networks. *Renew. Energy* **2019**, *133*, 422–432. [CrossRef]
28. Zou, F.; Zhang, H.; Sang, S.; Li, X.; He, W.; Liu, X. Bearing fault diagnosis based on combined multi-scale weighted entropy morphological filtering and bi-LSTM. *Appl. Intell.* **2021**, *51*, 6647–6664. [CrossRef]

29. Cao, L.; Qian, Z.; Zareipour, H.; Huang, Z.; Zhang, F. Fault Diagnosis of Wind Turbine Gearbox Based on Deep Bi-Directional Long Short-Term Memory Under Time-Varying Non-Stationary Operating Conditions. *IEEE Access* **2019**, *7*, 155219–155228. [CrossRef]
30. Huang, T.; Zhang, Q.; Tang, X.; Zhao, S.; Lu, X. A novel fault diagnosis method based on CNN and LSTM and its application in fault diagnosis for complex systems. *Artif. Intell. Rev.* **2022**, *55*, 1289–1315. [CrossRef]
31. Zhi, Z.; Liu, L.; Liu, D.; Hu, C. Fault Detection of the Harmonic Reducer Based on CNN-LSTM With a Novel Denoising Algorithm. *IEEE Sens. J.* **2021**, *22*, 2572–2581. [CrossRef]
32. Wang, T.; Zhang, L.; Wang, X. Fault Detection for Motor Drive Control System of Industrial Robots Using CNN-LSTM-based Observers. *CES Trans. Electr. Mach. Syst.* **2023**, *7*, 144–152. [CrossRef]
33. Chen, Y.; Yin, X.; Zhang, R.; Gao, F. Reinforced convolutional neural network fault diagnosis of industrial production systems. *Chem. Eng. Sci.* **2024**, *299*, 120466. [CrossRef]
34. Qiao, M.; Yan, S.; Tang, X.; Xu, C. Deep Convolutional and LSTM Recurrent Neural Networks for Rolling Bearing Fault Diagnosis Under Strong Noises and Variable Loads. *IEEE Access* **2020**, *8*, 66257–66269. [CrossRef]
35. Liu, X.; Chen, G. A Siamese CNN-BiLSTM-based method for unbalance few-shot fault diagnosis of rolling bearings. *Meas. Control* **2024**, *57*, 551–565.
36. Ullah, W.; Ullah, A.; Haq, I.U.; Muhammad, K.; Sajjad, M.; Baik, S.W. CNN features with bi-directional LSTM for real-time anomaly detection in surveillance networks. *Multimedia Tools Appl.* **2020**, *80*, 16979–16995. [CrossRef]
37. Sun, H.; Chen, M.; Weng, J.; Liu, Z.; Geng, G. Anomaly Detection for In-Vehicle Network Using CNN-LSTM With Attention Mechanism. *IEEE Trans. Veh. Technol.* **2021**, *70*, 10880–10893. [CrossRef]
38. Xia, T.; Song, Y.; Zheng, Y.; Pan, E.; Xi, L. An ensemble framework based on convolutional bi-directional LSTM with multiple time windows for remaining useful life estimation. *Comput. Ind.* **2020**, *115*, 103182. [CrossRef]
39. Huang, G.; Zhang, Y.; Ou, J. Transfer remaining useful life estimation of bearing using depth-wise separable convolution recurrent network. *Measurement* **2021**, *176*, 109090. [CrossRef]
40. Ilesanmi, A.E.; Ilesanmi, T.O. Methods for image denoising using convolutional neural network: A review. *Complex Intell. Syst.* **2021**, *7*, 2179–2198. [CrossRef]
41. Zhao, B.; Zhang, X.; Zhan, Z.; Pang, S. Deep multi-scale convolutional transfer learning network: A novel method for intelligent fault diagnosis of rolling bearings under variable working conditions and domains. *Neurocomputing* **2020**, *24*, 24–38. [CrossRef]
42. Lecun, Y.; Bottou, L.; Bengio, Y.; Haffner, P. Gradient-based learning applied to document recognition. *Proc. IEEE* **1998**, *86*, 2278–2324. [CrossRef]
43. Samek, W.; Montavon, G.; Lapuschkin, S.; Anders, C.J.; Muller, K.-R. Explaining Deep Neural Networks and Beyond: A Review of Methods and Applications. *Proc. IEEE* **2021**, *109*, 247–278. [CrossRef]
44. Huang, Y.; Zhang, J.; Liu, R.; Zhao, S. Improving Accuracy and Interpretability of CNN-Based Fault Diagnosis through an Attention Mechanism. *Processes* **2023**, *11*, 3233. [CrossRef]
45. Cui, Z.; Ke, R.; Pu, Z.; Wang, Y. Stacked bidirectional and unidirectional LSTM recurrent neural network for forecasting network-wide traffic state with missing values. *Transp. Res. Part C Emerg. Technol.* **2020**, *118*, 102674. [CrossRef]
46. Zhang, X.; Yang, J.; Yang, X. Residual Life Prediction of Rolling Bearings Based on a CEEMDAN Algorithm Fused with CNN–Attention-Based Bidirectional LSTM Modeling. *Processes* **2024**, *12*, 8. [CrossRef]
47. Zhang, A.; Li, S.; Cui, Y.; Yang, W.; Dong, R.; Hu, J. Limited Data Rolling Bearing Fault Diagnosis With Few-Shot Learning. *IEEE Access* **2019**, *7*, 2169–3536. [CrossRef]

processes

MDPI

Article

Modified Deng's Grey Relational Analysis Model for Panel Data and Its Applications in Assessing the Water Environment of Poyang Lake

Fanghong Jian [1], Jiangfeng Li [1,2], Xiaomei Liu [1,*], Qiong Wu [1] and Dan Zhong [3]

1 College of Science, Jiujiang University, 551, Qianjin St., Lianxi District, Jiujiang 332005, China; jfhrecoba@126.com (F.J.); jiangfengli2000@163.com (J.L.); penny5251@163.com (Q.W.)
2 Jiangxi Key Laboratory of Industrial Ecological Simulation and Environmental Health in Yangtze River Basin, Jiujiang University, 551, Qianjin St., Lianxi District, Jiujiang 332005, China
3 Jiujiang Ecological Environment Monitoring Center of Jiangxi Province, Jiujiang 332005, China; zdan01@126.com
* Correspondence: liuxiaomeijju@126.com

Abstract: Deng's grey relational analysis (GRA) model is widely used in clustering because of its simple mathematical mechanisms. For sample data of different dimensions, people have put forward different Deng's GRA models, including time series data, panel data, and panel time series data. The purpose of this paper is to improve the clustering accuracy of the existing Deng's GRA model for panel data in order to overcome some of its shortcomings. Firstly, the existing Deng's GRA model for panel data was tested based on the dataset LP1 of Robot Execution Failures. Then, according to the test results, the existing Deng's GRA model for panel data is modified by means of Taylor's formula, and the modified model is successfully validated by the dataset LP1 of Robot Execution Failures. Finally, as a practical application, the modified Deng's GRA model for panel data is applied to assess the water environment of Poyang Lake over the past five years. Compared with other cluster methods, the results of the case study show that the modified Deng's GRA model for panel data is applicable and also confirm the remarkable effectiveness of the Chinese government's water quality regulation in Poyang Lake. Therefore, the modified Deng's GRA model presented in this paper improves the clustering accuracy compared to the original model and can be applied well to the classification of data with a large dimension.

Keywords: Deng's grey relational analysis model; panel data; Robot Execution Failures Dataset; Poyang Lake; water environment

1. Introduction

Grey relational theory, first proposed by Professor Deng in 1984 [1], is a mathematical method to assess the similarity or closeness between samples by calculating the relational degree among them. Unlike the factor analysis method of statistics inference, it does not need a large number of samples, and so it provides an effective tool for analyzing the interactions between various factors within a system. The grey relational analysis (GRA) model, as an important concrete manifestation of grey relational theory, has been widely applied in various fields. For example, Zhang [2] applied GRA models to rank the importance of 22 factors in the process industry, pinpointing security inspection, risk identification, and security awareness as the most critical, so as to develop an intelligent monitoring system for key factors across subsystems, utilizing video surveillance and sensors for real-time safety management and accident prevention. Xu [3] utilized an enhanced GRA model to assess the influence of innovation strategies on marine industry clustering and transformation, revealing stronger ties between innovation investment and industry aggregation and contributing to marine economic sustainability. Chen [4] used an improved GRA model to develop a subcontractor selection model that integrates quality function deployment (QFD) and analytic hierarchy process (AHP) to enhance the objectivity and rationality of the selection process.

Since the GRA model was proposed 40 years ago, new patterns have been continuously proposed to adapt to wider applications. Overall, they can be divided into two types: One is univariate GRA models, which focus on the relationship between a single factor and the system's behavior, such as Deng's GRA model [5], the area GRA model [6], and the slope GRA model [7]. The other is multivariate GRA models, which consider the interactions between multiple factors within the system, such as the three-dimensional Deng's GRA model [5], the norm GRA model [8], the convex GRA model [9], the grid GRA model [10], the matrix GRA model [11], and the curvature GRA model [12]. But in these GRA models, Deng's GRA model is most widely applied due to its simple mathematical expression and easy implementation.

Deng's GRA model is the most basic model of grey relational theory, it constructs a mathematical model based on the difference sequence between the reference sequence and comparison sequences and calculates the relational coefficient to measure the relational degree between them. In practical applications, Deng's GRA model can effectively identify the interactions between various factors in the system, help researchers identify key factors, and optimize the system structure. For example, Zhou et al. [13] developed a backpropagation neural network model based on statistical analysis and Deng's GRA theory to predict the sulfur content in the COREX process. Zhang et al. [14] introduced a novel multivariate grey relational model based on spatial pyramid pooling for the analysis of time series data on different scales. Javed et al. [15] provided a new perspective for supplier evaluation and classification in multi-sourcing through the dynamic grey relational method. Xu et al. [16] utilized Deng's GRA model and PLS-SEM to analyze the integration level of China's digital economy and real economy. Li et al. [17] proposed a grey-adversary perceptual network to improve the performance of anomaly detection in surveillance videos. Overall, Deng's GRA model is an effective grey relational analysis tool that reveals the inherent connections between data sequences by quantifying their differences, providing a new perspective and method for solving practical problems.

However, as the sample data become relatively more complex, Deng's GRA model sometimes deviates from the actual facts. The literature [18] has pointed out that Deng's GRA model cannot distinguish between the normal state and the failure state in the Robot Execution Failures Dataset. The reason for this may be that Deng's GRA model is only constructed based on the raw difference sequence between the reference sequence and the comparison sequence. From the perspective of approximation theory, Deng's GRA model may not have sufficient depth in data mining. In fact, researchers have always improved Deng's GRA model to adapt to a wider range of applications. Liu et al. [19] proposed a new GRA model for measuring the relationships between inverse sequences. Huang et al. [20] proposed a new GRA model based on information differences for the performance evaluation of sanatoriums. However, the follow-up research on Deng's GRA model has changed towards other theoretical frameworks. That is to say, the mechanisms of the subsequent models proposed have significantly diverged from the foundational principles of Deng's original GRA model. In this way, the models become more complex, which can have a certain impact on the practical applications.

In view of the above facts, this paper will use Taylor's approximation theory to improve Deng's GRA model based on the original theoretical framework. Firstly, we use small-scale multivariate sample data to test the reliability of Deng's original GRA model for panel data, and then a possible method to modify Deng's original GRA model for panel data is provided according to the test results. The main work of this paper lies in the following aspects:

(1) A series of testing experiments were conducted on the performances of Deng's GRA model for panel data based on the dataset LP1 of Robot Execution Failures.
(2) A modified Deng's GRA model for panel data is presented.

(3) The water environment of Poyang Lake is assessed over past years using a modified Deng's GRA model for panel data.

The outline of this paper is organized as follows. Section 2 reviews Deng's original GRA model for panel data and tests the performance of the model based on the dataset LP1 of Robot Execution Failures. Section 3 presents a modification Deng's GRA model for panel data. Section 4 validates the modified Deng's GRA model for panel data by three numerical experiments. Section 5 applies the modified Deng's GRA model for panel data to assess the water environment of Poyang Lake. Section 6 makes conclusions.

2. Deng's GRA Model for Panel Data

2.1. Review of Deng's GRA Model for Panel Data

Deng's GRA model is constructed based on the difference sequence between the reference sequence and the comparison sequence and measures the degree of relation between sequences by calculating the relational coefficient. For simplicity in description, we denote

$$(a_{ij}^k)_{m\times n}=\begin{bmatrix} a_{11}^k & \cdots & a_{1n}^k \\ \cdots & \cdots & \cdots \\ a_{m1}^k & \cdots & a_{mn}^k \end{bmatrix}, k=0,1,2,3,\cdots L.$$

Assume $X_0 = (a_{ij}^0)_{m\times n}$ is the behavior matrix of system characteristics, $X_k = (a_{ij}^k)_{m\times n}(k \neq 0)$ is the behavior matrix of system factors, $1 \leqslant i \leqslant m$ represents the time dimension, and $1 \leqslant j \leqslant n$ represents the indicator dimension, L is the sample size.

Definition 1 ([5]). *Assume $X_0 = (a_{ij}^0)_{m\times n}$ is the behavior matrix of system characteristics, $X_k = (a_{ij}^k)_{m\times n}$ is the behavior matrix of system factors, and $k = 1,2,\cdots,L$, $\xi \in [0,1]$. Let $\varepsilon_{ij}^{0k} = \frac{\min\limits_{1\leqslant i\leqslant m}\min\limits_{1\leqslant j\leqslant n}\min\limits_{1\leqslant k\leqslant L}\left|a_{ij}^0-a_{ij}^k\right|+\xi\cdot\max\limits_{1\leqslant i\leqslant m}\max\limits_{1\leqslant j\leqslant n}\max\limits_{1\leqslant k\leqslant L}\left|a_{ij}^0-a_{ij}^k\right|}{\left|a_{ij}^0-a_{ij}^k\right|+\xi\cdot\max\limits_{1\leqslant i\leqslant m}\max\limits_{1\leqslant j\leqslant n}\max\limits_{1\leqslant k\leqslant L}\left|a_{ij}^0-a_{ij}^k\right|}$. Then, $\rho_{0k} = \frac{1}{mn}\sum\limits_{i=1}^{m}\sum\limits_{j=1}^{n}\varepsilon_{ij}^{0k}$ is defined as Deng's grey relational degree of X_k and X_0, and is also called Deng's original GRA model for panel data.*

Note that there is a distinguished coefficient ξ, the main function of which is to enhance the resolution, but in practical applications, ξ is commonly set to 0.5.

Obviously, Deng's GRA model for panel data offers a global perspective to analyze the incidence relationships for panel data. Meanwhile, it has several primary properties, including normality, closeness, symmetry, translation invariance, and indicator permutation. Particularly, the indicator permutation is very important, which implies that there is no need to pay attention to the arrangement among the indicators of panel data.

2.2. Performance Testing of Deng's GRA Model for Panel Data

As mentioned in the Introduction, the literature [18] points out that Deng's GRA model cannot distinguish between the normal state and the failure state in the Robot Execution Failures Dataset. In this section, based on the research presented in literature [18], we further discuss specific details of performance testing on the Robot Execution Failures Dataset, aiming to provide a comprehensive assessment for Deng's GRA model.

2.2.1. Robot Execution Failures Dataset

The Robot Execution Failures Dataset is a set of multivariate sample data from the UCI machine learning database. Detailed raw data can be obtained from the website (http://kdd.ics.uci.edu/databases/robotfailure/robotfailure.html), accessed on 23 April 1999. The Robot Execution Failures Dataset is generated by the fault monitoring of robots with six sensors, and there are a total of five sub-datasets. In this paper, the first sub dataset LP1 is adopted to perform experiments.

In dataset LP1, there are two states and a total of 88 samples, including 21 normal states and 67 failure states. Each sample is represented as a matrix of 15 by 6, including 6 indicators and 15 times of continuous acquisition. Specifically, columns 1–3 of the matrix are the evolution of force, and columns 4–6 are the evolution of torque.

According to the features of dataset LP1, we suppose that $X_0 = (a_{ij}^0)_{15\times 6}$ is the behavior matrix of system characteristics, where $1 \leqslant i \leqslant 15$ represents the time dimension, $1 \leqslant j \leqslant 6$ represents the indicator dimension, as shown in Table 1. At the same time, all of the 88 samples in dataset LP1 are considered as the behavior matrix of system factors, that is, $X_k = (a_{ij}^k)_{15\times 6}, k = 1, 2, \cdots, 88$.

Table 1. Behavior matrix of system characteristics X_0.

Time	F_x	F_y	F_z	T_x	T_y	T_z
1	−1	−1	63	−3	−1	0
2	−1	−1	63	−3	−1	0
3	−1	−1	63	−3	−1	0
4	−1	−1	63	−3	−1	0
5	−1	−1	63	−3	−1	0
6	−1	−1	63	−3	−1	0
7	−1	−1	63	−3	−1	0
8	−1	−1	63	−3	−1	0
9	−1	−1	63	−3	−1	0
10	−1	−1	63	−3	−1	0
11	−1	−1	63	−3	−1	0
12	−1	−1	63	−3	−1	0
13	−1	−1	63	−3	−1	0
14	−1	−1	63	−3	−1	0
15	−1	−1	63	−3	−1	0

2.2.2. Data Pre-Processing

Because Columns 1–3 and Columns 4–6 on the behavior matrix of the system are the evolution of force and torque respectively, the testing data would be pre-processed for dimensionless form before modeling. There are many methods for dimensionless data processing, but in grey relational theory, only the interval operator, initial value operator, and mean operator are often used. Furthermore, all features of the LP1 dataset are numeric and often continuous, and each sample was collected at regular time intervals within 315 ms. Therefore, based on the characteristics of the LP1 dataset, there is a subtle connection between these 88 samples in the time dimension. To overcome this problem, the interval operator is selected to pre-process the testing data. The interval operator is defined as follows.

Definition 2 ([5]). *Assume the behavior matrix of the system is represented as* $X_k = (x_{ij})_{m\times n}$, $1 \leqslant i \leqslant m$, $1 \leqslant j \leqslant n, k = 1, 2, \cdots, L$, *here L is the sample size. If D satisfies* $DX_k = Y_k = (y_{ij})_{m\times n}$ *and* $y_{ij} = \frac{x_{ij} - \min\limits_{1\leqslant k\leqslant L}\min\limits_{1\leqslant i\leqslant m} x_{ij}}{\max\limits_{1\leqslant k\leqslant L}\max\limits_{1\leqslant i\leqslant m} x_{ij} - \min\limits_{1\leqslant k\leqslant L}\min\limits_{1\leqslant i\leqslant m} x_{ij}}$, *then D is called the interval operator.*

2.2.3. Testing Results

As mentioned above, the relational degree order of Deng's GRA model for panel data does not change as the change of indicator arrangements. So there will be only 88 values of relational degree in the total 88 samples of dataset LP1. By programming in Python, the 88 values were obtained. For illustration, the maximum and the minimum of the relational degree of normal and failure states are listed in Table 2.

Table 2. Relational degree interval of normal and failure states offered by Deng's GRA model for panel data.

States	21 Normal States	67 Failure States
Interval of relational degree	[**0.9908**, 0.9996]	[0.6206, **0.9915**]

Table 2 shows that Deng's GRA model for panel data cannot fully distinguish normal and failure states. Because the maximum relational degree of failure states is greater than the minimum relational degree of normal states, there is no threshold between the relational degree of normal and failure states. In order to investigate the detailed reasons, the 88 relational degree values of Deng's GRA model for panel data are presented through a scatter plot, where green and red dots display normal and failure states, respectively, as shown in Figure 1.

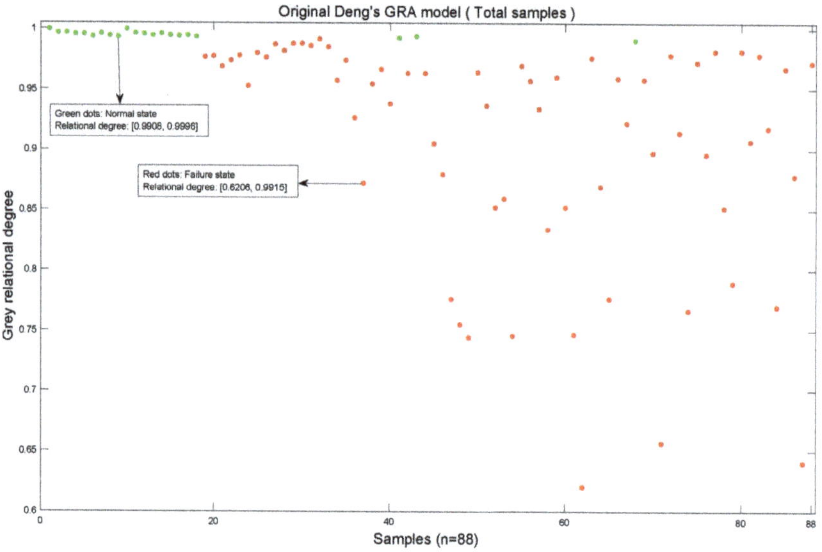

Figure 1. Scatter plot of 88 relational degrees calculated by Deng's GRA model for panel data.

Figure 1 shows that the majority of green dots are located at the top, and very few green dots are intersected with red dots. This indicates that very few normal states have not been distinguished. In other words, Deng's GRA model for panel data has demonstrated certain advantages in its ability to distinguish, but it still needs to be improved.

3. Modified Deng's GRA Model for Panel Data

From a mathematical perspective, Deng's original GRA model is constructed based on the raw difference sequence between the reference sequence and the comparison sequence. According to Taylor's formula, the approximation will inevitably result in significant errors. For example, let the reference sequence X_0 be a constant matrix with all elements equal to 0.5, the comparison sequence X_k is a binary matrix such that all elements can only be 0 or 1. That is,

$$X_0 = \begin{bmatrix} 0.5 & 0.5 & \cdots & 0.5 \\ \vdots & \vdots & \vdots & \vdots \\ 0.5 & 0.5 & \cdots & 0.5 \end{bmatrix}_{m\times n}, X_k = \begin{bmatrix} 0 \text{ or } 1 & 0 \text{ or } 1 & \cdots & 0 \text{ or } 1 \\ \vdots & \vdots & \vdots & \vdots \\ 0 \text{ or } 1 & 0 \text{ or } 1 & \cdots & 0 \text{ or } 1 \end{bmatrix}_{m\times n}, \forall k \in N^+.$$

According to Deng's original GRA model, for $\forall k \in N^{+}$, Deng's grey relational degree ρ_{0k} of X_k and X_0 will be 1. Obviously, this deviates from the facts and it is difficult to distinguish the samples.

Taylor approximation theory provides an approximation of a function as a polynomial sum derived from its derivatives at a single point. It is widely used in various fields of mathematics, physics, engineering, and computer science for its ability to solve problems that are otherwise difficult to handle analytically. In this paper, we adopted the concept of Taylor's formula and converted the derivative into the differential form to optimize Deng's GRA model. Therefore, Deng's GRA model is expanded to the first-order and second-order differences.

Consequently, Deng's GRA model for panel data is modified as follows.

Definition 3. *Assume* $X_0 = (a_{ij}^0)_{m\times n}$ *is the behavior matrix of system characteristics, and* $X_k = (a_{ij}^k)_{m\times n}$ *is a behavior matrix of system factors, where* $1 \leqslant i \leqslant m$ *represents the time dimension,* $1 \leqslant j \leqslant n$ *represents the indicator dimension,* $1 \leqslant k \leqslant L$ *is the sample size, and* $\xi \in [0,1]$. *Let*

$$r'_{0k} = \frac{\min\limits_{1\leqslant i\leqslant m}\min\limits_{1\leqslant j\leqslant n}\min\limits_{1\leqslant k\leqslant L}\left|a_{ij}^0-a_{ij}^k\right|+\xi\max\limits_{1\leqslant i\leqslant m}\max\limits_{1\leqslant j\leqslant n}\max\limits_{1\leqslant k\leqslant L}\left|a_{ij}^0-a_{ij}^k\right|}{\left|a_{ij}^0-a_{ij}^k\right|+\xi\max\limits_{1\leqslant i\leqslant m}\max\limits_{1\leqslant j\leqslant n}\max\limits_{1\leqslant k\leqslant L}\left|a_{ij}^0-a_{ij}^k\right|},$$

$$r''_{0k} = \frac{\min\limits_{1\leqslant i\leqslant m}\min\limits_{1\leqslant j\leqslant n-1}\min\limits_{1\leqslant k\leqslant L}|\Delta|+\xi\max\limits_{1\leqslant i\leqslant m}\max\limits_{1\leqslant j\leqslant n-1}\max\limits_{1\leqslant k\leqslant L}|\Delta|}{|\Delta|+\xi\max\limits_{1\leqslant i\leqslant m}\max\limits_{1\leqslant j\leqslant n-1}\max\limits_{1\leqslant k\leqslant L}|\Delta|}, \text{ here } \Delta = (a_{i,j+1}^0 - a_{ij}^0) - (a_{i,j+1}^k - a_{ij}^k).$$

$$r'''_{0k} = \frac{\min\limits_{1\leqslant i\leqslant m}\min\limits_{1\leqslant j\leqslant n-1}\min\limits_{1\leqslant k\leqslant L}\left|\Delta^2\right|+\xi\max\limits_{1\leqslant i\leqslant m}\max\limits_{1\leqslant j\leqslant n-1}\max\limits_{1\leqslant k\leqslant L}\left|\Delta^2\right|}{\left|\Delta^2\right|+\xi\max\limits_{1\leqslant i\leqslant m}\max\limits_{1\leqslant j\leqslant n-1}\max\limits_{1\leqslant k\leqslant L}\left|\Delta^2\right|},$$

here $\Delta^2 = (a_{i,j+2}^0 - 2a_{i,j+1}^0 + a_{ij}^0) - (a_{i,j+2}^k - 2a_{i,j+1}^k + a_{ij}^k)$. *Then,* $r_{0k} = \frac{1}{2}\cdot\frac{1}{mn}\sum_{i=1}^{m}\sum_{j=1}^{n} r'_{0k} + \frac{1}{4}\cdot \frac{1}{m(n-1)}\sum_{i=1}^{m}\sum_{j=1}^{n-1} r''_{0k} + \frac{1}{4}\cdot\frac{1}{m(n-2)}\sum_{i=1}^{m}\sum_{j=1}^{n-2} r'''_{0k}$ *is called the modified Deng's grey relational degree of* X_k *and* X_0 *and is also called the modified Deng's GRA model for panel data.*

Similarly, ξ is a distinguished coefficient, the function of which is to enhance resolution, and, as usual, $\xi = 0.5$. Furthermore, the grey relational degree defined by the modified Deng's GRA model for panel data still satisfies those primary properties, including normality, closeness, symmetry, translation invariance, and indicator permutation.

Next, we will continue to use the Robot Execution Failures Dataset adopted in Section 2 to test the performance of the modified Deng's GRA model for panel data.

4. Performance Testing of the Modified Deng's GRA Model for Panel Data

In this section, we will conduct a comprehensive assessment for the modified Deng's GRA model from three aspects of the Robot Execution Failures Dataset, including the total samples, the force samples, and the torque samples. As in Section 2, all the testing data are pre-processed in a dimensionless form before modeling using an interval operator.

4.1. Experiments Based on the Total Samples of the Robot Execution Failures Dataset

Based on the total samples of the Robot Execution Failures Dataset, by programming in Python, the 88 relational degree values of the modified Deng's GRA model for panel data were obtained. Table 3 lists the maximum and the minimum relational degree of normal and failure states. Similarly, the 88 values are presented through a scatter plot, where normal and failure states are still displayed by green and red dots, respectively, as shown in Figure 2.

Table 3. Relational degree interval calculated by the modified Deng's GRA model based on total samples.

States	21 Normal States	67 Failure States
Interval of relational degree	[**0.9933**, 0.9994]	[0.8010, **0.9887**]

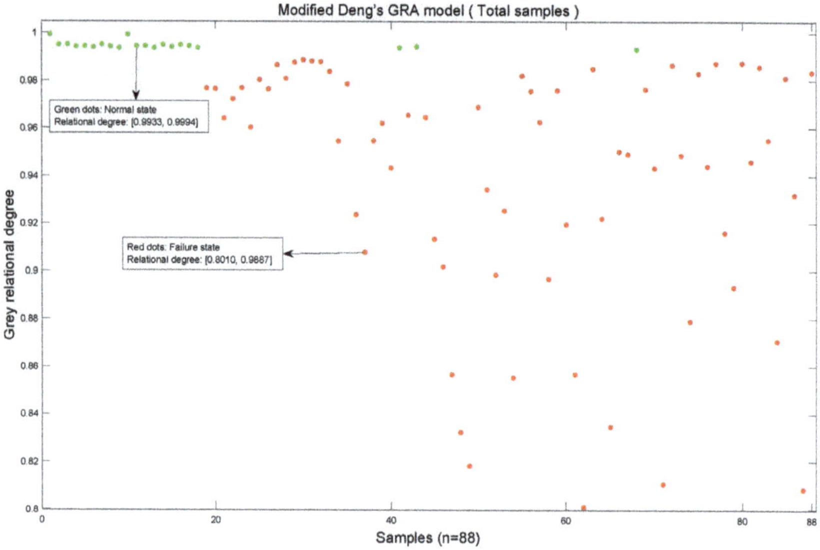

Figure 2. Scatter plot of 88 relational degrees provided by modified Deng's GRA model based on total samples.

Table 3 shows that the maximum relational degree of failure states is smaller than the minimum relational degree of normal states, and there is a significant gap in the relational degree interval of normal and failure states. Similarly, Figure 2 shows that the green and red dots are completely separated. Therefore, based on the total samples of the Robot Execution Failures Dataset, the modified Deng's GRA model for panel data can fully distinguish between normal and failure states.

4.2. Experiments Based on the Force Samples of Robot Execution Failures Dataset

In the same way, based on the force samples of the Robot Execution Failures Dataset, the 88 relational degree values of the modified Deng's GRA model for panel data were obtained by programming in Python. Table 4 lists the maximum and the minimum of the relational degree of normal and failure states. Meanwhile, Figure 3 displays the 88 relational degree values of normal and failure states by green and red dots, respectively.

Table 4. Relational degree interval was calculated by modifying Deng's GRA model for panel data based on force samples.

States	21 Normal States	67 Failure States
Interval of relational degree	[**0.9927**, 0.9995]	[0.7441, **0.9898**]

In Table 4, similar to Table 3, the maximum relational degree of failure states is smaller than the minimum relational degree of normal states, and there is also a significant gap in the relational degree interval of normal and failure states. Figure 3 also shows that the green and red dots are completely separated. Therefore, based only on the force samples of the Robot Execution Failures Dataset, the modified Deng's GRA model for panel data still can fully distinguish between normal states and failure states.

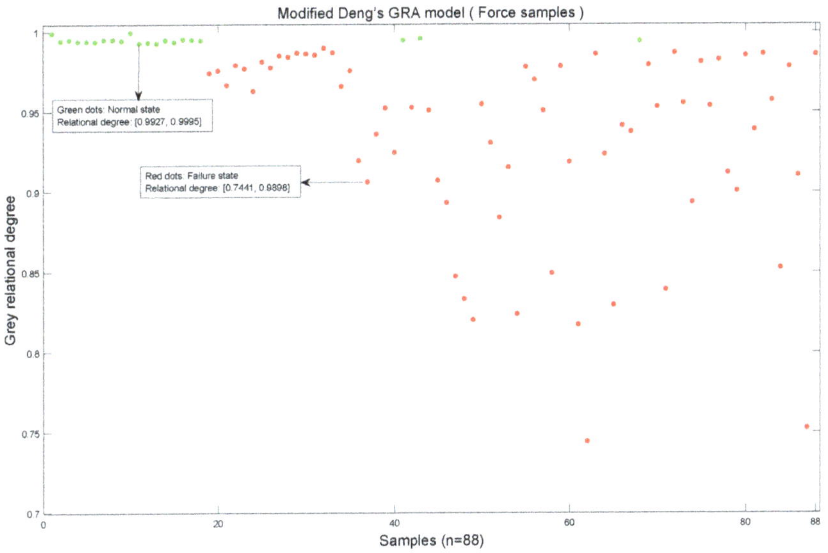

Figure 3. Scatter plot of 88 relational degrees provided by the modified Deng's GRA model based on force samples.

4.3. Experiments Based on the Torque Samples of the Robot Execution Failures Dataset

Similarly, based on the torque samples of the Robot Execution Failures Dataset, the 88 relational degree values of the modified Deng's GRA model for panel data were obtained by programming in Python. The maximum and the minimum relational degrees of normal and failure states are listed in Table 5. Meanwhile, the 88 values are presented in Figure 4, and the green and red dots represent the normal and failure states, respectively.

Table 5. Relational degree interval calculated by modifying Deng's GRA model for panel data based on torque samples.

States	21 Normal States	67 Failure States
Interval of relational degree	[**0.9868**, 0.9993]	[0.7019, **0.9862**]

Table 5 shows that the maximum relational degree of failure states is still smaller than the minimum relational degree of normal states, and there is a slight gap in the relational degree interval of normal and failure states. Nevertheless, Table 5 and Figure 4 still demonstrate that the modified Deng's GRA model for panel data can completely distinguish between normal and failure states based solely on the torque samples.

Therefore, all three test results indicate that the modified Deng's GRA model for panel data significantly improves the distinguishing ability of the original model and can be effectively performed by utilizing only a subset of the sample data. In summary, the performance of the modified Deng's GRA model for panel data is validated through three numerical experiments based on the Robot Execution Failures Dataset.

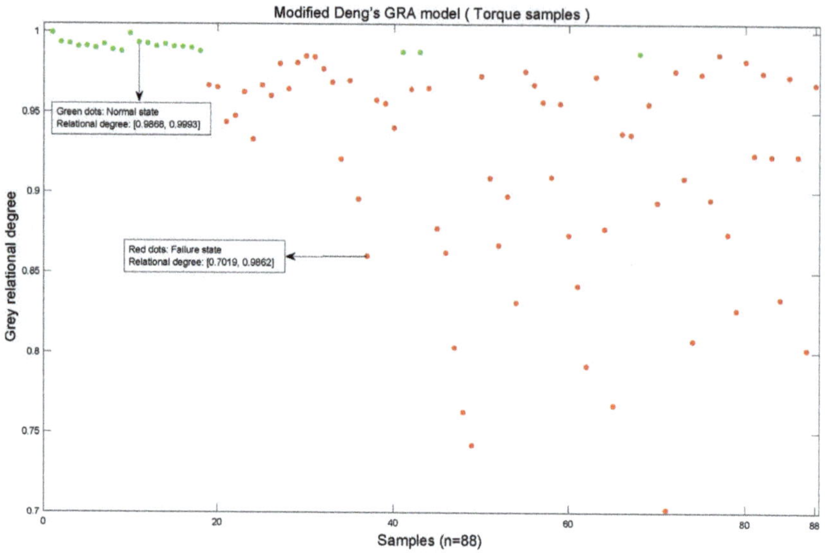

Figure 4. Scatter plot of 88 relational degree provided by modified Deng's GRA model based on torque samples.

4.4. Comparison and Analysis

In this section, to reveal the effectiveness of the modified Deng's GRA model for panel data, the classification results are compared with five other methods, including Deng's original GRA model, the Euclidean Norm GRA model [8], the matrix GRA model [21], the C-type GRA model [22], and the k-nearest neighbor (KNN) algorithm [23]. Similarly, the testing data are pre-processed in dimensionless form using an interval operator. By programming in Python, the metrics of overall performance for all models are listed in Table 6, where there are four classification evaluation metrics: accuracy, precision, recall (sensitivity), and F1 (score). These are quantitative measures used to assess the performance of a classification model and provide insights into the accuracy and reliability of the model's predictions.

Table 6. Classification results of the modified Deng's GRA model and other compared models.

Samples	Metrics	Deng's GRA	Euclidean Norm GRA	Matrix GRA	C-Type GRA	KNN	Modified Deng's GRA
Total samples	Accuracy	0.9773	1	1	1	1	1
	Precision	0.9524	1	1	1	1	1
	Recall	0.9524	1	1	1	1	1
	F1	0.9524	1	1	1	1	1
Force samples	Accuracy	1	1	1	1	1	1
	Precision	1	1	1	1	1	1
	Recall	1	1	1	1	1	1
	F1	1	1	1	1	1	1
Torque samples	Accuracy	0.9545	0.9773	1	1	0.9773	1
	Precision	0.9048	0.9524	1	1	0.9524	1
	Recall	0.9048	0.9524	1	1	0.9524	1
	F1	0.9048	0.9524	1	1	0.9524	1

Table 6 shows that Deng's original GRA model performs the worst, followed by the Euclidean norm GRA model and the KNN algorithm. By comparison, the modified Deng's GRA model, the matrix GRA model, and the C-type GRA model all perform well. This also demonstrates the effectiveness of the modified Deng's GRA model. In addition, the mathematical structure of the modified Deng's GRA model is simpler and easier to implement than that of the matrix GRA model and the C-type GRA model. Therefore, it can be deemed that the modified Deng's GRA model has more application prospects.

5. Case Study

In this section, as a practical application, the modified Deng's GRA model for panel data is applied to assess the water quality of Poyang Lake over the past five years. Poyang Lake, the largest freshwater lake in China, has abundant ecological resources and plays a significant role in economic and social spheres. Therefore, the Chinese government has always placed a high priority on the protection of water quality in Poyang Lake.

Among the various indicators of water quality, total phosphorus (TP) is the most crucial indicator and has a decisive influence on the health assessment of aquatic ecosystems. So TP is acknowledged as one of the pollutants that require stringent monitoring and control within water quality assessment and management practices. This case study will focus on the monthly total phosphorus data measured from eight monitoring sites in Poyang Lake District from 2019 to 2023. For confidentiality requirements, the eight monitoring sites are represented by symbols Site 1–8, respectively, and the detailed data can be found in Table A1.

According to the previous discussion in this paper, we suppose that $X_0 = (0)_{12\times 8}$ (which is zero matrix) is the behavior matrix of system characteristics, where $i(1 \leqslant i \leqslant 12)$ represents the time dimension and $j(1 \leqslant j \leqslant 8)$ represents the indicator dimension. The years are supposed as the samples for the behavior matrix of system factors, that is, $X_k = (a_{ij}^k)_{12\times 8}, k = 1, 2, \cdots, 5$.

Meanwhile, to verify the effectiveness of the modified Deng's GRA model for panel data, the results are compared with Deng's original GRA model and other GRA models for panel data, including the Euclidean Norm GRA model [8], the matrix GRA model [21], and the C-type GRA model [22]. As in Section 4, the testing data are pre-processed in dimensionless form using interval operator. By programming in Python, the relational degree values are all obtained, as depicted in Table 7.

Table 7 shows that the relational degree values differ significantly from each other, which is due to the fact that different GRA models for panel data have different functional expressions, and the range of relational degree is also different. Because the main purpose of GRA models is for cluster analysis based on the relational degree order, the numerical values of relational degree between various GRA models are meaningless. In other words, we merely need to compare the relational degree order of the five GRA models. Therefore, the relational degree order yielded by the five GRA models are displayed, as shown in Table 8.

Table 7. Relational degree values provided by five GRA models for panel data.

Years	Deng's GRA	Euclidean Norm GRA	Matrix GRA	C-Type GRA	Modified Deng's GRA
2019	0.6611	0.1888	0.7908	0.7418	0.6819
2020	0.6745	0.2208	0.8289	0.8011	0.7360
2021	0.6364	0.2089	0.8180	0.7855	0.7170
2022	0.6631	0.2305	0.8412	0.8182	0.7518
2023	0.6454	0.2118	0.8237	0.7933	0.7273

Table 8. Relational degree order yielded by five GRA models for panel data.

Models	Relational Degree Order
Deng's GRA	$r_{2020} \succ r_{2022} \succ r_{2019} \succ r_{2023} \succ r_{2021}$
Euclidean Norm GRA	$r_{2022} \succ r_{2020} \succ r_{2023} \succ r_{2021} \succ r_{2019}$
Matrix GRA	$r_{2022} \succ r_{2020} \succ r_{2023} \succ r_{2021} \succ r_{2019}$
C-type GRA	$r_{2022} \succ r_{2020} \succ r_{2023} \succ r_{2021} \succ r_{2019}$
Modified Deng's GRA	$r_{2022} \succ r_{2020} \succ r_{2023} \succ r_{2021} \succ r_{2019}$

Table 8 clearly shows that, among the order provided by the five GRA models, only Deng's original GRA model presents a different order from the others. This indicates that the modified Deng's GRA model for panel data is effective and significantly improves the clustering accuracy of the original model. Furthermore, the earlier the year appearing in the order, the lower the TP of Poyang Lake that year. Therefore, the data in Table 8 further confirm that the Chinese government has achieved significant results in reducing TP emissions from Poyang Lake and effectively protecting the water environment of Poyang Lake.

6. Conclusions

In this paper, a new method is provided to improve the clustering accuracy of the existing Deng's GRA model for panel data. Based on the numerical test results for Deng's original GRA model, it is found that the mechanism of Deng's original GRA model is reasonable, but it lacks in-depth modeling. To solve this problem, a modified Deng's GRA model is presented that can be applied to the cluster analysis of data with any number of dimensions. Three numerical experiments on the Robot Execution Failures Dataset have verified the significant improvement in clustering accuracy compared to Deng's original GRA model for panel data. Furthermore, a case study on the monthly TP data of Poyang Lake over the past five years is given to demonstrate the superiority of the modified model. By comparing it with other cluster methods, the results of the case study show that the modified Deng's GRA model for panel data is applicable and also confirm the remarkable effectiveness of the Chinese government's water quality regulation in Poyang Lake for many years. Due to the simplicity of the model, the modified Deng's GRA model for panel data presented in this paper can be more widely applied in the future research of grey relational theory and is expected to more effectively reveal the interactions between various factors in the system, helping researchers to identify key factors and optimize the system structure.

However, comparative analysis results show that the modified Deng's GRA model, the matrix GRA model, and the C-type GRA model all perform well. Based on the simple mathematical structure and easy implementation of the model, the modified Deng's GRA model is deemed effective and has more application prospects. There may be some potential shortcomings. Therefore, a more extensive simulation study with other clustering methods will be a good idea for our future work. On the other hand, the selection of data pre-processing methods is crucial for ensuring the accuracy of the model. Therefore, discussing the specific impact of different data pre-processing methods on the performance of various GRA models will also be meaningful research work for the future.

Author Contributions: Conceptualization, F.J. and X.L.; methodology, X.L.; software, F.J.; validation, F.J., J.L. and X.L.; formal analysis, F.J. and X.L.; investigation, F.J. and X.L.; resources, J.L., Q.W. and D.Z.; data curation, J.L., Q.W. and D.Z.; writing—original draft preparation, X.L.; writing—review and editing, X.L.; visualization, F.J.; supervision, J.L., Q.W. and D.Z.; project administration, F.J., J.L., X.L., Q.W. and D.Z.; funding acquisition, F.J., J.L., X.L., Q.W. and D.Z. All authors have read and agreed to the published version of the manuscript.

Funding: This research was funded by the National Natural Science Foundation of China (62341308), Natural Science Foundation of Jiangxi Province (20232BAB201020, 20224BAB201010), and Open Fund of Key Laboratory of Industrial Ecological Simulation and Environmental Health in Yangtze River Basin of Jiangxi Province (jj20212021).

Data Availability Statement: Details of dataset LP1 can be found from Robot Execution Failures Dataset linking to the website (http://kdd.ics.uci.edu/databases/robotfailure/robotfailure.html), and the data of the case study can be found in Appendix A.

Conflicts of Interest: The authors declare no conflicts of interest.

Appendix A

Table A1. Total phosphorus data in Poyang Lake District (Unit: mg/L).

Sample Time	Site 1	Site 2	Site 3	Site 4	Site 5	Site 6	Site 7	Site 8
2019/1/1	0.52	0.12	0.08	0.13	0.11	0.02	0.06	0.1
2019/2/1	0.08	0.03	0.04	0.13	0.09	0.01	0.04	0.05
2019/3/5	0.08	0.07	0.07	0.07	0.07	0.08	0.1	0.06
2019/4/1	0.04	0.03	0.04	0.04	0.06	0.04	0.04	0.04
2019/5/6	0.08	0.04	0.04	0.06	0.05	0.05	0.04	0.07
2019/6/4	0.04	0.02	0.03	0.03	0.06	0.03	0.06	0.03
2019/7/2	0.02	0.01	0.03	0.03	0.06	0.03	0.04	0.03
2019/8/1	0.03	0.02	0.02	0.02	0.01	0.02	0.03	0.03
2019/9/3	0.11	0.05	0.06	0.06	0.05	0.08	0.05	0.08
2019/10/8	0.04	0.08	0.03	0.03	0.03	0.02	0.04	0.02
2019/11/1	0.34	0.11	0.19	0.08	0.11	0.02	0.07	0.08
2019/12/4	0.07	0.07	0.06	0.06	0.06	0.02	0.09	0.07
2020/1/6	0.07	0.05	0.12	0.047	0.055	0.08	0.05	0.05
2020/2/12	0.02	0.1	0.02	0.068	0.03	0.08	0.08	0.08
2020/3/5	0.04	0.06	0.03	0.04	0.027	0.08	0.05	0.05
2020/4/1	0.04	0.04	0.05	0.05	0.05	0.04	0.03	0.06
2020/5/1	0.05	0.04	0.06	0.043	0.06	0.04	0.07	0.06
2020/6/1	0.05	0.072	0.05	0.05	0.047	0.06	0.07	0.04
2020/7/1	0.02	0.033	0.03	0.03	0.05	0.03	0.05	0.03
2020/8/1	0.03	0.025	0.017	0.017	0.027	0.03	0.027	0.03
2020/9/1	0.04	0.04	0.026	0.053	0.04	0.03	0.038	0.05
2020/10/1	0.04	0.037	0.033	0.05	0.042	0.04	0.05	0.097
2020/11/1	0.075	0.07	0.045	0.093	0.08	0.03	0.05	0.06
2020/12/1	0.09	0.048	0.052	0.06	0.07	0.03	0.12	0.08
2021/1/1	0.08	0.045	0.045	0.05	0.05	0.03	0.06	0.07
2021/2/1	0.12	0.038	0.04	0.05	0.06	0.03	0.075	0.07
2021/3/1	0.09	0.047	0.055	0.077	0.05	0.06	0.11	0.105
2021/4/1	0.04	0.041	0.051	0.058	0.06	0.05	0.05	0.055
2021/5/1	0.03	0.029	0.039	0.04	0.04	0.04	0.05	0.04
2021/6/1	0.03	0.022	0.037	0.03	0.04	0.03	0.062	0.04
2021/7/1	0.04	0.037	0.039	0.12	0.053	0.05	0.06	0.06
2021/8/1	0.108	0.032	0.041	0.05	0.125	0.03	0.07	0.04
2021/9/1	0.065	0.035	0.037	0.052	0.048	0.05	0.055	0.052
2021/10/1	0.08	0.048	0.065	0.077	0.06	0.05	0.065	0.077
2021/11/1	0.05	0.063	0.066	0.047	0.1	0.07	0.06	0.05
2021/12/1	0.07	0.05	0.048	0.065	0.065	0.07	0.08	0.07
2022/1/1	0.05	0.055	0.04	0.028	0.065	0.07	0.05	0.05
2022/2/1	0.05	0.048	0.013	0.04	0.062	0.05	0.055	0.05
2022/3/1	0.07	0.048	0.031	0.048	0.04	0.05	0.05	0.05
2022/4/1	0.04	0.042	0.045	0.048	0.062	0.04	0.04	0.05
2022/5/1	0.05	0.033	0.046	0.06	0.043	0.08	0.072	0.07
2022/6/1	0.05	0.023	0.039	0.04	0.047	0.03	0.065	0.05
2022/7/4	0.03	0.033	0.035	0.05	0.048	0.04	0.055	0.06
2022/8/1	0.07	0.076	0.046	0.07	0.07	0.04	0.062	0.07
2022/9/1	0.07	0.063	0.051	0.08	0.1	0.04	0.075	0.075
2022/10/1	0.085	0.04	0.044	0.05	0.08	0.04	0.06	0.005
2022/11/1	0.04	0.044	0.036	0.045	0.05	0.04	0.04	0.05
2022/12/1	0.1	0.052	0.05	0.03	0.1	0.041	0.04	0.04
2023/1/1	0.05	0.039	0.042	0.02	0.075	0.04	0.04	0.035
2023/2/1	0.04	0.065	0.062	0.08	0.08	0.04	0.04	0.02
2023/3/1	0.04	0.054	0.046	0.05	0.052	0.04	0.04	0.06
2023/4/1	0.14	0.045	0.053	0.05	0.08	0.04	0.08	0.09
2023/5/1	0.05	0.068	0.053	0.085	0.09	0.06	0.05	0.055
2023/6/1	0.05	0.073	0.044	0.06	0.045	0.05	0.065	0.045
2023/7/1	0.04	0.062	0.045	0.04	0.042	0.05	0.065	0.04
2023/8/1	0.05	0.074	0.037	0.045	0.048	0.16	0.05	0.042
2023/9/1	0.03	0.067	0.055	0.05	0.06	0.16	0.04	0.04
2023/10/1	0.05	0.044	0.052	0.045	0.04	0.04	0.06	0.05
2023/11/1	0.05	0.067	0.065	0.03	0.06	0.04	0.11	0.05
2023/12/1	0.04	0.058	0.045	0.04	0.048	0.04	0.05	0.04

References

1. Deng, J. The theory and methods of socio-economy grey system. *Soc. Sci. China* **1984**, *6*, 47–60.
2. Zhang, W.; Shi, H.; Zhao, Y.; Ding, T. Significance ranking and correlation identification of accident causes in process industry based on system thinking and statistical analysis. *Process Saf. Prog.* **2024**, *43*, 144–159. [CrossRef]
3. Xu, S.; Lu, Q.; Bin, B. Grey correlation analysis on the synergistic development between innovation-driven strategy and marine industrial agglomeration: Based on China's coastal provinces. *Grey Syst. Theory Appl.* **2022**, *12*, 269–289. [CrossRef]
4. Chen, X.; Ding, X.; Clark, A.; Wu, Y.; Feng, G.; Xiao, Y. A decision support model for subcontractor selection using a hybrid approach of QFD and AHP-improved grey correlation analysis. *Eng. Constr. Archit. Manag.* **2020**, *28*, 1780–1806. [CrossRef]
5. Liu, S.; Yang, Y.; Forrest, J. *Grey Data Analysis: Methods, Models and Applications*; Springer: London, UK, 2016; pp. 67–103.
6. Liu, X.; Ke, L.; Yu, J. An improved model to the absolute degree of grey incidence. *Math. Pract. Theory* **2018**, *48*, 16–22.
7. Liu, Z.; Dang, Y.; Xiao, Y. A new multivariate grey incidence model and its application to economic growth driving analysis. *J. Grey Syst.* **2018**, *30*, 116–133.
8. Gui, Y.; Xia, Y.; Deng, L. Grey incident grade in linear norm space. *J. Wuhan Univ. Technol.* **2004**, *28*, 399–412.
9. Wu, L.; Liu, S.; Yao, L.; Yan, S. Grey convex relational degree and its application to evaluate regional economic sustainability. *Sci. Ironical* **2013**, *20*, 44–49. [CrossRef]
10. Liu, Z.; Dang, Y.; Qian, W.; Zhou, W. Grey grid incidence model based on panel data. *Syst. Eng. Theory Pract.* **2014**, *34*, 991–996.
11. Liu, X.; Ke, L.; Yu, J. An improved model of three-dimensional absolute degree of grey incidence. *Stat. Decis.* **2018**, *34*, 20–24.
12. Wu, H.; Qu, Z. A grey incidence model for panel data based on the curvature of discrete surface. *J. Grey Syst.* **2022**, *34*, 75–87.
13. Zhou, H.; Tang, Z.; Wen, B.; Wang, S.; Yang, J.; Kou, M.; Wu, S. Application of statistical analysis, Deng's relevancy and BP neural network for predicting molten iron sulfur in COREX process. *Int. J. Chem. React. Eng.* **2020**, *18*, 20200122. [CrossRef]
14. Zhang, K.; Cui, L.; Yin, Y. A multivariate grey incidence model for different scale data based on spatial pyramid pooling. *J. Syst. Eng. Electron.* **2020**, *31*, 770–779.
15. Javed, S.; Gunasekaran, A.; Mahmoudi, A. DGRA: Multi-sourcing and supplier classification through dynamic grey relational analysis method. *Comput. Ind. Eng.* **2022**, *173*, 108674. [CrossRef]
16. Xu, G.; Lu, T.; Chen, X.; Liu, Y. The convergence level and influencing factors of China's digital economy and real economy based on grey model and PLS-SEM. *J. Intell. Fuzzy Syst.* **2022**, *42*, 1575–1605. [CrossRef]
17. Li, C.; Li, H.; Zhang, G. Grey-adversary perceptual network for anomaly detection. *Multimed. Tools Appl.* **2024**, *83*, 41273–41291. [CrossRef]
18. Liu, X.; YU, J. Grey incidence analysis models for matrix data and matrix sequences data. *J. Grey Syst.* **2019**, *31*, 59–70.
19. Liu, S. Negative grey relational model and measurement of the reverse incentive effect of fields medal. *Grey Syst.-Theory Appl.* **2023**, *13*, 1–13. [CrossRef]
20. Huang, J.; Dang, Y.; Wang, J.; Xue, Q. Novel Deng's grey development relation model based on information difference and its application in sanatorium performance evaluation. *Math. Probl. Eng.* **2020**, *6*, 3427040. [CrossRef]
21. Luo, D.; Zhang, H. Grey incidence analysis method for regional drought vulnerability. *J. North China Univ. Water Resour. Electr. Power (Nat. Sci. Ed.)* **2018**, *39*, 61–67.
22. Luo, D.; Li, J. Grey C-type correlation analysis of regional agricultural drought disaster risk based on panel data. *J. North China Univ. Water Resour. Electr. Power (Nat. Sci. Ed.)* **2020**, *41*, 47–53.
23. Uddin, S.; Haque, I.; Lu, H.; Moni, M.; Gide, E. Comparative performance analysis of K-nearest neighbor (KNN) algorithm and its different variants for disease prediction. *Sci. Rep.* **2024**, *12*, 6256.

processes

MDPI

Article

Local Path Planner for Mobile Robot Considering Future Positions of Obstacles

Xianhua Ou, Zhongnan You and Xiongxiong He *

College of Information Engineering, Zhejiang University of Technology, Hangzhou 310014, China; xhou@zjut.edu.cn (X.O.); youayou121@163.com (Z.Y.)
* Correspondence: hxx@zjut.edu.cn

Abstract: Local path planning is a necessary ability for mobile robot navigation, but existing planners are not sufficiently effective at dynamic obstacle avoidance. In this article, an improved timed elastic band (TEB) planner based on the requirements of mobile robot navigation in dynamic environments is proposed. The dynamic obstacle velocities and TEB poses are fully integrated through two-dimensional (2D) lidar and multi-obstacle tracking. First, background point filtering and clustering are performed on the lidar points to obtain obstacle clusters. Then, we calculate the data association matrix of the obstacle clusters of the current and previous frame so that the clusters can be matched. Thirdly, a Kalman filter is adopted to track clusters and obtain the optimal estimates of their velocities. Finally, the TEB poses and obstacle velocities are associated: we predict the obstacle position corresponding to the TEB pose through the detected obstacle velocity and add this constraint to the corresponding TEB pose vertex. Then, a pose sequence considering the future positions of obstacles is obtained through a graph optimization algorithm. Compared with the original TEB, our method reduces the total running time by 22.87%, reduces the running distance by 19.23%, and increases the success rate by 21.05%. Simulations and experiments indicate that the improved TEB enables robots to efficiently avoid dynamic obstacles and reach the goal as quickly as possible.

Keywords: dynamic obstacle avoidance; two-dimensional lidar; Kalman filter; TEB local planner

1. Introduction

Wheeled mobile robots are increasingly used in today's society in such forms as food delivery robots and transportation robots. In the autonomous navigation of robots, avoiding obstacles and safely reaching the target are the challenges they encounter, especially in the face of dynamic obstacles, and the robot must process a large amount of environmental information and make optimal control decisions. After the robot receives the reference path given by the global path planner [1,2], the local path planner needs to give the robot an optimal local path through sensors such as cameras, lidar, or ultrasonic devices so that the mobile robot can reach the target as soon as possible while avoiding obstacles on the path.

Obstacle detection plays a key role in the navigation of mobile robots in dynamic environments. This is the main way for mobile robots to receive obstacle motion information. Many recent object detection algorithms are vision-based [3,4]; these need a large amount of calculations, and vision-based object detection is greatly affected by light. Of course, some people use low-cost sensors such as infrared sensors for detecting obstacles [5]. Although the amount of calculation is small, there is too little information, so in this article, we weigh the pros and cons of using two-dimensional lidar to detect obstacle movement information [6–11]. We think that for robots, the data provided by 2D lidar can basically meet the robot's obstacle avoidance needs because for robot obstacle avoidance, we only need to consider obstacle detection rather than obstacle recognition.

Local planners include the dynamic window approach (DWA) and TEB. The DWA [12] mainly samples multiple groups of velocities in the linear and angular velocity space, simulates the trajectories formed by these velocities within a certain time period, evaluates these trajectories through the evaluation function, and selects the best trajectory. Its advantage is low computational complexity and strong real-time performance. The disadvantage is that it easily becomes stuck in local optimal, and it is suitable only for certain robots. The TEB [13–16] local planner's working principle is to insert N time intervals of T seconds between the start and target. Then, it finds the optimal pose sequence through the optimization algorithm and motion instructions to drive the robot to the target. In the Robot Operating System (ROS), TEB is an open-source package. The optimization method uses the general graph optimization (G2O) algorithm [17].

The two classic algorithms mentioned above can basically realize the local planning for most robots in most static environments, but these algorithms do not take the movement information of obstacles into consideration, and they all regard obstacles as static obstacles, which causes the planned trajectory to be only currently optimal, regardless of the future positions of the obstacles, so the obstacle avoidance effect in a dynamic environment is not good enough. However, robot obstacle avoidance is systematic work that not only needs path planners [18,19] but also needs to combine obstacle perception [20,21], robot control [22], trajectory tracking [23,24], and localization [25–27].

Therefore, on the basis of the TEB algorithm structure, we use 2D lidar to fully integrate dynamic obstacle velocities and finally plan a collision-free trajectory considering the future positions of obstacles. First, the robot receives a 2D point set through lidar and obtains obstacle point clusters through operations such as coordinate transformation, background point filtering, and obstacle point clustering. Then, a Kalman filter [28] is used to track obstacles to obtain obstacle velocities. Afterward, the velocities of obstacles are used to predict the future positions of the obstacles for various TEB poses; then, the robot optimizes the trajectory by the G2O method to obtain a collision-free trajectory.

The main contributions of this manuscript are as follows:

1. A multi-obstacle velocities detection algorithm through 2D lidar and Kalman filter is proposed.
2. An improved TEB local planner that combines the TEB poses and obstacle velocities is designed to improve the robot's dynamic obstacle avoidance efficiency.
3. A series of simulations and experiments are performed to verify the effectiveness of the proposed method.

2. Related Work

Path planning in response to dynamic obstacles is an open research topic that has been studied for decades. There are many obstacle avoidance algorithms for dynamic environments, but many of these algorithms were initially proposed for static environments and then adapted to dynamic environments, including artificial potential field (APF), DWA, and TEB. The APF [29] algorithm is a typical reactive method. It uses repulsion and attraction to plan the local trajectory of the robot. Attraction is positively correlated with the distance between the robot and the target, while repulsion is the opposite. This algorithm has a simple structure, but there are many defects. For example, it does not consider the kinematics of the robot, and it can easily fall into a local optimum; moreover, the algorithm does not take the future positions of obstacles into consideration. Methods based on velocity obstacle (VO) [30–33] are also noteworthy reactive methods that utilize the collision region principle to detect regions where collisions are possible for dynamic obstacle avoidance. Although these reactive methods are effective at obstacle avoidance, there is still considerable room for improvement. The application of the original TEB algorithm in dynamic scenes only continuously updates the local cost map without considering the statuses of obstacles. In this situation, if the speed of the robot or obstacle is too fast, the robot may not be able to successfully avoid obstacles.

There are some recent studies considering dynamic obstacles. Guo [34] proposed an obstacle avoidance algorithm suitable for two-wheeled differential-driven robots by using a stereo vision system to detect dynamic obstacles, estimating the motion state of dynamic obstacles through an extended Kalman filter, and then by using the velocities of obstacles to calculate dynamic risk regions and realize dynamic obstacle avoidance for the robot.

However, it can be applied to only the differential speed robot. For pedestrians, a special type of obstacle avoidance model has been proposed: Reddy [35] is a hybrid path planning method that takes into account some navigation habits of pedestrians walking, such as avoiding obstacles from the left side. Using these behaviors as a reference, the method combines the social force field with geometric methods to select the optimal path for the robot. According to experimental results, this algorithm outperforms baseline and individual methods. Smith [36] combines path planning with reinforcement learning and proposes a cooperative Markov decision process (Co-MDP) approach. This method utilizes an MDP to predict human behavior and integrates the prediction results with social rules and the state of the robot; the research aims to humanize the robot's movements, and the robot performs well in human–robot co-populated environments. Chen and Lin [37] provided a plugin of the cost map2d [38] layer in the framework of the ROS. On the basis of the obstacle layer and the obstacle expansion layer, the obstacle velocity layer was introduced. This layer passed the DeepSort [39] algorithm to detect dynamic obstacles, to further extract 2D lidar messages, and to generate moving obstacles and the range of future moments through the k-means clustering algorithm; this method provides better trajectory planning for mobile robots to avoid obstacles in dynamic environments. Dong [40] proposed an algorithm that combines two-dimensional lidar and the TEB algorithm. Simulation and experiments have proven the ability of the system in real-time detection and obstacle tracking and show that it has a certain robustness. Wang [41] set up a virtual target (VT) on the global path and tracked the VT and tracked the global path when the robot was moving in the dynamic environment to achieve safe and stable operation of the robot in a dynamic environment. In this article, we also propose a global path replanning strategy to address the situation of untrackable areas in dense environments. Through a large number of experiments, it is proven that this method can reach the target faster, safer, and more smoothly than can other methods.

3. Methodology

In this section, we first introduce how to use 2D lidar for obstacle detection, and then, we describe the details of obstacle tracking when using the Kalman filter and the calculation of the interframe data association matrix. Finally, we introduce how to use obstacle velocities and TEB pose information to predict the obstacle positions of different poses in the future as well as a method for associating TEB pose vertices and obstacle edges.

3.1. Obstacle Detection

3.1.1. Clustering

A point set obtained by 2D lidar is based on the polar coordinate system with the lidar as the origin and consists of the polar diameter and polar angle; the expression is (θ, r). However, this form of data does not satisfy our requirements for obstacle velocity detection, so we need to transform it into the coordinate system (x, y) with the map as the origin. Therefore, this algorithm relies more on the accuracy of the real-time localization algorithm of the robot. The formula to transform the coordinates from (θ, r) to (x, y) is as follows:

$$\begin{bmatrix} x \\ y \end{bmatrix} = \begin{bmatrix} \cos\theta_r & -\sin\theta_r \\ \sin\theta_r & \cos\theta_r \end{bmatrix} \times \begin{bmatrix} \cos\theta \\ \sin\theta \end{bmatrix} \times r + \begin{bmatrix} x_r \\ y_r \end{bmatrix} \tag{1}$$

where (x_r, y_r, θ_r) is the robot's global pose in the map, (x_r, y_r) are the coordinates and θ_r is the orientation of the robot, (x, y) are the point coordinates after transformation, while (θ, r) are the polar coordinate points before.

After obtaining the global point cloud, these points need to be clustered. This paper uses the conditional Euclidean clustering algorithm to cluster the point set and uses the known inflated global grid map as a mask to filter out the background points; this clustering results in K obstacle clusters. Figure 1 presents the schematic diagram of background point clustering. The original Euclidean has only one hyperparameter: that is, ϵ, which is the neighborhood range. The setting of this hyperparameter will depend on the width of the mobile robot, the pedestrian's step length, and the resolution of the lidar. Conditional Euclidean clustering adds a hyperparameter based on the original algorithm: that, is the background grid map after inflation. Whenever the clustering algorithm searches for a point, in addition to judging how the point cloud is distanced from the neighboring points, the background map will also be used to determine whether the point is a noise point. After the points are clustered, K clusters will be obtained. In addition, the length and width of each cluster needs to be calculated to facilitate the subsequent calculation of the intersection over union (IOU).

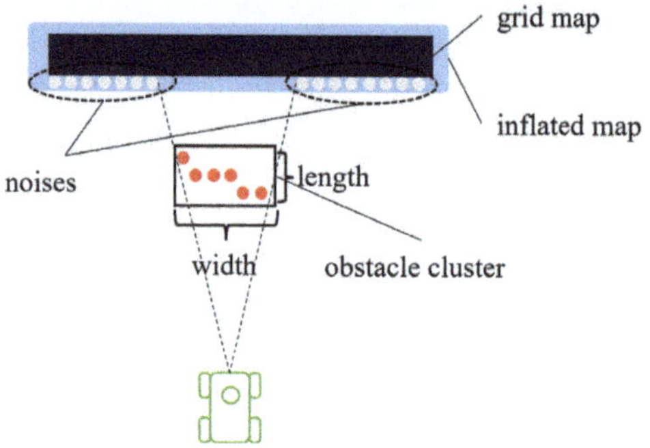

Figure 1. Schematic diagram of background point clustering. The points in the figure are the filtered obstacle points: the white points are the noise points, the black rectangle is the obstacle, and the blue area is the inflated obstacle area.

3.1.2. Data Association Matrix

Calculation of the data association matrix is crucial in multi-obstacle tracking because it is necessary to ensure that the clusters of the current frame corresponds to the clusters of the previous frame. The calculation method of the data association matrix in this article is introduced below. First, the expression of the obstacle cluster center set obtained by clustering is as follows:

$$P^{pre} \triangleq \{(px_i^{pre}, py_i^{pre}), i \in [1, M]\} \tag{2}$$

$$P^{now} \triangleq \{(px_j^{now}, py_j^{now}), j \in [1, K]\} \tag{3}$$

where P^{pre} is the cluster set in the previous frame, P^{now} is the cluster set in the current frame, and M and K are the total number of point cloud cluster sets. Then, the data association matrix mat is weighted by two matrices: a distance difference matrix mat_{dist} and an IOU matrix mat_{IOU}. The specific expansion formula is as follows:

$$mat = \alpha \times mat_{dist} + \beta \times mat_{IOU} \tag{4}$$

$$mat_{dist} = \begin{bmatrix} d(p_1^{pre}, p_1^{now})^{-1} & \cdots & d(p_1^{pre}, p_K^{now})^{-1} \\ \vdots & \ddots & \vdots \\ d(p_M^{pre}, p_1^{now})^{-1} & \cdots & d(p_M^{pre}, p_K^{now})^{-1} \end{bmatrix} \tag{5}$$

$$mat_{IOU} = \begin{bmatrix} IOU(p_1^{pre}, p_1^{now}) & \cdots & IOU(p_1^{pre}, p_K^{now}) \\ \vdots & \ddots & \vdots \\ IOU(p_M^{pre}, p_1^{now}) & \cdots & IOU(p_M^{pre}, p_K^{now}) \end{bmatrix} \tag{6}$$

The expansion of the data in the matrix mat_{dist} is as follows (that is, the distance between the center of the i-th cluster in the previous frame and the j-th cluster in the current frame):

$$d(p_i^{pre}, p_j^{now}) = \sqrt{(x_i^{pre} - x_j^{now})^2 + (y_i^{pre} - y_j^{now})^2} \tag{7}$$

The specific meaning of the data in the matrix mat_{IOU} is the intersection and union ratio between the current cluster and the previous frame. The IOU is the intersection ratio and the union of the binding boxes of the two clusters. Finally, we get a value between (0, 1).

In addition, in order to eliminate the dimensional influence between different features and to improve the matching accuracy, the two matrices need to be normalized: the original data are linearly mapped to the [0, 1] interval, and the mapping function is:

$$x' = (x - min)/(max - min) \tag{8}$$

where x are the original data, x' are the normalized data, min is the minimum value in the matrix, and max is the maximum value in the matrix.

3.1.3. Kalman Filter Tracking

The pairing relationship of clusters between frames can be obtained through the data association matrix, and then, the obstacles are tracked through Kalman filtering to obtain their motion information. The state equation of the obstacles is:

$$X_t = AX_{t-1} + W_{t-1} \tag{9}$$

where $X_t = [x, y, v^x, v^y, a^x, a^y]^T$ is a six-dimensional state vector that includes position, velocity, and acceleration in the x- and y-directions. X_{t-1} is the state at the previous moment, and W_{t-1} is the process noise, for which the probability distribution is unknown. A is the state transition matrix; the expansion of this matrix is as follows:

$$A = \begin{bmatrix} 1 & 0 & dt & 0 & dt^2/2 & 0 \\ 0 & 1 & 0 & dt & 0 & dt^2/2 \\ 0 & 0 & 1 & 0 & dt & dt^2/2 \\ 0 & 0 & 0 & 1 & 0 & dt \\ 0 & 0 & 0 & 0 & 1 & 0 \\ 0 & 0 & 0 & 0 & 0 & 1 \end{bmatrix} \tag{10}$$

where dt is the observation interval: that is, the time interval between frames. The observation matrix for dynamic obstacles is:

$$Z_t = HX_t + V_t \tag{11}$$

where $Z_t = [x, y, v^x, v^y]^T$ is the observation value and includes the coordinates and velocities of the obstacle in the x- and y-directions. Observations come from inter-frame clusters: from such clusters we obtain the positions of the obstacles. Referring to the idea of differentiation, v^x and v^y can be obtained. As long as the sampling interval dt is small enough, the obstacles in the two sampling intervals can be seen to move at a uniform speed: that is $v_t^x = (x_t - x_{t-1})/dt$. Similarly, v_t^y, V_t is observation noise. The observation matrix H is:

$$H = \begin{bmatrix} 1 & 0 & 0 & 0 \\ 0 & 1 & 0 & 0 \\ 0 & 0 & 1 & 0 \\ 0 & 0 & 0 & 1 \end{bmatrix} \tag{12}$$

The posterior estimate $\hat{X}_t$ is calculated through the Kalman gain from the following equation:

$$\hat{X}_t = \hat{X}_t^- + K_t(Z_t - H\hat{X}_t^-) \tag{13}$$

The term $\hat{X}_t$ is the optimal motion estimates of obstacles. Following, we introduce how to associate motion information with the original TEB algorithm to achieve local planning and dynamic obstacle avoidance.

3.2. Improved TEB Algorithm

The TEB algorithm is a classic local path planning algorithm. Based on the elastic band (EB) algorithm, it adds various constraints such as time, velocities, obstacles, and path and robot kinematic constraints. It imagines the ideal path as a elastic band and treats all external constraints as external forces that cause the elastic band to deform. The original TEB algorithm obtains the location information of obstacles through a local cost map. Therefore, the information obtained by the algorithm does not include the motion information of the obstacles. This will cause the algorithm to perform poorly in an environment with dynamic obstacles, mainly due to the possibilities that there is no time to avoid obstacles and that the obstacle avoidance path is not optimal. Therefore, in order to optimize the operating effect of the algorithm in a dynamic environment, this article optimizes the path by associating the TEB poses and the obstacle positions at the corresponding future time. This article introduces a new constraint function:

$$f_{fpobs} = e(-d_{min,j}, -r_{o_{min}}, \epsilon, S, n) \tag{14}$$

where $d_{min,j}$ is the distance between the TEB pose and the obstacle at the corresponding time, $r_{o_{min}}$ is the minimum distance between the obstacle and the pose point, and the specific expansion formula of the e function is as follows:

$$e(x, x_r, \epsilon, S, n) \cong \begin{cases} \left(\frac{x-(x_r-\epsilon)}{S}\right)^n & x > x_r - \epsilon \\ 0 & else \end{cases} \tag{15}$$

where x_r is the limit value, and S, n, and ϵ affect the accuracy of the approximation and are scaling, polynomial order, and a small displacement around the approximation value, respectively.

The obstacle avoidance effects of the improved TEB algorithm when facing obstacles in different directions are shown in Figure 2. As can be seen from Figure 2a, when it is detected that the obstacle is moving from left to right, the improved TEB algorithm plans an obstacle avoidance path to avoid the obstacle to the left. The same is true for the obstacle avoidance in Figure 2b. As can be seen from the schematic diagram, when it detects that the direction of movement of an obstacle is towards the robot from top to bottom, the algorithm makes a decision to avoid obstacles in advance instead of waiting until the robot enters the area affected by the obstacle.

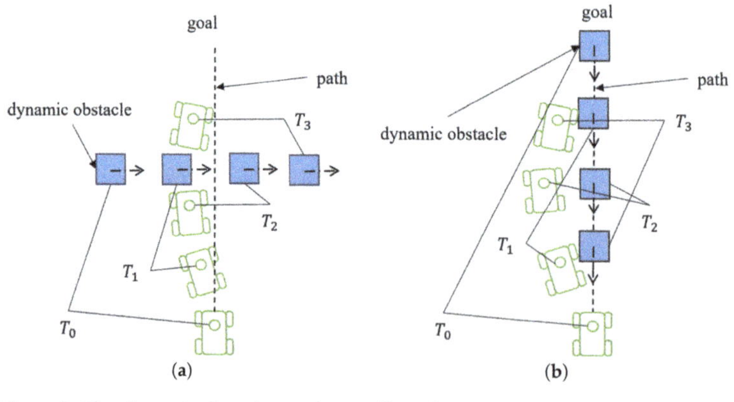

Figure 2. The dynamic obstacle avoidance effect of improved TEB: The green car represents the position of the mobile robot at different times, and the blue rectangle represents the position of the one obstacle at different times. The arrow represents the speed and direction of the obstacle. The robot's target is at the top of the figure. The black dotted line is the ideal motion trajectory of the robot, with $T_0 - T_3$ indicating time 0 to time 3. (**a**) Obstacle from the side. (**b**) Obstacle from the front.

Figure 3 is a schematic diagram of the proposed obstacle position in the future. The position of the obstacle is related to the corresponding TEB pose, where the position of the j-th obstacle is $P_j = [x_j, y_j]^T$, as follows:

$$P_j = P_0 + \sum_{i=0}^{j} \left(V + dt_i^{i+1} + \frac{A \times (dt_i^{i+1})^2}{2} \right), j \in [1, k] \quad (16)$$

where $V = [v_x, v_y]^T$, $A = [a_x, a_y]^T$ are the speed and acceleration, respectively, of the obstacle detected at the current moment, and dt_i^{i+1} is the time interval between the i-th and $(i+1)$-th TEB poses.

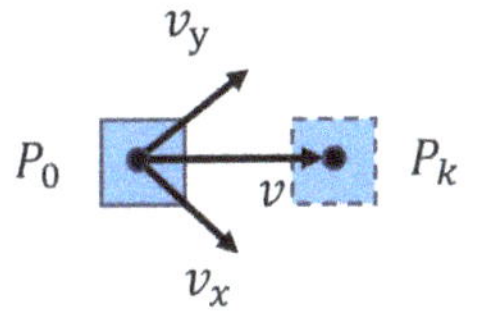

Figure 3. Schematic diagram of the obstacle's future position. The solid square area in the figure is the obstacle information detected at the current moment. It has component velocities in the x- and y-directions. The dotted line area is the k-th TEB pose.

Figure 4 shows the details of dynamic obstacle avoidance achieved by the improved TEB algorithm in Figure 2. It can be clearly seen from Figure 4a that at time T_0, the robot has planned a path to avoid obstacles to the left, because at this time, the obstacle corresponding to the second TEB pose is on the ideal path of the robot, so the TEB pose at this position will be stretched to the left. Due to the characteristics of the TEB poses, the TEB poses adjacent to it will also be affected and stretched to the left. The robot will move to the left and avoid obstacles. The same applies to the poses in Figure 4b.

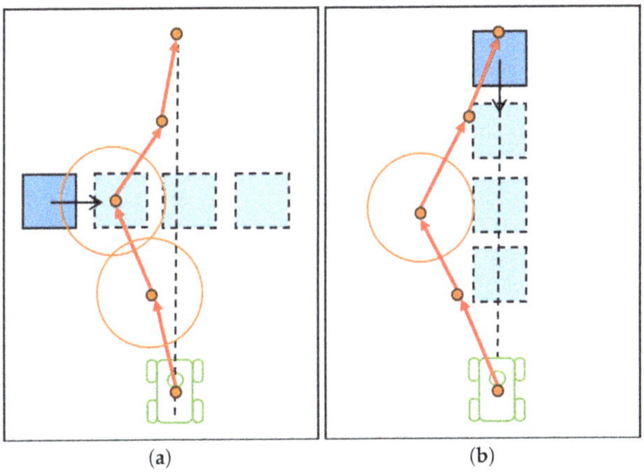

Figure 4. Details of improved TEB. This figure presents the details at time T_0. The local pose sequence of the robot in the figure consists of four red arrows, and the blue square is the obstalce at T_0 while the light-blue squares are the future positions of the obstalce at different times. And the green car represents the robot. The specific calculation process can be referenced in Figure 3 and Equation (16).

It can be seen from the above that the improved TEB algorithm can combine the TEB poses with the future positions of obstacles and plan a local path that integrates the obstacle velocities, improving the obstacle avoidance efficiency of the mobile robot when facing dynamic obstacles. In order to verify the proposed algorithm's effectiveness, this article will verify the algorithm on simulation and experimental platforms.

4. Simulations and Experiments

4.1. Simulations

The simulation environment has good repeatability and controllable conditions. To evaluate the robustness and efficacy of the proposed approach in a dynamic environment, we use Gazebo and the ROS for simulations (the simulation environment is shown in Figure 5) and choose an Ackermann type of car as the mobile robot. First, we map the environment in a simulation environment without dynamic obstacles, and then, we add corresponding obstacles for simulation testing experiments. We use rosbag to extract the simulation data and then export the relevant topic information in the .bag file into a .csv file for analysis. We verify the navigation effect of the proposed method for a single obstacle with various velocities and directions and in a complex environment with multiple obstacles. For the first two simulations, to highlight our improvements over the original TEB local planner, we compare the navigation effect of the improved TEB with the effect of the original. In order to demonstrate the advantages of the proposed method over other local planners, we also compare the improved TEB with a model predictive control (MPC) local planner and a state lattice planner in the third simulation.

(**a**) (**b**) (**c**)

Figure 5. Simulation environment: (**a**) the simulation environment, which is a cross corridor, (**b**) the car used in the simulation, (**c**) the pedestrian acting as a dynamic obstacle in the simulation. The blue area in the figure represents the area that the lidar can scan.

4.1.1. Obstacle Ahead of the Robot

The first set of simulations is used to verify the obstacle avoidance effect of the method for a single obstacle with different velocities coming from the front. The simulation environment is shown in the first row of Figure 6. The start position of the pedestrian is (12, 0), and the pedestrian moves at a speed of 0.8 m/s, 1.0 m/s, and 1.2 m/s, respectively, in the three simulations. From the comparison of Figure 6d,g, it can be seen that the distance between the robot and the pedestrian is approximately 5 m at this time, but the TEB poses of the original algorithm are still in the reference path, while for our method, the robot predicts that in the future, a pedestrian will affect its trajectory, so advanced obstacle avoidance actions are taken.

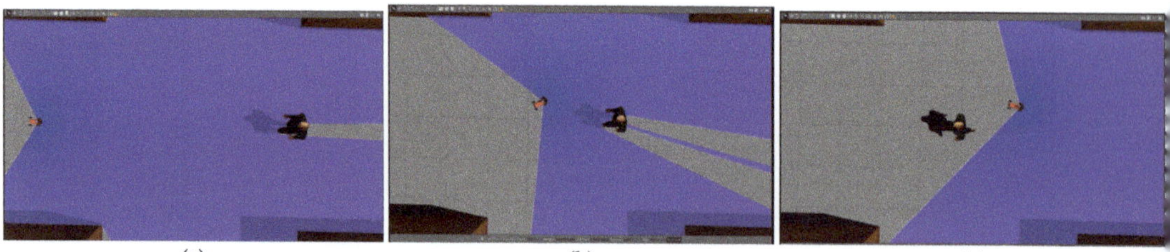

(**a**) (**b**) (**c**)

Figure 6. *Cont.*

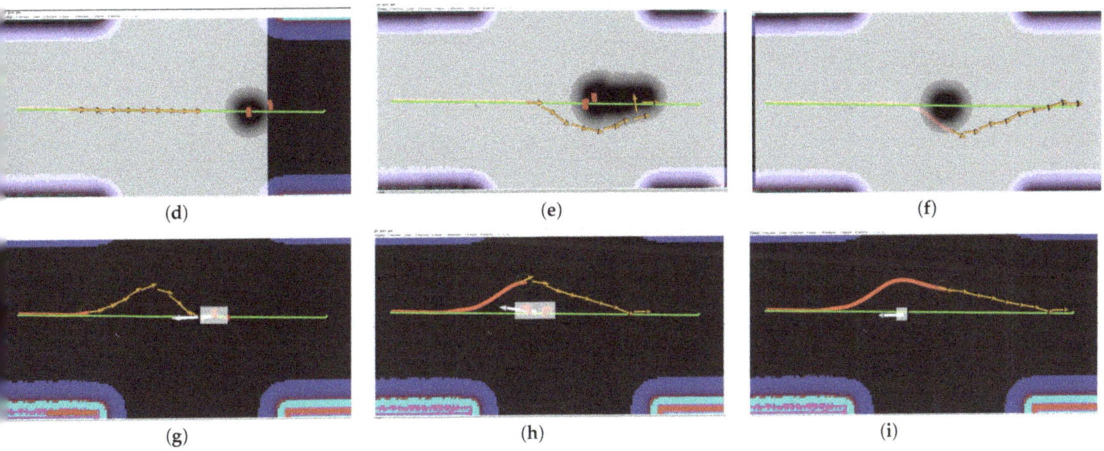

Figure 6. Simulation processions: (a–c) Gazebo snapshots of the robot's avoidance maneuver when confronted with a pedestrian moving at a speed of 1.0 m/s. (d–i) Robot Visualization Tools (rviz) screenshots at different times for the original algorithm and our method; these figures display the robot's global reference path, local path, history path, TEB poses, etc. In these figures, the yellow arrows represent the TEB poses; they can also be seen to be the positions that the robot will reach in the future. The red point cloud is the obstacle point cloud, the green path is the global reference path, the red trajectory represents the robot's history path, and the white binding box is the detected obstacle.

From Figure 7, we can observe that the proposed method can perform obstacle avoidance behavior when the robot is at a considerable distance from the obstacle. When the robot takes obstacle avoidance actions, the distance between the robot and the obstacle is called the reaction distance. We computed the relationship between the reaction distance and the obstacle velocity and formed Table 1. From Table 1, we can conclude that for the proposed method, the reaction distance is determined by the detected obstacle velocity, while for the original TEB, the reaction distance does not change much, or it even decreases in the case of encountering a faster obstacle—the robot may react too late to avoid obstacles. Therefore, it is necessary to take obstacle avoidance actions in advance. This simulation proves that the improved TEB can control the reaction distance to avoid obstacles in advance according to the velocities of obstacles.

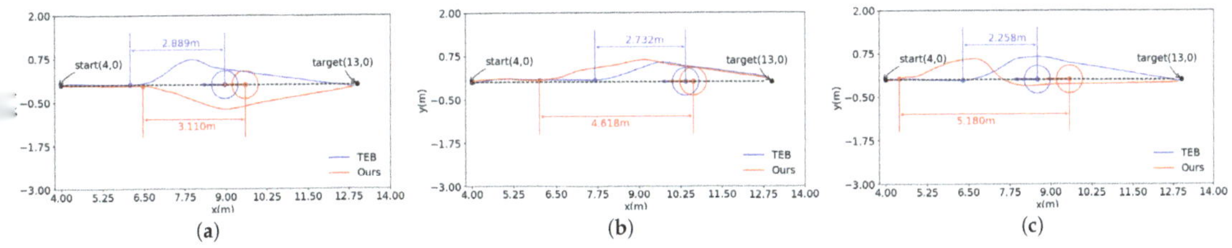

Figure 7. (a–c) Trajectories of the robot for pedestrian speeds of 0.8 m/s, 1.0 m/s, and 1.2 m/s, where the blue line is the original TEB, the red line is our method, and the circles are the positions of the pedestrian when the robot performs obstacle avoidance behavior under different algorithms. The position of the robot at this time is represented by red and blue solid dots, and the reaction distance is calculated and marked with numbers with different colors in the figure.

Table 1. Relationship table between reaction distance and pedestrian velocity.

Pedestrian Velocity	TEB	Ours
0.8 m/s	2.998 m	3.110 m
1.0 m/s	2.732 m	4.618 m
1.2 m/s	2.258 m	5.180 m

4.1.2. Obstacle Approaching from the Side of the Robot

The second simulation changes the direction of the pedestrian. The simulation environment is shown in Figure 8. The start and target of the robot remain unchanged, but the ideal path conflicts with pedestrians, forcing the robot to avoid them. We observe the obstacle avoidance effect of the robot. From the first row and the second row in Figure 8, we can clearly see that the robot runs at low efficiency under the original TEB. However, from Figure 8g–l, it can be seen that in the improved TEB, under the influence of the risk area, the robot adopts the strategy of avoiding obstacles to the left.

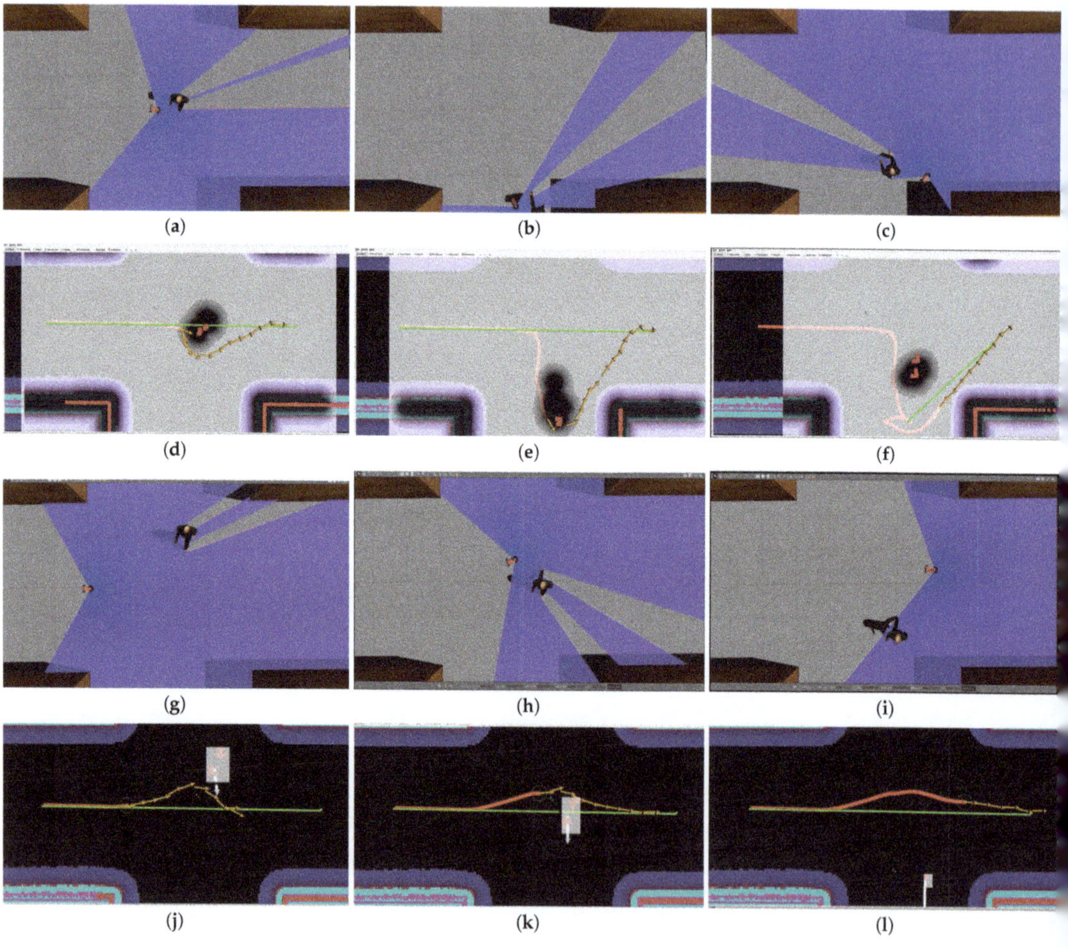

Figure 8. **(a–c,g–i)** Snapshots of the robot using the original TEB and our method to avoid side pedestrians. **(d–f,j–l)** The rviz screenshots during the operation of the original TEB and our method, respectively, in which the yellow arrows are the TEB poses.

From the experimental effect in Figure 9a, we can see that, although the pedestrian does not block the robot at this time, the obstacle risk area affects the TEB poses. Therefore, the robot decides in advance to go in the opposite direction of the area where pedestrians are walking. The difference between Figure 9a,b lies in the initial position of the pedestrian when the robot starts. The position y of the pedestrian in Figure 9b is smaller than that in Figure 9a. The robot does not need to make too much of a left-turn deflection so that obstacles can be avoided using our method, and the original TEB also appears to move around obstacles. In Figure 9c, the direction of the pedestrian is from the negative direction of y to the positive direction. In the improved algorithm, when the position of the pedestrian is far away from the ideal trajectory of the robot, the robot takes a left turn to avoid obstacles in advance. However, the original TEB takes obstacle avoidance measures only when pedestrians almost collide with robots. The obstacle avoidance efficiency is low, and collisions are prone to occur. This set of simulations proves that the robot will choose the obstacle avoidance direction based on the direction of the pedestrians.

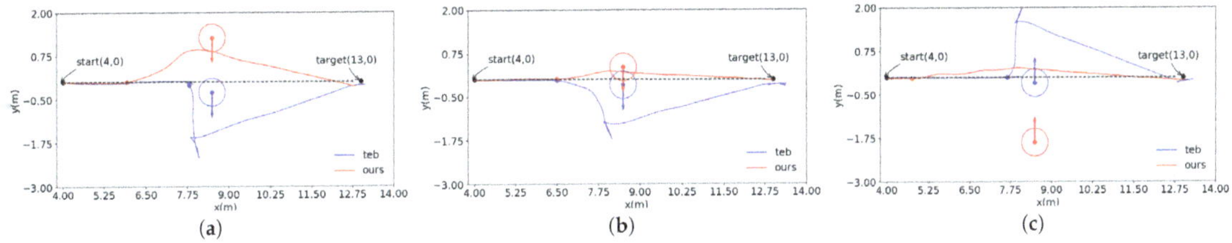

Figure 9. (**a**–**c**) Experimental results of the robot in three different situations. The three different situations here are aimed at different moments when the three robots are started. The blue line is the original algorithm, the red line is the proposed algorithm, and the circle represents the positions of the pedestrian when the robot is performing obstacle avoidance behavior. At this time, the position of the robot is marked with red and blue solid circles.

4.1.3. Navigation in Complex Environments

The third set of simulation experiments allowed the robot to run in a complex, multidynamic obstacle scene, as shown in Figure 10. Four pedestrians moved at a speed of 0.8 m/s. We compare our method with the original TEB, the MPC local planner, and the state lattice planner in this section. For the four local planners, we conducted 50 simulation experiments. For the robot, it is considered successful if it reaches the target within the specified time, and it is considered to have failed if it collides or gets stuck on the path. Through the rosbag file, we performed some processing on the data to calculate the success rates, total running distances, and times of the four methods to form Table 2. We can see from Table 2 that the improved TEB not only improves the success rate of dynamic obstacle avoidance, but it also increases the average speed of the robot on the global path and reduces the running distance. Through this simulation, we found that the improved TEB is not only better than the original TEB, but it is also more effective than other advanced local planners.

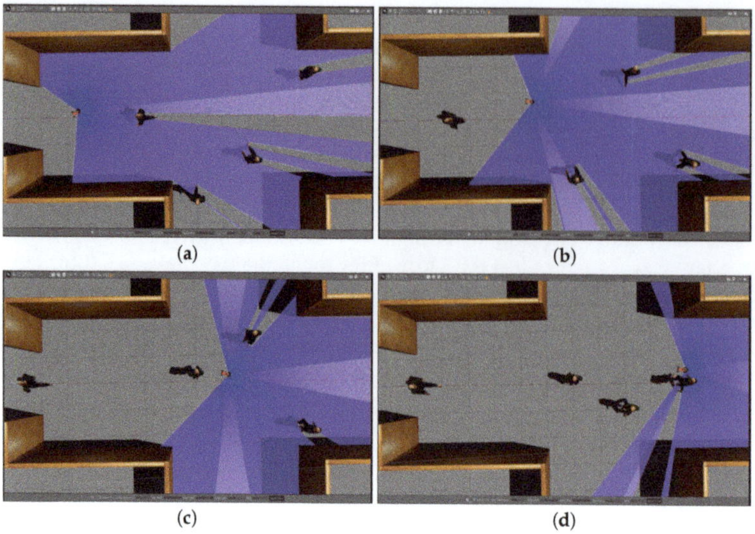

Figure 10. Snapshots of the robot using the improved method to avoid obstacles in a complex, dynamic simulation environment. (a–d) the different time during the simulation.

Table 2. The success rates, running distances, and total times of the four methods.

Algorithm	Success	Running Distance	Time
TEB local planner	75%	18.61 m	21.81 s
MPC local planner	80%	17.54 m	21.62 s
State lattice	90%	16.33 m	18.82 s
Ours	95%	15.03 m	16.82 s

4.2. Experiments

In order to verify the efficiency of the proposed method on an actual robot platform, we designed a mobile robot experimental platform. The platform includes a two-dimensional lidar for sensing the environment, and the lidar is connected to the upper computer: a Jetson Nano. The chassis is controlled by an STM32 to control the motor movement of the robot. Our method works under ROS-Melodic. The real environment experiment must be used to verify whether the robot uses obstacle avoidance operations for oncoming pedestrians in advance. As shown in Figure 11, the start of the given robot is $(0, 0)$, and the target starts at $(8, 0)$. When the robot starts, a person moves slowly toward the robot as a dynamic obstacle.

Figure 11. (a) the front view of the robot platform, (b) the side view.

From Figure 12a–d in the original algorithm, when the pedestrian moves in front of the robot, since the position of the pedestrian does not affect the first few TEB poses, the robot still runs on the ideal trajectory. Only when the position of the pedestrian reaches a position close enough to the robot does the robot identify the obstacle and devise a trajectory to circumvent it. Because it is too close to the obstacle, there is an operation of turning in place. If the pedestrian does not notice the robot at this time, it may collide with the pedestrian. From Figure 12e–h, we can see that in the improved method, the robot performs maneuvering operations in advance when facing oncoming pedestrians: the position of the pedestrian in Figure 12g is the same as in Figure 12c, and the distance between the pedestrian and the robot is approximately the same or even greater, but the robot has already started to avoid obstacles in advance, so even if the pedestrian maintains his/her own trajectory at this time, the route of the robot can perfectly stagger with the pedestrian.

From the trajectory path diagram in Figure 13, we can see the difference between the two methods of advance obstacle avoidance more intuitively. The reaction distance of the original TEB is only approximately 1 m, while the reaction distance of the improved TEB algorithm is as high as 3.5 m. The total running path of the robot in the original TEB is longer.

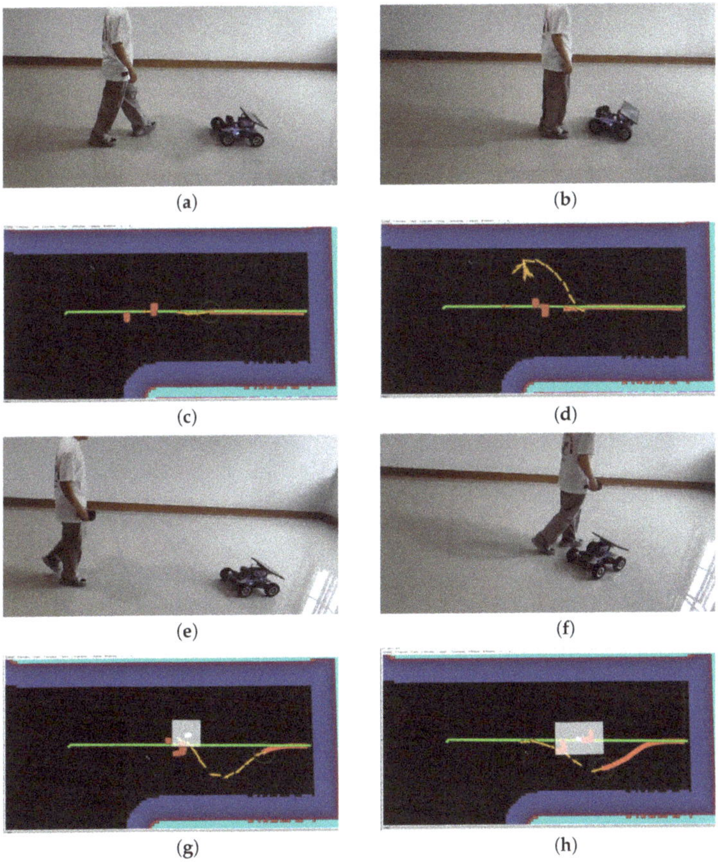

Figure 12. Photos of the robot facing oncoming pedestrians and the corresponding rviz screenshots. The yellow arrow indicates the TEB poses, and the red point cloud is the pedestrian scanned by the 2D lidar: (**a**–**d**) the original TEB, and (**e**–**h**) our method.

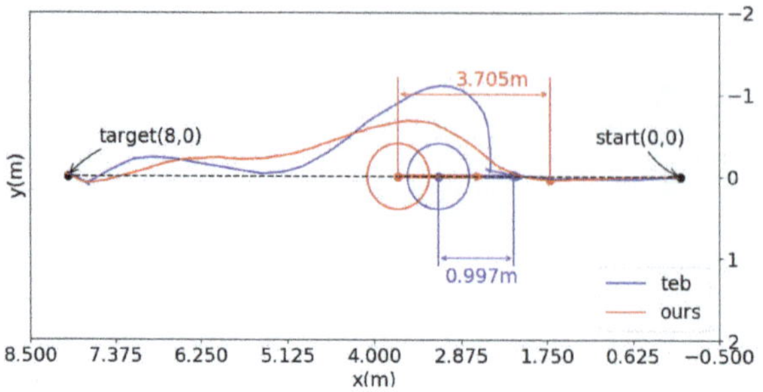

Figure 13. Trajectories of two methods.

4.3. Discussion

Simulations and experiments show that the combination of obstacle velocities and TEB poses can improve the efficiency of dynamic obstacle avoidance. The first simulation verified that the improved TEB can plan different obstacle avoidance paths for obstacles with different velocities, the robot can avoid obstacle in advance based on the velocities of obstacles, and the reaction distance increases as the speed of the obstacles increases, which can ensure the safety of the robot while moving. The second simulation showed that our method can make different obstacle avoidance strategies for obstacles from different directions, which improves the efficiency of obstacle avoidance. The third simulation demonstrated the effectiveness of our method in environments with multiple dynamic obstacles, and we compared the improved TEB with other advanced local planners, including an MPC local planner and a state lattice planner. The comparative experiments with other planners showed that the improved TEB reduces the total running time and running distance and increases the success rate, proving that the improved TEB can indeed improve the efficiency of local path planning in dynamic environments. Finally, the robot platform experiment verified that our method can run on a robot platform and is better than the original TEB. Simulations and experiments prove the contribution of this article and the necessity of our research.

5. Conclusions

Due to the uncertainty of robot movement, dynamic obstacle avoidance poses many challenges to robot navigation. The traditional method of avoiding static obstacles is to set the inflation radius of obstacles or to increase the update frequency of the local cost map. However, the positions of dynamic obstacles change. These methods do not robustly consider the velocities of obstacles. If the speed of the obstacle is too fast, the robot may not have enough time to avoid the obstacle when it is too close to the obstacle. In this manuscript, we propose a new dynamic obstacle avoidance method that detects and estimates the velocities of dynamic obstacles through 2D lidar and integrates the TEB local planner. We conduct simulations and real experiments to demonstrate that the proposed method can perform advanced avoidance according to the velocity of obstacles, but this algorithm is more dependent on the accuracy of the localization algorithm. In the future, we plan to conduct further research on the uncertainty of the directions of dynamic obstacles.

Author Contributions: Methodology, X.O., Z.Y. and X.H.; Writing—original draft, Z.Y.; Supervision, X.O. and X.H. All authors have read and agreed to the published version of the manuscript.

Funding: This work was supported in part by the National Natural Science Foundation of China (Nos. 62233016 and 62303418) and the Zhejiang Provincial Natural Science Foundation of China (Nos. LQ22F030016 and LQ23F010024).

Data Availability Statement: The data presented in this study are available on request from the corresponding author.

Conflicts of Interest: The authors declare no conflicts of interest.

References

1. Qi, J.; Yang, H.; Sun, H. MOD-RRT*: A sampling-based algorithm for robot path planning in dynamic environment. *IEEE Trans. Ind. Electron.* **2020**, *68*, 7244–7251. [CrossRef]
2. Zhang, Y.; Wang, S. LSPP: A novel path planning algorithm based on perceiving line segment feature. *IEEE Sens. J.* **2021**, *22*, 720–731. [CrossRef]
3. Kim, B.; Pineau, J. Socially adaptive path planning in human environments using inverse reinforcement learning. *Int. J. Soc. Robot.* **2016**, *8*, 51–66. [CrossRef]
4. Sun, T.; Pan, W.; Wang, Y.; Liu, Y. Region of interest constrained negative obstacle detection and tracking with a stereo camera. *IEEE Sens. J.* **2022**, *22*, 3616–3625. [CrossRef]
5. Almasri, M.M.; Alajlan, A.M.; Elleithy, K.M. Trajectory planning and collision avoidance algorithm for mobile robotics system. *IEEE Sens. J.* **2016**, *16*, 5021–5028. [CrossRef]
6. Kobayashi, Y.; Sugimoto, T.; Tanaka, K.; Shimomura, Y.; Arjonilla Garcia, F.J.; Kim, C.H.; Yabushita, H.; Toda, T. Robot navigation based on predicting of human interaction and its reproducible evaluation in a densely crowded environment. *Int. J. Soc. Robot.* **2022**, *14*, 373–387 . [CrossRef]
7. Duong, H.T.; Suh, Y.S. Human gait tracking for normal people and walker users using a 2D LiDAR. *IEEE Sens. J.* **2020**, *20*, 6191–6199. [CrossRef]
8. Lee, H.; Lee, H.; Shin, D.; Yi, K. Moving objects tracking based on geometric model-free approach with particle filter using automotive LiDAR. *IEEE Trans. Intell. Transp. Syst.* **2022**, *23*, 17863–17872. [CrossRef]
9. Wei, C.; Wang, Y.; Shen, Z.; Xiao, D.; Bai, X.; Chen, H. AUQ–ADMM algorithm-based peer-to-peer trading strategy in large-scale interconnected microgrid systems considering carbon trading. *IEEE Syst. J.* **2023**, *17*, 6248–6259. [CrossRef]
10. Chen, Q.; Li, Y.; Hong, Y.; Shi, H. Prescribed-Time Robust Repetitive Learning Control for PMSM Servo Systems. *IEEE Trans. Ind. Electron.* **2024**, *in press*. [CrossRef]
11. Shim, I.; Choi, D.G.; Shin, S.; Kweon, I.S. Multi lidar system for fast obstacle detection. In Proceedings of the 2012 9th International Conference on Ubiquitous Robots and Ambient Intelligence (URAI), Daejeon, Republic of Korea, 26–28 November 2012; IEEE: Piscataway, NJ, USA, 2012; pp. 557–558.
12. Fox, D.; Burgard, W.; Thrun, S. The dynamic window approach to collision avoidance. *IEEE Robot. Autom. Mag.* **1997**, *4*, 23–33. [CrossRef]
13. Rösmann, C.; Feiten, W.; Wösch, T.; Hoffmann, F.; Bertram, T. Trajectory modification considering dynamic constraints of autonomous robots. In Proceedings of the ROBOTIK 2012: 7th German Conference on Robotics, VDE, Munich, Germany, 21–22 May 2012; pp. 1–6.
14. Rösmann, C.; Hoffmann, F.; Bertram, T. Planning of multiple robot trajectories in distinctive topologies. In Proceedings of the 2015 European Conference on Mobile Robots (ECMR), Lincoln, UK, 2–4 September 2015; IEEE: Piscataway, NJ, USA, 2015; pp. 1–6.
15. Rösmann, C.; Hoffmann, F.; Bertram, T. Kinodynamic trajectory optimization and control for car-like robots. In Proceedings of the 2017 IEEE/RSJ International Conference on Intelligent Robots and Systems (IROS), Vancouver, BC, Canada, 24–28 September 2017; IEEE: Piscataway, NJ, USA, 2017; pp. 5681–5686.
16. Rösmann, C.; Oeljeklaus, M.; Hoffmann, F.; Bertram, T. Online trajectory prediction and planning for social robot navigation. In Proceedings of the 2017 IEEE International Conference on Advanced Intelligent Mechatronics (AIM), Munich, Germany, 3–7 July 2017; IEEE: Piscataway, NJ, USA, 2017; pp. 1255–1260.
17. Rösmann, C.; Feiten, W.; Wösch, T.; Hoffmann, F.; Bertram, T. Efficient trajectory optimization using a sparse model. In Proceedings of the 2013 European Conference on Mobile Robots, Barcelona, Spain, 25–27 September 2013; IEEE: Piscataway, NJ, USA, 2013; pp. 138–143.
18. Hwang, C.L.; Huang, H.H. Experimental validation of a car-like automated guided vehicle with trajectory tracking, obstacle avoidance, and target approach. In Proceedings of the IECON 2017—43rd Annual Conference of the IEEE Industrial Electronics Society, Beijing, China, 29 October–1 November 2017; IEEE: Piscataway, NJ, USA, 2017; pp. 2858–2863.
19. Ding, Z.; Liu, J.; Chi, W.; Wang, J.; Chen, G.; Sun, L. PRTIRL based socially adaptive path planning for mobile robots. *Int. J. Soc. Robot.* **2023**, *15*, 129–142. [CrossRef]
20. Zhou, Y.; Tuzel, O. Voxelnet: End-to-end learning for point cloud based 3D object detection. In Proceedings of the IEEE Conference on Computer Vision and Pattern Recognition, Salt Lake City, UT, USA, 19–23 June 2018; pp. 4490–4499.
21. Ren, W.; Wang, X.; Tian, J.; Tang, Y.; Chan, A.B. Tracking-by-counting: Using network flows on crowd density maps for tracking multiple targets. *IEEE Trans. Image Process.* **2020**, *30*, 1439–1452. [CrossRef]
22. Li, C.; Guo, S.; Guo, J. Study on obstacle avoidance strategy using multiple ultrasonic sensors for spherical underwater robots. *IEEE Sens. J.* **2022**, *22*, 24458–24470. [CrossRef]
23. Jha, B.; Turetsky, V.; Shima, T. Robust path tracking by a Dubins ground vehicle. *IEEE Trans. Control Syst. Technol.* **2018**, *27*, 2614–2621. [CrossRef]
24. Sun, C.; Zhang, X.; Zhou, Q.; Tian, Y. A model predictive controller with switched tracking error for autonomous vehicle path tracking. *IEEE Access* **2019**, *7*, 53103–53114. [CrossRef]
25. Qu, P.; Li, S.; Zhang, J.; Duan, Z.; Mei, K. A Low-cost and Robust Mapping and Relocalization Method Base on Lidar Inertial Odometry. In Proceedings of the 2021 IEEE International Conference on Real-time Computing and Robotics (RCAR), Xining, China, 15–19 July 2021; IEEE: Piscataway, NJ, USA, 2021; pp. 257–262.

26. Wang, E.; Chen, D.; Fu, T.; Ma, L. A Robot Relocalization Method Based on Laser and Visual Features. In Proceedings of the 2022 IEEE 11th Data Driven Control and Learning Systems Conference (DDCLS), Chengdu, China, 3–5 August 2022; IEEE: Piscataway, NJ, USA, 2022; pp. 519–524.
27. Chen, X.; Vizzo, I.; Läbe, T.; Behley, J.; Stachniss, C. Range image-based LiDAR localization for autonomous vehicles. In Proceedings of the 2021 IEEE International Conference on Robotics and Automation (ICRA), Xian, China, 30 May–5 June 2021; IEEE: Piscataway, NJ, USA, 2021; pp. 5802–5808.
28. Welch, G.; Bishop, G. An Introduction to the Kalman Filter. 1995. Available online: https://perso.crans.org/club-krobot/doc/kalman.pdf (accessed on 20 April 2024).
29. Huang, Y.; Ding, H.; Zhang, Y.; Wang, H.; Cao, D.; Xu, N.; Hu, C. A motion planning and tracking framework for autonomous vehicles based on artificial potential field elaborated resistance network approach. *IEEE Trans. Ind. Electron.* **2019**, *67*, 1376–1386. [CrossRef]
30. Fiorini, P.; Shiller, Z. Motion planning in dynamic environments using velocity obstacles. *Int. J. Robot. Res.* **1998**, *17*, 760–772. [CrossRef]
31. Bharath, G.; Singh, A.; Kaushik, M.; Krishna, K.; Manocha, D. Prvo: Probabilistic reciprocal velocity obstacle for multi robot navigation under uncertainty. In Proceedings of the 2017 IEEE/RSJ International Conference on Intelligent Robots and Systems (IROS), Vancouver, BC, Canada, 24–28 September 2017; pp. 24–28.
32. Van den Berg, J.; Lin, M.; Manocha, D. Reciprocal velocity obstacles for real-time multi-agent navigation. In Proceedings of the 2008 IEEE International Conference on Robotics and Automation, Pasadena, CA, USA, 19–23 May 2008; IEEE: Piscataway, NJ, USA, 2008; pp. 1928–1935.
33. Guo, K.; Wang, D.; Fan, T.; Pan, J. VR-ORCA: Variable responsibility optimal reciprocal collision avoidance. *IEEE Robot. Autom. Lett.* **2021**, *6*, 4520–4527. [CrossRef]
34. Guo, B.; Guo, N.; Cen, Z. Obstacle avoidance with dynamic avoidance risk region for mobile robots in dynamic environments. *IEEE Robot. Autom. Lett.* **2022**, *7*, 5850–5857. [CrossRef]
35. Reddy, A.K.; Malviya, V.; Kala, R. Social cues in the autonomous navigation of indoor mobile robots. *Int. J. Soc. Robot.* **2021**, *13*, 1335–1358. [CrossRef]
36. Smith, T.; Chen, Y.; Hewitt, N.; Hu, B.; Gu, Y. Socially aware robot obstacle avoidance considering human intention and preferences. *Int. J. Soc. Robot.* **2021**, *15*, 661–678. [CrossRef]
37. Chen, C.S.; Lin, S.Y. Costmap generation based on dynamic obstacle detection and velocity obstacle estimation for autonomous mobile robot. In Proceedings of the 2021 21st International Conference on Control, Automation and Systems (ICCAS), Jeju, Republic of Korea, 12–15 October 2021; IEEE: Piscataway, NJ, USA, 2021; pp. 1963–1968.
38. Lu, D.V.; Hershberger, D.; Smart, W.D. Layered costmaps for context-sensitive navigation. In Proceedings of the 2014 IEEE/RSJ International Conference on Intelligent Robots and Systems, Chicago, IL, USA, 14–18 September 2014; IEEE: Piscataway, NJ, USA, 2014; pp. 709–715.
39. Wojke, N.; Bewley, A.; Paulus, D. Simple online and realtime tracking with a deep association metric. In Proceedings of the 2017 IEEE International Conference on Image Processing (ICIP), Beijing, China, 17–20 September 2017; IEEE: Piscataway, NJ, USA, 2017; pp. 3645–3649.
40. Dong, H.; Weng, C.Y.; Guo, C.; Yu, H.; Chen, I.M. Real-time avoidance strategy of dynamic obstacles via half model-free detection and tracking with 2d lidar for mobile robots. *IEEE/ASME Trans. Mechatronics* **2020**, *26*, 2215–2225. [CrossRef]
41. Wang, C.; Chen, X.; Li, C.; Song, R.; Li, Y.; Meng, M.Q.H. Chase and track: Toward safe and smooth trajectory planning for robotic navigation in dynamic environments. *IEEE Trans. Ind. Electron.* **2022**, *70*, 604–613. [CrossRef]

processes

MDPI

Article

Data-Driven-Based Intelligent Alarm Method of Ultra-Supercritical Thermal Power Units

Xingfan Zhang, Lanhui Ye, Cheng Zhang and Chun Wei *

College of Information Engineering, Zhejiang University of Technology, Hangzhou 310014, China; 2112103148@zjut.edu.cn (X.Z.); 221122030410@zjut.edu.cn (L.Y.); 211123030146@zjut.edu.cn (C.Z.)
* Correspondence: chunwei18@zjut.edu.cn

Abstract: In order to ensure the safe operation of the ultra-supercritical thermal power units (USCTPUs), this paper proposes an intelligent alarm method to enhance the performance of the alarm system. Firstly, addressing the issues of slow response and high missed alarm rate (*MAR*) in traditional alarm systems, a threshold optimization method is proposed by integrating kernel density estimation (KDE) and convolution optimization algorithm (COA). Based on the traditional approach, the expected detection delay (*EDD*) indicator is introduced to better evaluate the response speed of the alarm system. By considering the false alarm rate (*FAR*), and *EDD*, a threshold optimization objective function is constructed, and the COA is employed to obtain the optimal alarm threshold. Secondly, to address the problem of excessive nuisance alarms, this paper reduces the number of nuisance alarms by introducing an adaptive delay factor into the existing system. Finally, simulation results demonstrate that the proposed method significantly reduces the *MAR* and *EDD*, improves the response speed and performance of the alarm system, and effectively reduces the number of nuisance alarms, thereby enhancing the quality of the alarms.

Keywords: USCTPUs; intelligent alarm; false alarm rate; missed alarm rate; convolution optimization

1. Introduction

Compared to traditional thermal power units, ultra-supercritical thermal power units (USCTPUs) exhibit higher thermal conversion efficiency and lower coal consumption, which can significantly reduce the pollution emissions of power plants, thereby enhancing the energy-saving and environmental protection effects of thermal power plants [1]. However, due to the complex system structure and increased coupling among equipment, USCTPUs are prone to experiencing abnormal operating conditions. The occurrence of malfunctions during operation not only reduces the production efficiency and economic benefits of the power plant, but can also lead to serious safety accidents [2]. In order to mitigate the adverse effects of malfunctions and prevent unit shutdown or equipment damage caused by malfunctions, it is necessary to utilize an alarm system to monitor critical equipment and parameters.

In the current industrial power generation field, alarm systems are widely used due to their critical role. However, the alarm system of USCTPUs faces issues such as slow system response, missed alarms, and excessive nuisance alarms. The main cause of slow system response and increased missed alarms is the improper setting of alarm thresholds. When the alarm threshold set by the system is large, it delays the generation of alarm signals, resulting in a slow system response and an increase in missed alarms, thus increasing the risk of equipment damage due to abnormal operating conditions. The excessive number of nuisance alarms is primarily due to the lack of a delay factor in traditional alarm systems. Due to the complexity of USCTPUs, there are strong correlations among alarm variables. The repeated occurrence of alarm signals for a specific monitoring parameter may be influenced by other parameters and does not necessarily indicate a fault. This leads to a significant number of nuisance alarms in the alarm information, which severely interferes with the judgment of operators [3,4]. Therefore, an efficient and intelligent alarm system is crucial for ensuring the secure and steady running of USCTPUs.

Currently, the commonly used method for optimizing alarm systems is to optimize and design alarm thresholds. The optimization and design methods for alarm thresholds can be categorized into model-based [5], experience-based [6], and data-driven approaches [7,8]. Reference [9] proposes a method for designing alarm thresholds applicable to condensate water pumps in thermal power plants and constructs a model estimator based on Bayesian filtering to maximize the absolute cumulative error of model parameters, thus obtaining the alarm thresholds, and the effectiveness of the proposed method is verified through experiments. In [10], expert knowledge is combined with data featuring reliability characteristics to build an optimized model for alarm threshold based on interval evidence inference, and the effectiveness of this method is demonstrated through experimentation. Unlike the limitations of the above modeling and knowledge-based approaches, data-driven methods do not require the establishment of complex mechanistic models or extensive empirical knowledge; they only necessitate the analysis of alarm data, making them simple and efficient, and widely applied in industrial fields. In [11], a multivariable alarm threshold optimization method was proposed to address the issue of high false alarm rate (*FAR*) resulting from unreasonable alarm threshold settings, and simulation results demonstrate that this method effectively reduces *FAR*. Reference [12] introduces a variable alarm threshold optimization method based on correlation and cluster analysis, establishing an objective function concerning *FAR* and missed alarm rate (*MAR*) with variable weight coefficients, and employs a numerical optimization algorithm (NOA) to solve the objective function. Reference [13] uses standardized Euclidean distance to cluster and group the variables, calculates the variable weights by entropy weighting method, establishes the threshold optimization objective function with variable weights, and finally uses ant colony optimization algorithm to solve the optimal alarm threshold, and the experimental results show that this method can effectively reduce the *FAR*. Reference [14] initially groups parameter variables, calculates original weights, establishes an optimization objective function concerning *FAR*, *MAR*, and original weights, and then utilizes particle swarm optimization algorithm (PSO) to solve the objective function; experimental results indicate that the proposed method reduces the number of alarms by 37.8% compared to the initial situation. Reference [15] proposes a dynamic variable alarm threshold optimization algorithm suitable for non-stationary processes under different operating conditions, significantly reducing *FAR* and ensuring the reliability of the alarm system. Reference [16] presents a process alarm threshold optimization method based on satisfaction optimization, constructing a weighted objective function for *FAR* and *MAR*, and solving the optimal threshold through NOA, further enhancing the alarm system performance. Reference [17] utilizes non-parametric statistical methods based on windowing and testing to adjust the Hotelling T^2 statistic and Q statistic, thereby generating second-level control limits for alarm thresholds, and the experimental results demonstrate that using these two methods can effectively reduce the *FAR*. Reference [18] addresses the lack of the ability in traditional alarm threshold optimization methods to adjust thresholds online, offering an adaptive optimization method for simplifying multi-layer Bayesian network alarm thresholds based on active transfer entropy, achieving a reduction in the sum of the average *FAR* from 28.5% to 19.7%. The aforementioned studies have further optimized alarm thresholds, to some extent reducing *FAR*; however, they overlook the issue that simply reducing *FAR* may lead to an increase in *MAR*, posing significant safety risks to production. Additionally, they neglect the problem of slow response in the alarm system. Furthermore, due to the absence of delay factors in traditional alarm systems, there is an issue of excessive nuisance alarms, preventing substantial improvement in the performance of alarm systems.

To address the issues of slow response speed and high *MAR* in traditional alarm systems, an optimization method for alarm thresholds is proposed in this paper. Firstly, the data under both normal and abnormal operating conditions were fitted into probability density functions (PDF) using the kernel density estimation (KDE) approach. Then, based on the traditional threshold optimization methods, the detection delay expectation (*EDD*) to measure the agility of the response to the alarm system is considered, and the threshold optimization objective function is established with the *FAR*, *MAR*, and *EDD* as the optimization evaluation indexes. The convolution optimization algorithm (COA) is used to solve the problem, and the optimal alarm threshold is obtained. Furthermore, to address the issue of a high number of nuisance alarms, an adaptive alarm delay method is proposed. Based on the analysis of the alarm time interval data, an adaptive delay factor is introduced into the existing system to eliminate nuisance alarms. Finally, the effectiveness of the proposed methods is validated using historical operational data from 1000 MW and 660 MW USCTPU. The results demonstrate that the proposed method significantly reduces the *MAR* and *EDD*, improves the response speed and performance of the alarm system, effectively reduces the number of nuisance alarms, and enhances the quality of alarms.

The rest of the paper is organized as follows: Section 2 briefly introduces the relevant theoretical knowledge of KDE and COA. Section 3 presents the proposed alarm threshold optimization method and adaptive alarm delay method in detail. Section 4 applies the alarm threshold optimization method to 1000 MW and 660 MW USCTPU for case analysis and conducts an instance analysis of the adaptive alarm delay method to verify the performance of the proposed methods. Section 5 provides a summary of the entire paper.

2. Fundamental Theories

2.1. KDE

KDE is developed based on the statistical histogram, which differs from the traditional histogram estimation which yields a continuous PDF. In practical applications, KDE can estimate unknown variables by performing weighted averaging over fixed windows of data, thereby revealing the probability density distribution pattern. The estimated density function minimizes the integrated mean square error between the estimated and true density functions of the variables. Additionally, KDE allows for the study of distribution characteristics based on the data, making it suitable for data with arbitrary distribution properties. Therefore, selecting KDE to compute the PDF of the data in this paper facilitates the calculation of probability expressions for performance metrics [19]. It can be expressed as follows:

$$\hat{H}(x) = \frac{1}{mu}\sum_{p=1}^{m} K\left(\frac{x - x_p}{u}\right) \quad (1)$$

where x is the estimation point, x_p is the monitoring data point, m is the data length, $K(\,)$ represents the kernel function, u is the width of the window, and $\hat{H}(x)$ is the estimated value of the probability density function of the current monitoring data point. In KDE, the kernel function and the window width collectively decide on the precision of KDE. In this paper, we choose the Gaussian kernel function, and the window width u is obtained by minimizing the mean integrated square error (*MISE*) [20]. The definition of *MISE* is as follows:

$$\begin{aligned} MISE[\hat{H}(x)] &= E\left(\int [\hat{H}(x) - H(x)]^2 dx\right) \\ &= (mu)^{-1}\left[\int K^2(s)ds\right] + \frac{\left[\int s^2K(s)ds\right]^2}{4}\cdot\int\left[\frac{d^2}{dx}H(x)\right]^2 dx\cdot u^4 + o\left(\frac{1}{mu} + u^4\right) \end{aligned} \quad (2)$$

where $H(x)$ is the real worth of the probability density function at the monitoring point, and $o()$ denotes a higher-order infinitesimal term. After neglecting the infinitesimal term in (2), the asymptotic mean integrated square error (*AMISE*) is obtained [21], and it can be expressed as follows:

$$AMISE[\hat{H}(x)] = (mu)^{-1}\left[\int K^2(s)ds\right] + \frac{\left[\int s^2K(s)ds\right]^2}{4}\cdot\int\left[\frac{d^2}{dx}H(x)\right]^2 dx\cdot u^4 \quad (3)$$

When the derivative of *AMISE* is equal to 0, its value reaches the minimum. The optimal window width u_b is given by the following [22]:

$$u_b = \sqrt[5]{\frac{\int K^2(s)ds}{m[\int s^2K(s)ds]^2 \cdot \int [\frac{d^2}{dx}H(x)]^2 dx}} \tag{4}$$

Finally, employing the Silverman's rule of thumb [23] to simplify the calculation of (4), the u_b can be expressed as follows:

$$u_b = S \times (\frac{4}{3} \times \frac{1}{m})^{\frac{1}{5}} \tag{5}$$

where S is the standard deviation of the data.

2.2. COA

2.2.1. Convolutional Search Process

Step 1: Update the longitudinal convolution location. Firstly, define a longitudinal convolution as follows [24]:

$$X_1^t = X^t * [2 \times rand(k,1) - I_a] \tag{6}$$

where *rand*(k,1) is a $k \times 1$ matrix whose elements are random numbers between 0 and 1, k is the size of the convolution kernel, I_a is a $k \times 1$ matrix whose elements are all one, t is the current iteration number, * denotes convolution operation, X^t is the location vector of the population in generation t, and X_1^t is the location vector of the population in generation t after updating the longitudinal convolution location.

Then, compare the *fitness* values of X_t and X_1^t, and optimally replace the larger individual location in X_t as shown below:

$$X_b^t = \begin{cases} X_{1b}^t, \ fitness_{X_{1b}^t} < fitness_{X_b^t} \\ X_b^t, \ fitness_{X_{1b}^t} \geq fitness_{X_b^t} \end{cases} \tag{7}$$

where X_b^t is the location of the b-th individual, X_{1b}^t is the location of the b-th individual after updating the longitudinal convolution location, and *fitness* is an adaptation function.

Step 2: Update the horizontal convolution location. Firstly, define a horizontal convolution as follows:

$$X_2^t = X^t * [2 \times rand(1,k) - I_c] \tag{8}$$

where *rand*(1,k) is a $1 \times k$ matrix whose elements are random numbers between 0 and 1, I_c is a $1 \times k$ matrix whose elements are all one, and X_2^t is the location vector of the population in generation t after updating the horizontal convolution location.

$$X_b^t = \begin{cases} X_{2b}^t, \ fitness_{X_{2b}^t} < fitness_{X_b^t} \\ X_b^t, \ fitness_{X_{2b}^t} \geq fitness_{X_b^t} \end{cases} \tag{9}$$

where X_{2b}^t is the b-th location of the individual after updating the horizontal convolution location.

Step 3: update the matrix convolution location. Firstly, define a matrix convolution as follows:

$$X_3^t = X^t * [2 \times rand(k,k) - I_n] \tag{10}$$

where *rand*(k,k) is a $k \times k$ matrix whose elements are random numbers between 0 and 1, I_n is a $k \times k$ matrix whose elements are all one, and X_3^t is the location vector for the population of the t-th generation after updating the matrix convolution location.

Then, compare the *fitness* values of X^t and X_3^t, and optimally replace the larger individual location in X^t as shown below:

$$X_b^t = \begin{cases} X_{3b}^t, & fitness_{X_{3b}^t} < fitness_{X_b^t} \\ X_b^t, & fitness_{X_{3b}^t} \geq fitness_{X_b^t} \end{cases} \tag{11}$$

where X_{3b}^t is the location of the b-th individual after updating the matrix convolution location.

Step 4: Update the integrated location. Add X_1^t, X_2^t, and X_3^t together according to proportional weights, as shown below:

$$X_4^t = \frac{e_1 \times X_1^t + e_2 \times X_2^t + e_3 \times X_3^t}{e_1 + e_2 + e_3} \tag{12}$$

where X_4^t is the location vector for the population of the t-th generation after updating the integrated location, and e_1, e_2, and e_3 are random numbers between 0 and 1.

Then, compare the *fitness* values of X^t and X_4^t, and optimally replace the larger individual location in X^t as shown below.

$$X_b^t = \begin{cases} X_{4b}^t, & fitness_{X_{4b}^t} < fitness_{X_b^t} \\ X_b^t, & fitness_{X_{4b}^t} \geq fitness_{X_b^t} \end{cases} \tag{13}$$

where X_{4b}^t is the location of the b-th individual after updating the integrated location.

2.2.2. Result Enhancement

Select the best solution X_{be}^t from X^t, and apply Gaussian mutation with non-inertial weight to X_{be}^t in order to avoid falling into local optima during the iteration, as shown below:

$$\hat{X}_{be}^t = \left[1 - \left(\frac{t}{t_{\max}}\right)^2\right] \cdot X^t + randn \cdot X_{be}^t \tag{14}$$

where $\hat{X}_{be}^t$ is the best solution after enhanced perturbation, $t_{\max}$ is the maximum number of iterations, and *randn* is a random number. Finally, compare $\hat{X}_{be}^t$ and X_{be}^t, and replace the individual location of X_{be}^t on a merit basis:

$$X_{be}^t = \begin{cases} \hat{X}_{be}^t, & fitness_{\hat{X}_{be}^t} < fitness_{X_{be}^t} \\ X_{be}^t, & fitness_{\hat{X}_{be}^t} \geq fitness_{X_{be}^t} \end{cases} \tag{15}$$

3. Proposed Intelligent Alarm Method

3.1. Calculation of Performance Optimization Metric

For a univariate alarm system, one of the most critical performance metrics is accuracy, which can be evaluated using metrics such as *FAR* and *MAR* [25]. *FAR* refers to the probability of the system issuing an alarm signal when the process variable is operating normally. A high *FAR* can increase the number of alarms, significantly impacting the judgment and efficiency of the operators. *MAR* denotes the probability that the alarm system fails to respond with an alarm when the process variable experiences an anomaly. A high *MAR* can increase the potential risk of equipment failure, posing serious safety hazards to production. Therefore, the index value of *FAR* and *MAR* can reflect the performance of the designed alarm system. In alarm systems, *FAR* and *MAR* are inversely related, and this contradictory relationship is reflected in the receiver operating characteristic (ROC) curve, as shown in Figure 1. Assuming a process variable $x(t)$ with known PDF for normal and abnormal operating conditions, PDF curve distributions for normal and fault data can be obtained, as shown in Figure 2. Here, the solid line represents the PDF curve distribution of normal operating condition data $\hat{H}_1(x)$, while the dashed line represents the PDF curve distribution of abnormal operating condition data $\hat{H}_2(x)$.

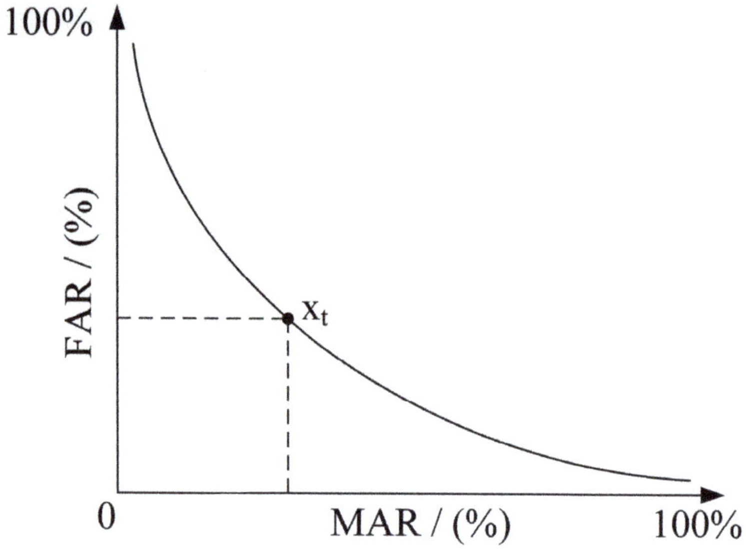

Figure 1. ROC curve.

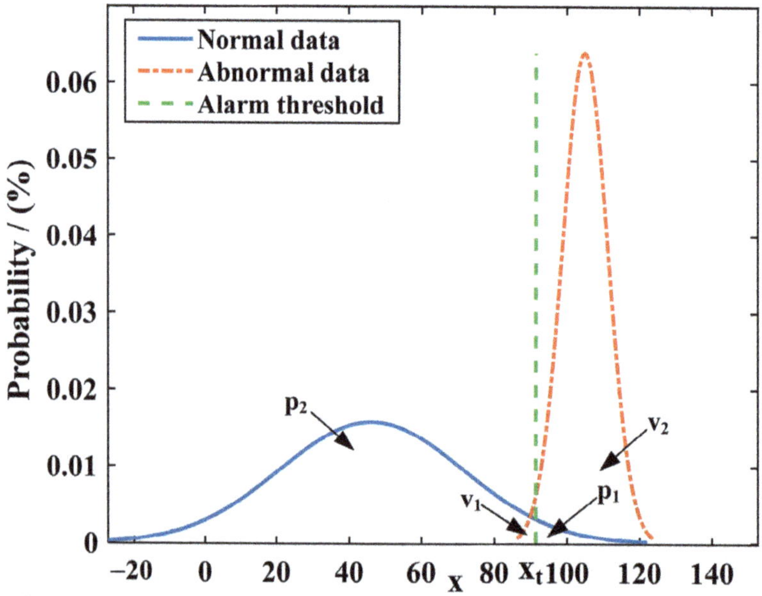

Figure 2. Distribution of PDF of process variables.

According to Figure 2, we can calculate p_1, p_2, v_1, and v_2, where $p_1 + p_2 = v_1 + v_2 = 1$. It can be expressed as follows:

$$\begin{cases} p_1 = \int_{x_t}^{+\infty} \hat{H}_1(x)dx \\ p_2 = \int_{-\infty}^{x_t} \hat{H}_1(x)dx \end{cases} \tag{16}$$

$$\begin{cases} v_1 = \int_{-\infty}^{x_t} \hat{H}_2(x)dx \\ v_2 = \int_{x_t}^{+\infty} \hat{H}_2(x)dx \end{cases} \tag{17}$$

Based on the definitions of *FAR* and *MAR* [25], the *FAR* and *MAR* can be expressed as follows:

$$FAR = p_1 = \int_{x_t}^{+\infty} \hat{H}_1(x)dx \tag{18}$$

$$MAR = v_1 = \int_{-\infty}^{x_t} \hat{H}_2(x)dx \tag{19}$$

Apart from accuracy, agility is another important performance metric for alarm systems. Agility is a measure of the response speed of the alarm system, typically characterized by *EDD*. Detection delay refers to the time difference between the occurrence of an anomaly and the triggering of the alarm state by the alarm system. A smaller value indicates a higher sensitivity and faster response speed of the alarm system. Therefore, minimizing the *EDD* as much as possible is essential to ensure the stable and reliable operation of USCTPUs. Assuming that the moment of failure occurrence is t_s during operation, and an alarm signal is sent out after h moments, with the signal being sent at time t_a, the detection delay at h moments can be expressed as follows:

$$\begin{cases} p(DD=0) = p\{t_{al} = t_s\} = v_2 \\ p(DD=1) = p\{t_{unal} = t_s\} \cdot p\{t_{al} = t_{s+1}\} = v_1 v_2 \\ \vdots \\ p(DD=h) = p\{t_{unal} = t_s\} \ldots p\{t_{unal} = t_{s+h-1}\} \cdot p\{t_{al} = t_{s+h}\} = v_2 v_1^h \end{cases} \tag{20}$$

where t_{al} is the time when an alarm signal is issued, and t_{unal} is the time when no alarm signal is issued.

Next, find the *EDD* [25], calculated as follows:

$$E(DD) = \sum_{h=0}^{\infty} h \cdot p(DD=h) = \frac{v_1}{v_2} = \frac{MAR}{1-MAR} \tag{21}$$

3.2. Design of Alarm Threshold Optimization Method

In order to find the optimal alarm threshold, the proposed threshold optimization method in this paper takes into account the *FAR*, *MAR*, and *EDD*. With the *FAR*, *MAR*, and *EDD* as optimization indicators, an alarm threshold optimization objective function is established, and the objective is to minimize the objective function using the COA for optimization calculation, thereby obtaining the optimal alarm threshold x_c. The flowchart of the threshold optimization algorithm is shown in Figure 3, the pseudocode is shown in Algorithm 1, and the objective function can be represented as follows:

$$F(x) = o_1 \frac{FAR}{FAR^*} + o_2 \frac{MAR}{MAR^*} + o_3 \frac{EDD}{EDD^*} \tag{22}$$

where o_1, o_2, and o_3 are the design weighting coefficients, and *FAR**, *MAR**, and *EDD** are the maximum acceptable *FAR*, *MAR*, and *EDD*.

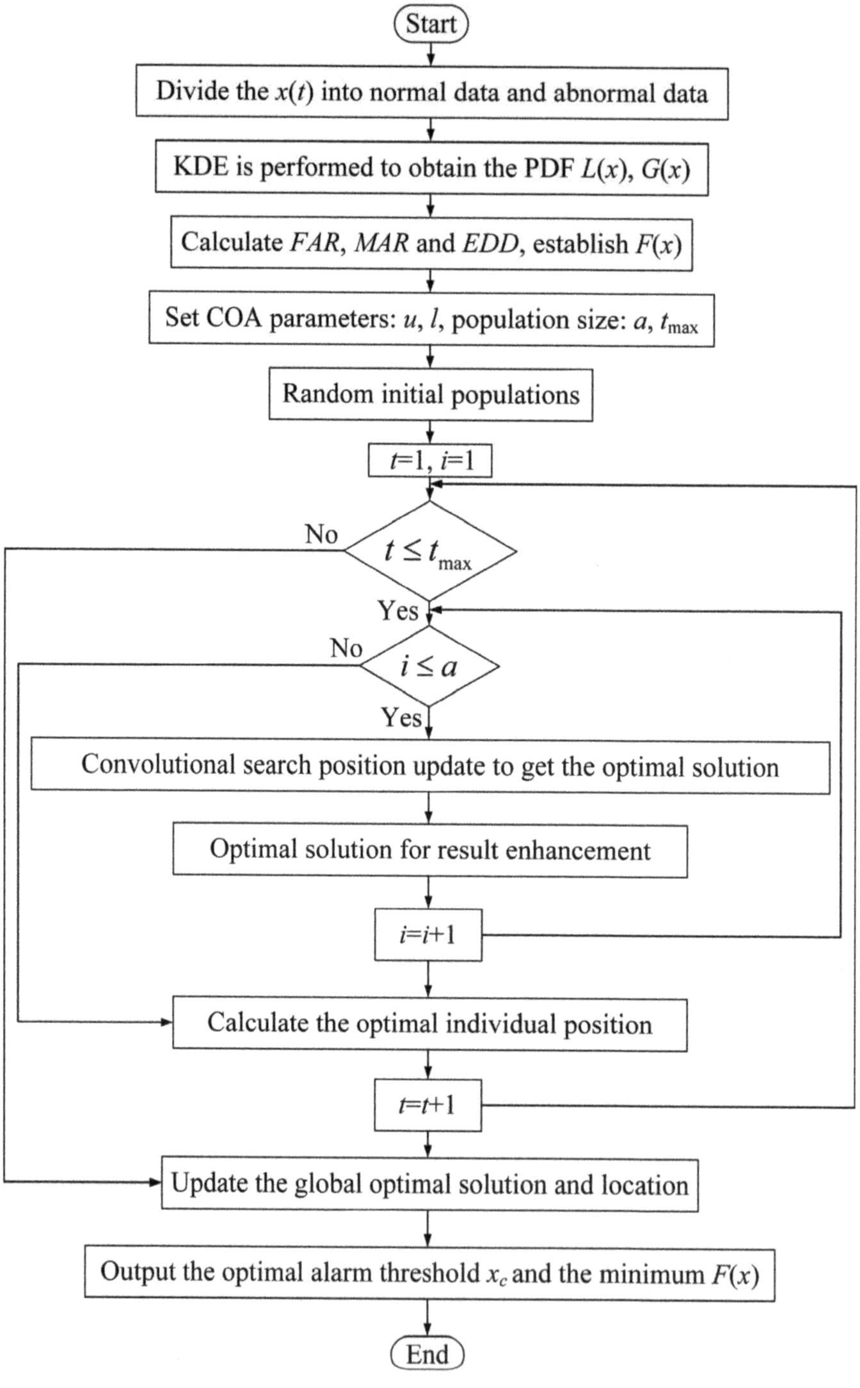

Figure 3. Flow chart of alarm threshold optimization.

Algorithm 1: The pseudocode of the alarm threshold optimization algorithm.

Algorithm 1: Threshold optimization algorithm

Input: Process variable $x(t)$
Parameter: x_t, dim, u_b, l_b, t_{max}, a
Output: FAR_{be}, MAR_{be}, EDD_{be}, x_c

```
normal data = find x(t) ≤ x_t, abnormal data = find x(t) > x_t
normal PDF = PDF(normal data), abnormal PDF = PDF(abnormal data),
  Function F(x) = fun(x)
    define o_1, o_2, o_3, FAR*, MAR*, and EDD*
    compute FAR, MAR, EDD
    construct F(x)
  end
Initial populations
repeat
  for t = 1:t_max
    for i = 1:a
        vertical convolution position update
        horizontal convolution position update
        square convolution position update
        comprehensive location update
    end for
    calculate the best position
    for j = 1:dim
        solution quality enhancement
    end for
    update the global best location
  end for
until Convergence
return x_c, FAR_be, MAR_be, EDD_be
```

3.3. *Design of Alarm Adaptive Delay Algorithm*

The principle of the alarm delay method is that the system starts alarming when a variable exceeds d consecutive alarm signals, and the system stops alarming when there are d consecutive non-alarm signals. The alarm adaptive delay method designed in this paper can dynamically adjust the size of the delay factor d based on the actual alarm data and the mean and variance of the alarm duration. The steps of the alarm adaptive delay method are as follows:

Step 1: Based on the original alarm threshold, the alarm interval data E processed by the delay factor d can be obtained, where $E = [E_1, E_2, E_3, \ldots, E_r]$, and E_i is the time interval of the ith alarm ($i = 1,2,\ldots,r$).

Step 2: Based on the alarm interval data E, calculate the total mean E_{avg} and the standard deviation S_E of E.

$$E_{avg} = \frac{\sum_{i=1}^{r} E_i}{r} \tag{23}$$

$$S_E = \sqrt{\frac{\sum_{i=1}^{r} (E_i - E_{avg})^2}{r-1}} \tag{24}$$

Step 3: Based on the E_{avg} and S_E, calculate the coefficient of variation CV_d for the alarm interval data processed by the delay factor d [26].

$$CV_d = \frac{S_E}{E_{avg}} \tag{25}$$

Step 4: Set the standard coefficient of variation CV_{st} to one, and compare the calculated CV_d with the CV_{st}. If $CV_d > CV_{st}$, update the delay factor $d = d + CV_d$, and return to step 1; if $CV_d < CV_{st}$, end the algorithm process, and obtain the final adaptive delay factor.

4. Experimental Analysis

4.1. Performance Validation of COA

In order to fully evaluate the performance of the COA, this paper compared it with seven intelligent optimization algorithms, including the subtraction-based averaging optimization (SABO) [27], optical microscope algorithm (OMA) [28], rimmed ice optimization (RIME) [29], as well as the classical genetic algorithm (GA), particle swarm optimization (PSO), whale optimization algorithm (WOA), and sparrow search optimization algorithm (SSA). The experimental dataset consisted of 12 test functions from the CEC 2017 benchmark. To make sure that the experiments are fair, the population size for all the algorithms was set to 20, the maximum number of iterations was set to 500, and the dimensionality was set to 100. Moreover, each algorithm was run five times separately, and the average value was taken as the final result to reduce random errors.

To visually observe the performance of the COA and the aforementioned seven optimization algorithms during the iteration process, an analysis of the iterative convergence curves of the eight optimization algorithms on test functions was conducted, as shown in Figure 4. Specifically, results from six challenging functions were selected for demonstration.

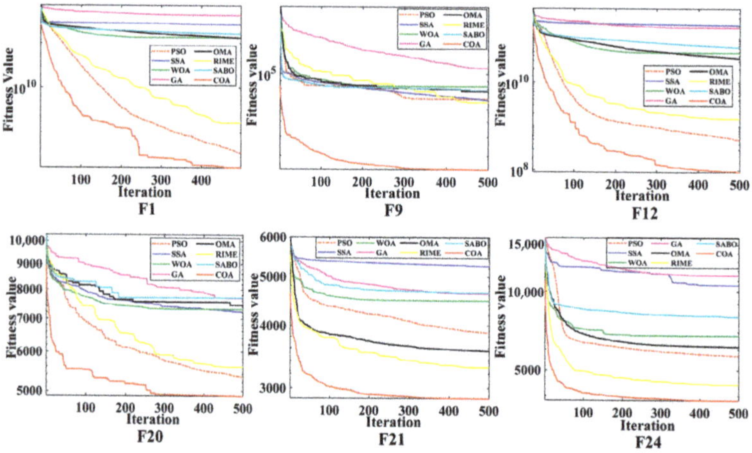

Figure 4. Iterative convergence curves for six test functions.

Figure 4 clearly demonstrates the variations in the *fitness* values during the optimization iteration process of the COA, SABO, OMA, RIME, GA, PSO, WOA, and SSA algorithms. From the convergence curves in Figure 4, it is evident that the COA consistently reaches the optimal value at the fastest rate during the iteration process of each test function, maintaining a leading convergence speed compared to the other seven algorithms. Furthermore, when reaching the maximum number of iterations, the *fitness* value of the COA is significantly lower than that of the other seven algorithms, indicating that the convergence accuracy of the COA exceeds that of the other algorithms. It can be observed that the COA possesses superior optimization capabilities, as well as stronger abilities in both local and global search.

Furthermore, to further validate the performance of the COA, the Friedman test was employed to compare the statistical differences between each group of results. The rank values were compared to evaluate the performance of the different algorithms, where smaller rank values indicate higher performance [30]. The Friedman rank and ranking results for each algorithm on the test functions are presented in Table 1.

Table 1. Friedman rank sum ranking of 12 algorithms on different test functions.

Function	COA	PSO	SSA	WOA	GA	OMA	RIME	SABO	COA Ranking
F1: Shifted and Rotated Bent Cigar Function	1	2	7	4.8	8	4.2	3	6	1
F2: Shifted and Rotated Sums of Different Power Functions	1.2	1.8	7.2	5.2	7.8	4	3	5.8	1
F3: Shifted and Rotated Zakharov Function	1	4.6	4.2	6.6	7.2	3.2	2	7.2	1
F5: Shifted and Rotated Rastrigin's Function	1	2.2	6.4	5.6	8	4.4	2.8	5.6	1
F8: Shifted and Rotated Lunacek BiRastrigin Function	1.2	3.2	7	5.6	7.8	3.8	1.8	5.6	1
F9: Shifted and Rotated Levy Function	1	5	2.6	5.8	7.8	5.4	2.6	5.8	1
F12: Hybrid Function 2 (N = 3)	1	2	7	4.8	8	4.2	3	6	1
F16: Hybrid Function 6 (N = 4)	1.4	1.6	8	6.2	6.8	4	3	5	1
F20: Hybrid Function 6 (N = 6)	1.4	1.6	5.2	6	6.8	6.2	3	5.8	1
F21: Composition Function 1 (N = 3)	1	4	7.4	5.8	6.6	2.8	2.2	6.2	1
F24: Composition Function 4 (N = 4)	1	3.6	7.4	5.2	7.6	3.8	2	5.4	1
F30: Composition Function 10 (N = 3)	1	2	7.2	4.4	7.8	4.8	3	5.8	1
Average	1.1	3	7	5.5	7.5	4.2	2.6	5.9	1

From Table 1, it can be observed that in the Friedman test conducted on each test function, the COA consistently achieved the smallest rank. The average rankings of the 8 algorithms were COA, RIME, PSO, OMA, WOA, SABO, SSA, and GA. The COA ranked first, indicating that the COA holds a statistical and performance advantage over the other algorithms.

4.2. Performance Analysis of Threshold Optimization

4.2.1. Data Validation of 1000 MW USCTPU

To verify the intelligent alarm capability of the proposed alarm threshold optimization method during the normal operation of the unit, experimental data from a 1000 MW-level secondary reheating ultra-supercritical DC boiler system in a certain thermal power plant were collected. The data are the variations in a high-pressure feed pump BV winding temperature over time, with an initial alarm threshold set at 99.78. The sample data curve is shown in Figure 5.

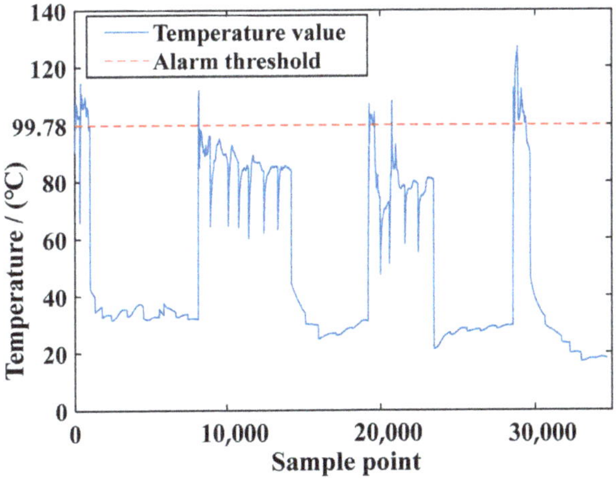

Figure 5. High-pressure feed pump BV winding temperature data curve.

Firstly, the process values of the BV winding temperature data of the high-pressure feedwater pump were separately collected under normal and abnormal working conditions based on the alarm threshold, resulting in two groups of data. Then, according to Equation (5), the optimal window widths for the two groups of data were calculated as

u_{b1} = 3.4 and u_{b2} = 1.4, respectively. Subsequently, Equation (1) was used to fit the PDF curves of the normal and abnormal states, as shown in Figure 6.

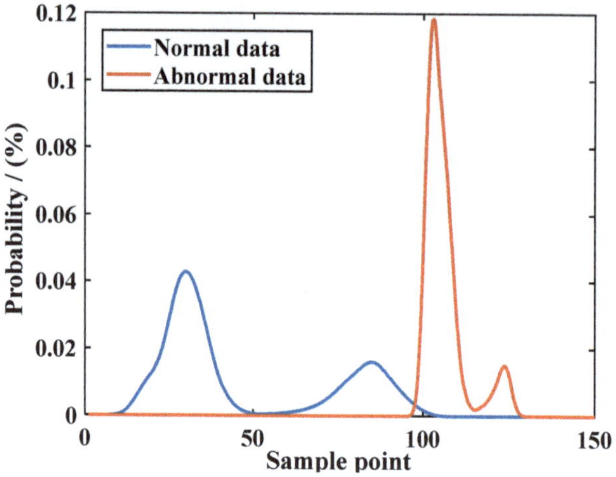

Figure 6. PDF curve distribution.

According to Figure 6, the PDFs of the normal and abnormal state data can be obtained, denoted as $L(x)$ and $R(x)$, respectively. It can be expressed as follows:

$$L(x) = \frac{1}{32753 \times 3.4} \sum_{q=1}^{32753} \frac{1}{\sqrt{2\pi}} e^{-\left(\frac{x-x_q}{\sqrt{2}\times 3.4}\right)^2} \tag{26}$$

$$R(x) = \frac{1}{1918 \times 1.4} \sum_{d=1}^{1918} \frac{1}{\sqrt{2\pi}} e^{-\left(\frac{x-x_d}{\sqrt{2}\times 1.4}\right)^2} \tag{27}$$

Based on Equations (26) and (27), we can further derive the expressions for the *FAR*, *MAR*, and *EDD*, as expressed below

$$FAR = \int_{x_b}^{+\infty} \frac{1}{32753 \times 3.4} \sum_{q=1}^{32753} \frac{1}{\sqrt{2\pi}} e^{-\left(\frac{x-x_q}{\sqrt{2}\times 3.4}\right)^2} dx \tag{28}$$

$$MAR = \int_{-\infty}^{x_b} \frac{1}{1918 \times 1.4} \sum_{d=1}^{1918} \frac{1}{\sqrt{2\pi}} e^{-\left(\frac{x-x_d}{\sqrt{2}\times 1.4}\right)^2} dx \tag{29}$$

$$EDD = \frac{\int_{-\infty}^{x_c} \frac{1}{1918\times 1.4} \sum_{d=1}^{1918} \frac{1}{\sqrt{2\pi}} e^{-\left(\frac{x-x_d}{\sqrt{2}\times 1.4}\right)^2} dx}{1 - \int_{-\infty}^{x_c} \frac{1}{1918\times 1.4} \sum_{d=1}^{1918} \frac{1}{\sqrt{2\pi}} e^{-\left(\frac{x-x_d}{\sqrt{2}\times 1.4}\right)^2} dx} \tag{30}$$

Next, by substituting the expressions for the *FAR*, *MAR*, and *EDD* into Equation (22), we obtain the objective function $F(x)$ for optimizing the alarm threshold, where the FAR^* = 0.02, MAR^* = 0.02, and EDD^* = 0.04. The optimization is performed using the COA; at this point, the variable search range is set to [95, 105], and the *fitness* value curve during the optimization process with increasing iteration numbers as well as the curve showing the variation in the objective function values under different alarm thresholds, are illustrated in Figure 7.

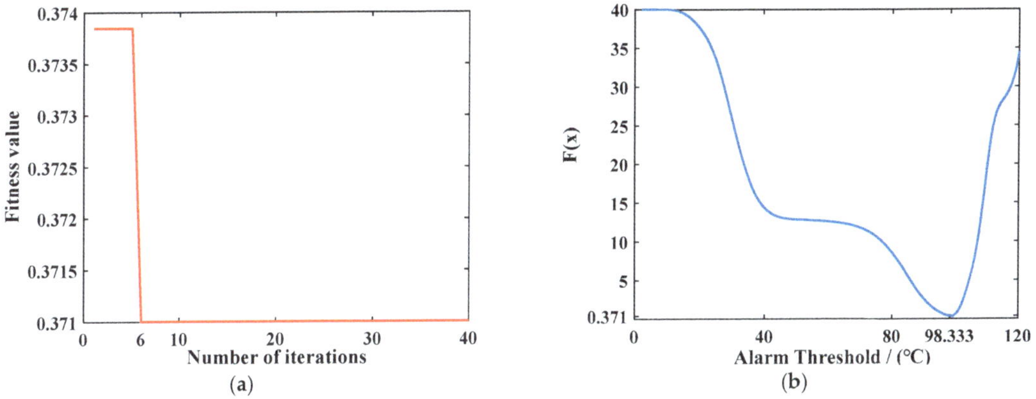

Figure 7. (**a**) COA iterative convergence curves; (**b**) curve of change in $F(x)$.

According to Figure 7a, it can be observed that when using the COA to optimize the objective function, the minimum value of the function is reached after 6 iterations, with a value of 0.371, which remains unchanged in the subsequent iterations. Additionally, as shown in Figure 7b, the optimal alarm threshold corresponding to the minimum function value is $x_c = 98.333$. Furthermore, based on Equations (28)–(30), the optimal values for the *FAR*, *MAR*, and *EDD* are calculated to be *FAR* = 0.0079, *MAR* = 0.0068, and *EDD* = 0.0069, respectively.

Finally, in order to better validate the effectiveness of the proposed threshold optimization method in this paper, the threshold optimization method proposed in this paper is compared with the 3σ method commonly used in the industry, NOA threshold optimization method, PSO threshold optimization method, and the optimization index results of the original system alarm threshold. The results are shown in Table 2.

Table 2. Comparison of indicator results.

Index	Method	*FAR*	*MAR*	*EDD*	Alarm Threshold
1	Original system	0.0046	0.0423	0.044	99.78
2	3σ	0.0029	0.0538	0.0568	106.11
3	KDE + NOA	0.0073	0.01	0.0101	98.5888
4	KDE + PSO	0.0072	0.0103	0.0104	98.6119
5	KDE + COA	0.0079	0.0098	0.0069	98.3333

From the results for the first and fifth methods in Table 2, it is evident that compared to the original system before optimization, although the proposed threshold optimization method in this paper has resulted in a slight increase in the *FAR* by only 0.0033, the obtained *FAR* remains within an acceptable range. Furthermore, the original values of the *MAR* and *EDD* exceeded the maximum acceptable range, posing safety hazards to production and reducing the response speed of the alarm system. However, after optimization using the proposed method, both the *MAR* and *EDD* are reduced to approximately one-sixth of their original values, and the optimized results fall within the maximum acceptable range. This achievement demonstrates that the proposed threshold optimization method significantly reduces the *MAR* and *EDD* without significantly increasing the *FAR*, thereby effectively mitigating the safety hazards associated with excessive false negatives and improving the response speed and performance of the alarm system. Although using the 3σ method can greatly reduce the *FAR*, the *MAR* and *EDD* are much higher than the threshold optimization method in this paper. Additionally, from the results for the third, fourth, and fifth methods in Table 2, it can be inferred that the utilization of the COA method in this paper further reduces the *MAR* and *EDD*, indicating that the COA method employed in the optimization process exhibits superior performance.

4.2.2. Data Validation of 660 MW USCTPU

To validate the applicability and scalability of the threshold method on the different types of USCTPUs, experimental data from the 660 MW ultra-supercritical variable-pressure direct current boiler system of a thermal power plant were collected. The data depict the variation in temperature over time in the non-drive section of the high-pressure feedwater pump A. The initial alarm threshold is set at 62, and the sample data curve is shown in Figure 8.

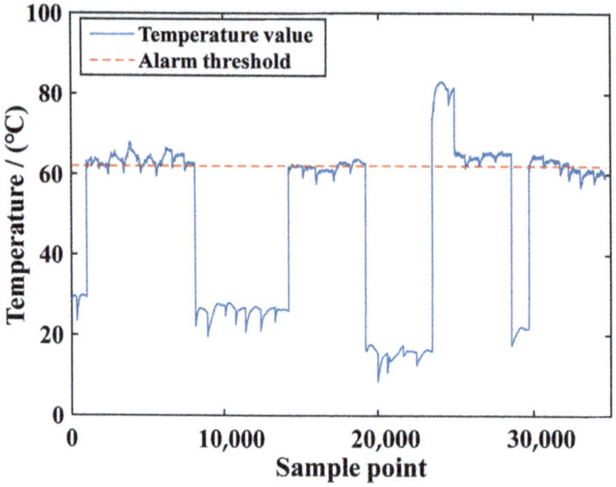

Figure 8. High-pressure feed pump A motor non-driven section temperature data curve.

To validate the effectiveness of the threshold optimization method proposed in this paper on the 660 MW USCTPU, the optimization indicators resulting from the threshold optimization method proposed in this paper are compared with those under the original system alarm thresholds, where the *FAR** = 0.02, *MAR** = 0.02, and *EDD** = 0.04. The final results are shown in Table 3.

Table 3. The performance indexes of the threshold optimization method are compared with the original system and the traditional threshold optimization method.

Index	Method	*FAR*	*MAR*	*EDD*	Alarm Threshold
1	Original system	0.12189	0.074827	0.080879	62
2	KDE + COA	0.15697	0.015809	0.016063	61.2357
3	3σ	0.038	0.1538	0.1818	65.46
4	KDE + NOA	0.155	0.055	0.058	61.465
5	KDE + PSO	0.1546	0.067	0.072	61.33

From Table 3, it can be observed that compared to the results of the original system before optimization, the threshold optimization method proposed in this paper increases the *FAR* by 0.035, but it remains within the maximum acceptable range without affecting the safe operation. However, both the *MAR* and *EDD* exceeded the maximum acceptable range before optimization, posing safety hazards to production. After optimization using the method proposed in this paper, the *MAR* and *EDD* are significantly reduced and are within the maximum acceptable range. Compared with the 3σ method, the *MAR* and *EDD* optimized by the threshold method proposed in this paper are 1/10 of the 3σ method, indicating that using the threshold method can improve the response speed of the alarm system. Additionally, from the results for the second, fourth, and fifth methods in Table 3, it can be inferred that the utilization of the COA method in this paper further reduces the *MAR* and *EDD*, indicating that the COA method employed in the optimization process exhibits superior performance. This demonstrates that the proposed threshold optimization method also exhibits a superior performance on the 660 MW USCTPU.

4.2.3. Performance Verification of the Proposed Threshold Optimization Method under System Fault Conditions

To verify the robustness of the threshold method proposed in this paper, it is applied when the system experiences faults. Fault data from a 1000 MW-level secondary reheating ultra-supercritical direct current boiler system were collected. The fault observed was an elevated temperature in the winding 1 of the high-pressure feedwater pump AV. The sample data operational curve is shown in Figure 9.

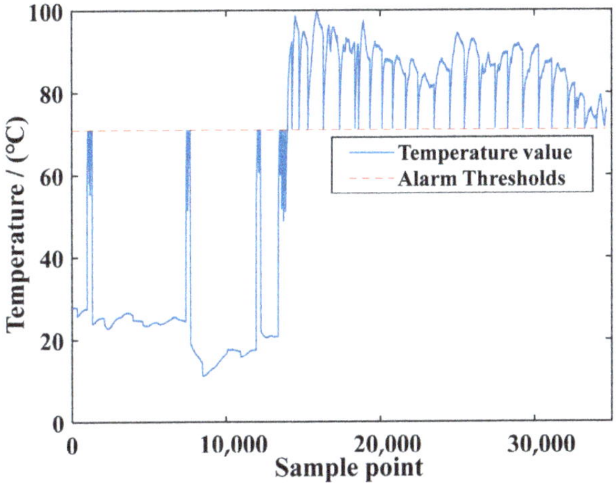

Figure 9. High-pressure feed pump AV winding temperature operating curve.

From Figure 9, it is evident that the system experienced a fault at sample point 13,967, after which the temperature remained above the alarm threshold set in the original system. Subsequently, the threshold optimization method proposed in this paper was applied to optimize the alarm thresholds of the original system. The optimized results were then compared with the optimization indicator results under the original system's alarm thresholds, where the $FAR^* = 0.02$, $MAR^* = 0.002$, and $EDD^* = 0.002$. The comparative results are presented in Table 4.

Table 4. Comparison of performance indicators under fault conditions.

Index	Method	*FAR*	*MAR*	*EDD*	Alarm Threshold
1	Original system	0.0094	0.004	0.004	71
2	KDE + COA	0.0129	0.00146	0.00146	70.4312
3	3σ	0.005	0.035	0.036	75.472
4	KDE + NOA	0.011	0.018	0.018	70.465
5	KDE + PSO	0.012	0.0174	0.018	71.3324

From Table 4, it can be observed that after optimizing the alarm thresholds using the method proposed in this paper under fault conditions, the *MAR* and *EDD* are significantly reduced and controlled within the safety range. Although there is an increase in the *FAR*, it is also within the safety range. Overall, the threshold optimization method proposed in this paper demonstrates good performance even under fault conditions. As shown by the second, third, fourth, and fifth methods in Table 4, the results obtained in this section are similar to those in Sections 4.2.1 and 4.2.3. This indicates that when the system fails, compared with traditional threshold design methods, the proposed threshold method is the optimal solution.

4.2.4. Real-Time Analysis of the Proposed Threshold Optimization Method

To validate the effectiveness of the threshold optimization method on real-time operational data, the algorithm was deployed onto a user interface within an online platform. The threshold optimization algorithm was integrated as a functional module on the user interface, enabling all functionalities of the optimization algorithm. Real-time operational data stored in the Distributed Control System (DCS) were inputted into the threshold optimization module through the database on the online platform to obtain the optimized performance metric results. This integration ensured seamless connectivity with existing alarm systems. The results before and after the optimization were visualized on the platform's graphical interface.

The real-time operational data of the high-pressure main steam temperature from a 1000 MW-level secondary reheating ultra-supercritical direct current boiler system were taken as the research subject. The sampling frequency was set at 1 s, and 9.5 h of real-time operational data were collected. Among them, the alarm threshold is 556 °C. The real-time operational data curve is depicted in Figure 10.

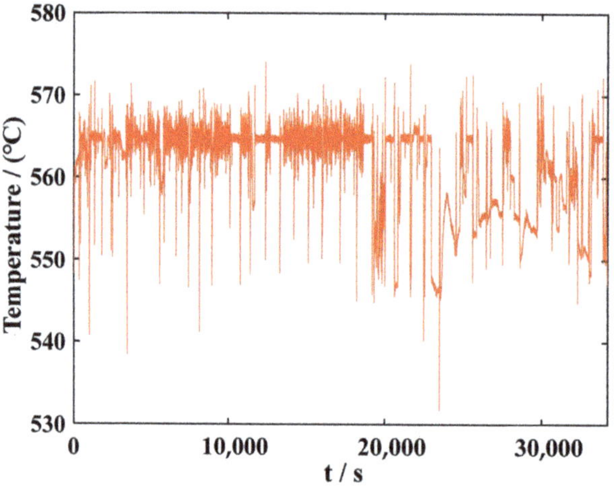

Figure 10. Real-time operation curve of high-pressure main steam temperature.

The performance comparison results of the alarm threshold before and after optimization using the threshold optimization algorithm module proposed in this article are displayed in the graphical configuration of the online platform, as shown in Figure 11. From Figure 11, it can be observed that during real-time operation, the threshold optimization algorithm proposed in this paper still achieves a further reduction in the *MAR* and *EDD* without significantly increasing the *FAR*. This demonstrates the effectiveness of the threshold optimization algorithm proposed in this paper during the real-time operation of the unit.

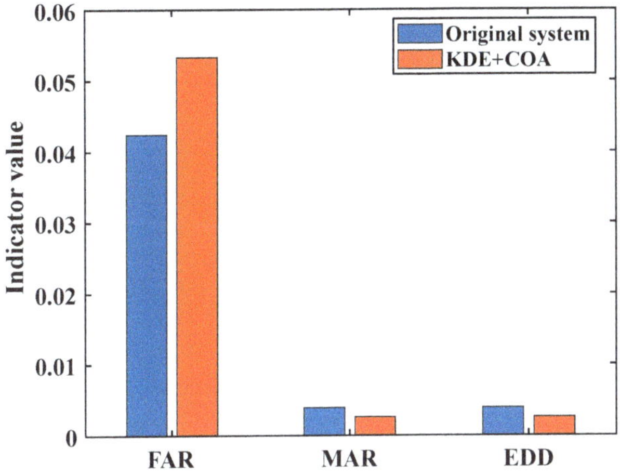

Figure 11. Comparison of the performance indicators before and after the lower threshold optimization in real-time running.

4.3. *Performance Analysis of Alarm Adaptive Delay Algorithm*

4.3.1. Data Verification of 1000 MW USCTPU

To validate the suppression capability of nuisance alarms during the normal operation of a 1000 WM USCTPU using the proposed adaptive alarm delay method, the reheat desuperheating water flow data in a 1000 MW-level secondary reheat ultra-supercritical DC boiler system were selected as the research subject, with its operational curve depicted in Figure 12. It can be observed from Figure 12 that the flow data contain numerous sample values fluctuating around the alarm threshold. Whenever the sample values of the flow data fluctuate below the alarm threshold, an alarm is triggered, leading to repetitive alarm signals, known as nuisance alarms. These alarms are often not caused by variable anomalies but rather by strong correlations among alarm variables, with the repeated occurrence of alarm signals potentially influenced by other parameters rather than indicating a malfunction. Upon statistical analysis, the original number of alarms reached 1647.

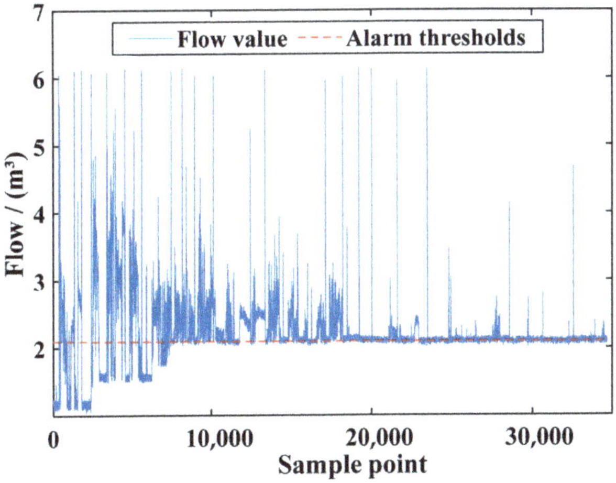

Figure 12. Operating data of reheat desuperheating water flow.

Firstly, based on the pre-set alarm threshold and the original alarm data, the time interval data for each alarm are obtained. Then, the alarm system is optimized using an alarm adaptive delay method according to the alarm time interval data. The changes in the coefficient of variation CV and the delay factor d are shown in Figure 13.

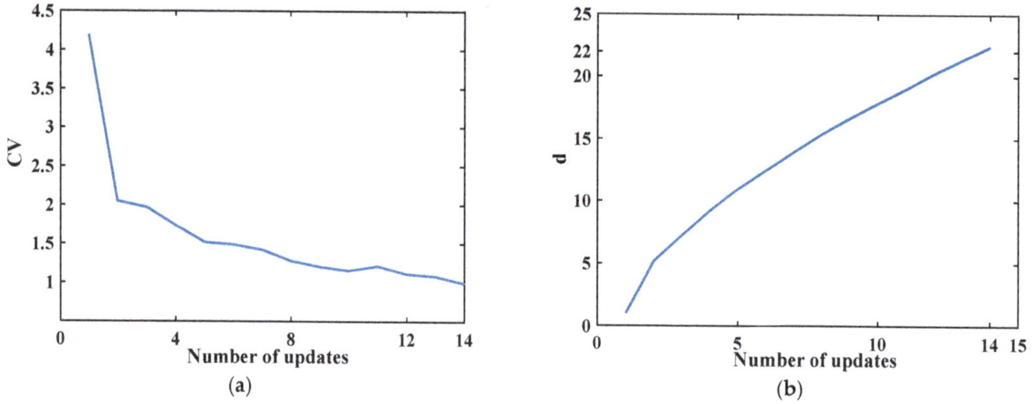

Figure 13. (**a**) Update of the coefficient of variation CV; (**b**) updated results for the delay factor d.

From Figure 13a, it can be seen that the value of the CV is 0.994 after the 14th update, which is less than the CV_{st}. This indicates that the setting of the delay factor is reasonable at this point, and the algorithm meets the stopping update condition. Figure 13b shows that the optimal delay factor after the optimization using the alarm adaptive delay method is d = 22.

In order to visually demonstrate the effect of setting the delay factor on suppressing nuisance alarms, a comparison is made between the alarm signal data before and after setting the delay factor. The alarm signal values are recorded as 1 when an alarm occurs, and 0 otherwise. The results are shown in Figure 14.

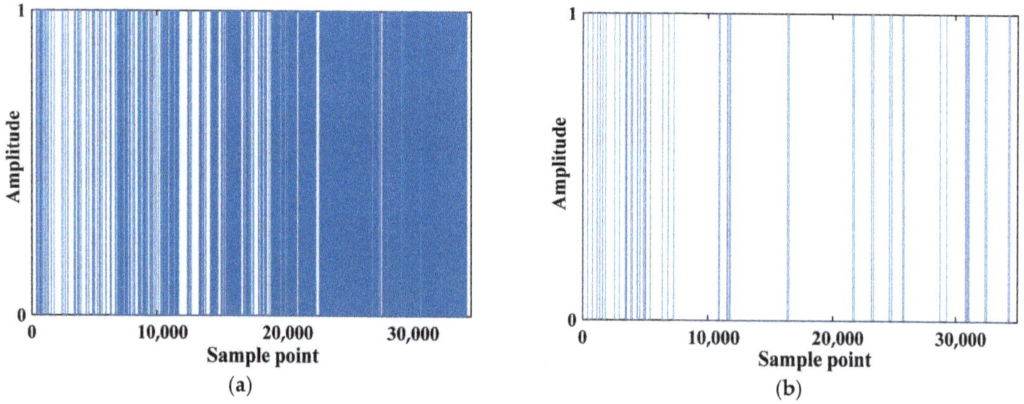

Figure 14. (**a**) Alarm signal data before setting delay factor; (**b**) alarm signal data after setting delay factor.

As shown in Figure 14a, without setting an adaptive delay factor in the traditional alarm system, the number of alarms is 1647, severely affecting the operation of the staff. However, in Figure 14b, due to the reasonable setting of the adaptive delay factor, the number of alarms is reduced to 27, which is 1/61 of the original number of alarms. Therefore, adding an appropriate delay factor in the alarm system can significantly reduce the number of nuisance alarms and improve the quality of the alarm signals.

4.3.2. Data Verification of 660 MW USCTPU

The expansion data of the gas exhaust expansion section in the 660 MW ultra-supercritical direct current boiler system are used as the research object to verify the effectiveness and universality of the proposed adaptive delayed alarm method on the different types of USCTPU. The operational curves are depicted in Figure 15.

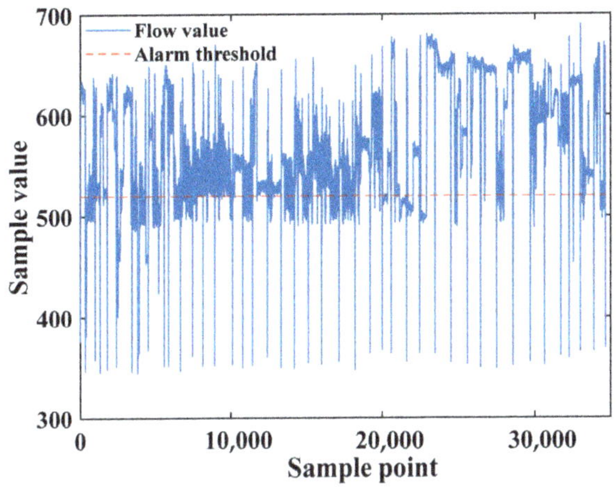

Figure 15. Data of gas exhaust expansion section.

The alarm signal data before and after setting the delay factor are compared, and the results are shown in Figure 16. As shown in Figure 16a, the number of alarms without setting the delay factor is 1121, while after setting the adaptive delay factor, the number of alarms decreases to 41. It can be seen that the adaptive delay alarm method proposed in this paper can greatly reduce the number of disturbance alarms. This also verifies the effectiveness and universality of the proposed delay alarm method on the different types of ultra-supercritical thermal power units.

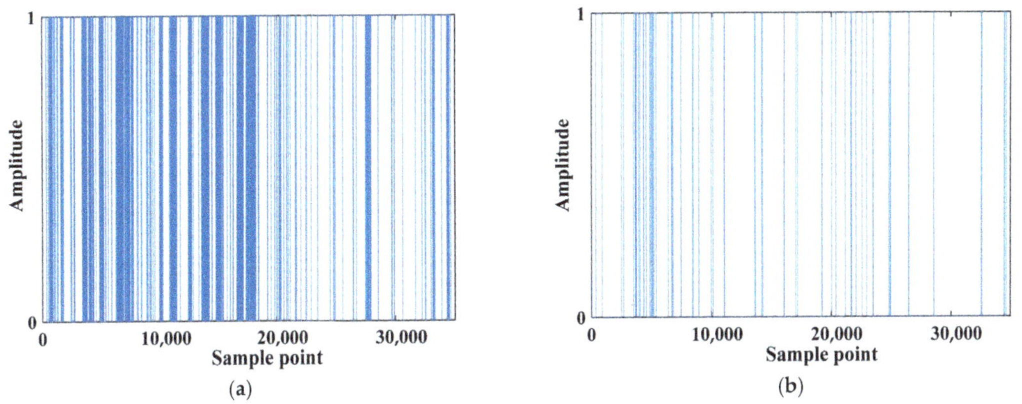

Figure 16. (**a**) Alarm signal data before adding adaptive delay factor; (**b**) alarm signal data after adding adaptive delay factor.

4.3.3. Performance Verification under System Fault

To further investigate the effectiveness and reliability of the proposed adaptive delay alarm method when the system experiences faults, data from a 1000 MW USCTPU experiencing a fault are used as the research subject. The fault scenario involves the RSV2 opening too slowly, and the operational curve during the fault occurrence is depicted in Figure 17.

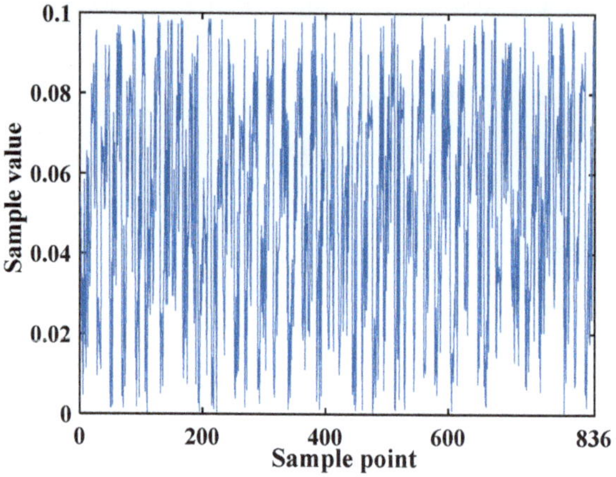

Figure 17. RSV2 opening speed curve when a fault occurs.

From Figure 17, it can be observed that during the fault occurrence, the opening speed of RSV2 is too slow, below 0.1, resulting in 836 alarms. Subsequently, when applying the adaptive delay alarm method during the fault occurrence, the generated alarm signals are shown in Figure 18. It can be seen from Figure 18 that the number of alarms after setting the adaptive delay factor is 118. Compared to the original system, the number of alarms has been significantly reduced. Therefore, it can be concluded that under fault conditions, the adaptive delay alarm method proposed in this paper still performs well, further validating the effectiveness and reliability of the proposed adaptive delay alarm method.

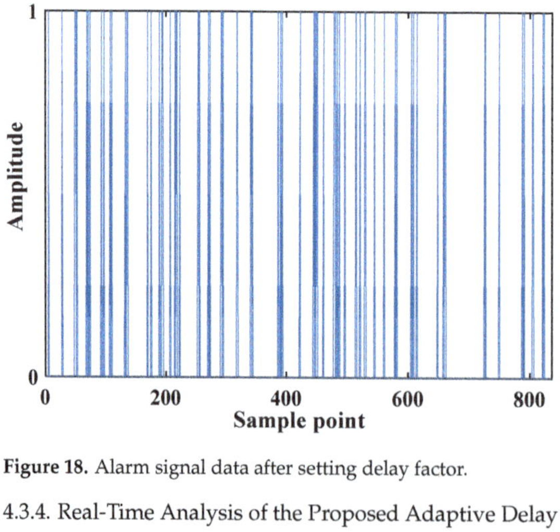

Figure 18. Alarm signal data after setting delay factor.

4.3.4. Real-Time Analysis of the Proposed Adaptive Delay Alarm Method

To verify the effectiveness of the proposed adaptive delay alarm method validated on real-time operational data, the adaptive delay alarm method proposed in this paper is deployed onto an online platform. The adaptive delay alarm algorithm and the alarm signal generation algorithm's code are integrated into different modules, replacing the entire functionality of the algorithms. Real-time operational data of a 1000 MW USCTPU DCS, stored in the online platform's database, are utilized, with the measurement point being the high-pressure main steam flow rate. The original alarm threshold is set to 250. The sampling period is 1 s, and 4 h of real-time operational data are captured as the test data. The real-time operational curve is depicted in Figure 19.

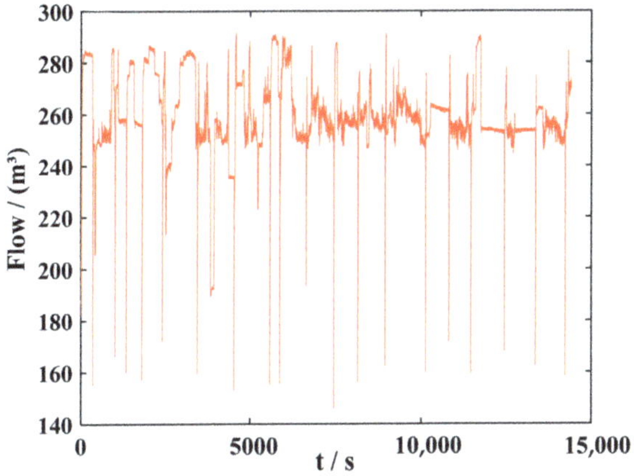

Figure 19. Real-time operation data curve of high-pressure main steam flow rate.

The real-time operational data are inputted into the alarm signal generation module, and based on the alarm threshold, the number of alarms is determined. After statistical analysis, it is found that a total of 184 alarms occurred within the 4 h period. Subsequently, the real-time operational data are processed using the adaptive delay alarm module. The processed alarm signals are depicted in Figure 20. From Figure 20, it can be observed that after applying the delay factor, the number of alarms reduces to 27. This indicates that the adaptive delay alarm method proposed in this paper exhibits a superior performance even during real-time system operation, effectively reducing the number of alarms.

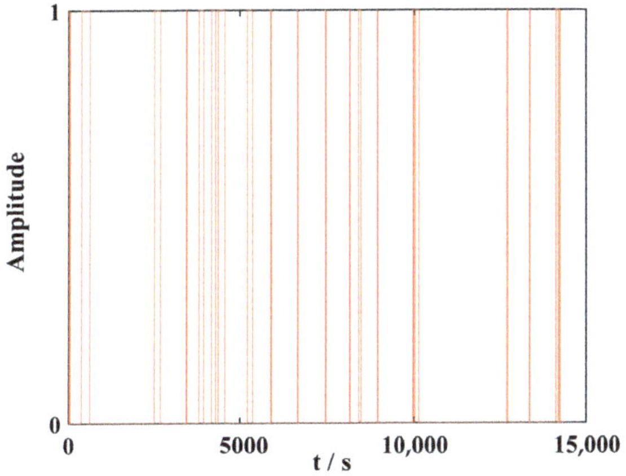

Figure 20. Real-time alarm data after setting delayed alarms.

5. Conclusions

This paper has proposed an intelligent alarm method, which includes the optimization method of alarm threshold based on KDE and COA. Unlike traditional methods, this method has taken the *MAR* as the optimization objective, aiming to significantly reduce the *MAR* while controlling the *FAR* within a reasonable range. It also considers the *EDD* of the alarm system and combines the *FAR*, *MAR*, and *EDD* to construct the objective function for threshold optimization. Finally, the COA has been used to optimize and solve the objective function, obtaining the optimized alarm threshold. The results have shown that the proposed threshold optimization method can significantly reduce the *MAR* and *EDD*, almost without increasing the *FAR*, thus eliminating the safety hazards that a high *MAR* may bring and improving the safety of USCTPUs.

Furthermore, to address the issue of a high number of nuisance alarms, this paper has introduced an alarm adaptive delay method. Initially, based on the pre-set alarm threshold and original alarm data, the time intervals between each alarm are obtained. Subsequently, the delay factor is adaptively updated based on the mean and variance of these alarm time intervals. The results have indicated that the inclusion of the delay factor can significantly reduce the number of nuisance alarms and enhance the quality of alarms.

The intelligent alarm method proposed in this paper is designed for single-variable alarm systems, but in practical production operations, multi-variable alarm situations are more common. Once an abnormality occurs in one variable, it may cause simultaneous abnormalities in other related variables, leading to alarm surges, which are highly detrimental to safe production operations. Therefore, future work will focus on studying the correlation of variables in multi-variable alarm systems. We will determine the topology of variables by analyzing causal relationships among them, thereby more accurately identifying the source of alarms.

Author Contributions: Conceptualization, X.Z.; methodology, X.Z.; software, L.Y.; validation, X.Z. and C.W.; formal analysis, X.Z.; investigation, C.Z.; resources, C.W.; data curation, L.Y.; writing—original draft preparation, X.Z.; writing—review and editing, C.Z.; visualization, C.Z.; supervision, C.W.; project administration, C.W.; funding acquisition, C.W. All authors have read and agreed to the published version of the manuscript.

Funding: This research was funded by the Key R&D Program Project of Zhejiang Province under Grants No. 2022C01206.

Data Availability Statement: Some or all data, models, or codes generated or used during the study are available from the corresponding author by request.

Conflicts of Interest: The authors declare no conflicts of interest.

References

1. Wei, C.; Wang, Y.; Shen, Z.; Xiao, D.; Bai, X.; Chen, H. AUQ-ADMM algorithm-based peer-to-peer trading strategy in large-scale interconnected microgrid systems considering carbon trading. *IEEE Syst. J.* **2023**, *17*, 6248–6259. [CrossRef]
2. He, K.X.; Wang, T.; Zhang, F.K.; Jin, X. Anomaly detection and early warning via a novel multiblock-based method with applications to thermal power plants. *Measurement* **2022**, *193*, 110979. [CrossRef]
3. Lai, S.; Chen, T. A method for pattern mining in multiple alarm flood sequences. *Chem. Eng. Res. Des.* **2017**, *117*, 831–839. [CrossRef]
4. Hu, W.; Chen, T.; Shah, S.L. Discovering Association Rules of Mode-Dependent Alarms From Alarm and Event Logs. *IEEE Trans. Control Syst. Technol.* **2018**, *26*, 971–983. [CrossRef]
5. Jablonski, A.; Barszcz, T.; Bielecka, M. Modeling of Probability Distribution Functions for Automatic Threshold Calculation in Condition Monitoring Systems. *Measurement* **2013**, *46*, 727–738. [CrossRef]
6. Fricker, R.D.; Banschbach, D. Optimizing Biosurveillance Systems that Use Threshold-based Event Detection Methods. *IEEE Trans. Control Syst. Technol.* **2012**, *13*, 117–128. [CrossRef]
7. Fricker, R.D.; Banschbach, D. A correlation consistency based multivariate alarm thresholds optimization approach. *ISA Trans.* **2016**, *65*, 37–43.
8. Aslansefat, K.; Gogani, M.B.; Kabir, S.; Shoorehdeli, M.A.; Yari, M. Performance evaluation and design for variable threshold alarm systems through semi-Markov process. *ISA Trans.* **2020**, *97*, 282–295. [CrossRef]

9. Xiong, W.; Wang, J.; Chen, K. Multivariate Alarm Systems for Time-Varying Processes Using Bayesian Filters With Applications to Electrical Pumps. *IEEE Trans. Ind. Inform.* **2018**, *14*, 504–513. [CrossRef]
10. Zhi, Z.; Tao, L.; Guan, H.; Si, L.; Gai, L.; Wei, H. A Fault Detection Method Based on Data Reliability and Interval Evidence Reasoning. *Acta Autom. Sin.* **2020**, *46*, 2628–2637.
11. Han, L.; Gao, H.; Xu, Y.; Zhu, Q. Combining FAP, MAP and correlation analysis for multivariate alarm thresholds optimization in industrial process. *J. Loss Prev. Proc.* **2016**, *40*, 471–478. [CrossRef]
12. Zhang, G.; Wang, Z. Correlation Degree and Clustering Analysis-Based Alarm Threshold Optimization. *Processes* **2018**, *10*, 224. [CrossRef]
13. Tian, W.; Zhang, G.; Liang, H. Alarm clustering analysis and ACO based multi-variable alarms thresholds optimization in chemical processes. *Process Saf. Environ. Prot.* **2018**, *113*, 132–140. [CrossRef]
14. Zhang, G.; Wang, Z.; Mei, H. Sensitivity clustering and ROC curve based alarm threshold optimization. *Process Saf. Environ. Prot.* **2018**, *141*, 471–478. [CrossRef]
15. Zhao, Y.; Zhao, C.H. Dynamic multivariate threshold optimization and alarming for nonstationary processes subject to varying conditions. *Control Eng. Pract.* **2018**, *124*, 105180. [CrossRef]
16. Cai, Y.; Hong, L. Satisfaction Optimization Based Self-adaptive Adjustment of Process Alarm Thresholds. *Control Eng. China* **2019**, *26*, 417–422.
17. Hashim, H.; Clifford, E.; Ryan, P. False Alarm Moderation for Performance Monitoring in Industrial Water Distribution Systems. *Adv. Eng. Inform.* **2022**, *52*, 101592. [CrossRef]
18. Luo, Y.; Gopaluni, B.; Cao, L.; Wang, Y.; Cheng, J. Adaptive online optimization of alarm thresholds using multilayer Bayesian networks and active transfer entropy. *Control Eng. Pract.* **2023**, *137*, 105534. [CrossRef]
19. Wang, Y.; Ding, Y.; Shahrampour, S. TAKDE: Temporal Adaptive Kernel Density Estimator for Real-Time Dynamic Density Estimation. *IEEE Trans. Pattern Anal.* **2023**, *45*, 13831–13843. [CrossRef]
20. Shi, X.; Gu, L.; Li, X.; Jiang, T.; Gao, T. Automated spectral transfer learning strategy for semi-supervised regression on Chlorophyll-a retrievals with Sentinel-2 imagery. *Int. J. Digit. Earth* **2024**, *17*, 2313856. [CrossRef]
21. Hu, B.; Li, Y.; Yang, H.; Wang, H. Wind speed model based on kernel density estimation and its application in reliability assessment of generating systems. *J. Mod. Power Syst. Clean Energy* **2017**, *5*, 220–227. [CrossRef]
22. Wei, S.; Liu, X.; Shi, H.; Gan, J. Remaining Useful Life Prediction of High-Dimensional Kernel Density Estimation With Adaptive Relative Density Window Width Considering Multisource Information Fusion. *IEEE Sens. J.* **2024**, *24*, 6543–6568. [CrossRef]
23. Yun, S.; Zanetti, R.; Jones, B. Kernel-based ensemble gaussian mixture filtering for orbit determination with sparse data. *Adv. Space Res.* **2022**, *69*, 4179–4197. [CrossRef]
24. Chen, K.; Wei, S.; Zhang, J. Intelligent Optimization Algorithm Based on Two-dimensional Convolution Operation. *J. Armored Forces* **2023**, *2*, 102–108.
25. Zeng, Z.; Tan, W.; Zhou, R. Performance assessment for generalized delay-times in alarm configuration. *J. Process Control* **2017**, *57*, 80–101. [CrossRef]
26. Yu, H.; Gao, M.; Zhang, H.; Chen, J.; Liu, J.; Yue, G. Dynamic Modeling of Key Operating Parameters for Supercritical Circulating Fluidized Bed Units based on Data-Knowledge-Driven Method. *J. Therm. Sci.* **2024**. [CrossRef]
27. Trojovsky, P.; Dehghani, M. Subtraction-Average-Based Optimizer: A New Swarm-Inspired Metaheuristic Algorithm for Solving Optimization Problems. *Biomimetics* **2023**, *8*, 149. [CrossRef] [PubMed]
28. Cheng, M.; Sholeh, M. Optical microscope algorithm: A new metaheuristic inspired by microscope magnification for solving engineering optimization problems. *Knowl.-Based Syst.* **2023**, *279*, 110939. [CrossRef]
29. Su, H.; Zhao, D.; Heidari, A.; Liu, L.; Zhang, X.; Mafarja, M.; Chen, H.L. RIME: A physics-based optimization. *Neurocomputing* **2023**, *532*, 183–214. [CrossRef]
30. Derrac, J.; García, S.; Molina, D.; Herrera, F. A practical tutorial on the use of nonparametric statistical tests as a methodology for comparing evolutionary and swarm intelligence algorithms. *Swarm Evol. Comput.* **2011**, *1*, 3–18. [CrossRef]

processes

MDPI

Article

Rotating Machinery Fault Diagnosis under Time–Varying Speed Conditions Based on Adaptive Identification of Order Structure

Xinnan Yu [1], Xiaowang Chen [1,*], Minggang Du [2], Yang Yang [2] and Zhipeng Feng [1]

1 School of Mechanical Engineering, University of Science and Technology Beijing, Beijing 100083, China
2 China North Vehicle Research Institute, Beijing 100072, China
* Correspondence: chenxw@ustb.edu.cn

Abstract: Rotating machinery fault diagnosis is of key significance for ensuring safe and efficient operation of various industrial equipment. However, under nonstationary operating conditions, the fault–induced characteristic frequencies are often time–varying. Conventional Fourier spectrum analysis is not suitable for revealing time–varying details, and nonstationary fault feature extraction methods are still in desperate need. Order spectrum can reveal the rotational–speed–related time–varying frequency components as spectral peaks in order domain, thus facilitating fault feature extraction under time–varying speed conditions. However, the speed–unrelated frequency components are still nonstationary after angular–domain resampling, thus causing wide–band features and interferences in the order spectrum. To overcome such a drawback, this work proposes a rotating machinery fault diagnosis method based on adaptive separation of time–varying components and order feature extraction. Firstly, the rotational speed is estimated by the multi–order probabilistic approach (MOPA), thus eliminating the inconvenience of installing measurement equipment. Secondly, adaptive separation of the time–varying frequency component is achieved through time–varying filtering and surrogate test. It effectively eliminates interference from irrelevant components and noise. Finally, a high–resolution order spectrum is constructed based on the average amplitude envelope of each mono–component. It does not involve Fourier transform or angular–domain resampling, thus avoiding spectral leakage and resampling errors. By identifying the fault–related spectral peaks in the constructed order spectrum, accurate fault diagnosis can be achieved. The Rényi entropy values of the proposed order spectrum are significantly lower than those of the traditional order spectrum. This result verifies the effective energy concentration and high resolution of the proposed order spectrum. The results of both numerical simulation and lab experiments confirm the effectiveness of the proposed method in accurately presenting the time–varying frequency components for rotating machinery diagnosing faults.

Keywords: signal processing; fault diagnosis; nonstationary; order analysis; order structure

1. Introduction

Rotating machineries are prone to fault under harsh environment and complex operating conditions. Undetected failures may deteriorate the accuracy and/or efficiency of the equipment, or even cause safety accidents. Thus, rotating machinery fault diagnosis is of key significance. Fault feature extraction from the measured signals, e.g., vibration, sound, and current, enables accurate fault detection and diagnosis [1–6], yet most efforts are based on the assumption of fixed operating conditions. Rotating machinery fault diagnosis under nonstationary operating conditions still deserves deep research.

In engineering practice, rotating machinery often operates under time–varying speed and/or load conditions, which are referred to as nonstationary operating conditions. Consequently, the frequency and/or amplitude of vibration signals undergo corresponding changes with varying operational conditions, disabling effective fault feature extraction in the frequency domain. To address this issue, researchers have tried to map the time–varying frequency details into a constant one [7–10]. Such an approach transforms the three–dimensional time–frequency features into two–dimensional spectral features, enabling more intuitive and convenient identification of nonstationary characteristics. It is capable of handling time–varying and multicomponent signals; thus, it is promising for rotating machinery fault diagnosis under nonstationary operating conditions.

Order tracking (OT) realizes this concept by resorting time–domain signals according to fixed angular increments, effectively converting time–varying frequency into constant frequency [7]. OT methods can be broadly classified into three categories: hardware order tracking (HOT), computed order tracking (COT), and tacholess order tracking (TOT) [8,9]. HOT adjusts the sampling frequency proportionally to the rotational speed via hardware control. It offers real–time angular–domain sampling [10], yet delays and errors may occur during rapid speed changes [11]. Additionally, the reliance on extra hardware components increases system complexity and cost. COT resamples the measured time–domain signal into an angular domain on a digital platform [12], and the order spectrum is generated by performing Fourier transform on the angular–domain signal. However, the angular–domain resampling needs to be guided by the precise instantaneous rotational speed. The acquisition of rotational speed typically relies on an extra tachometer, which can be available in real applications. To eliminate the need for an external tachometer, researchers have proposed the TOT method. It aims at estimating the instantaneous speed based on the signal itself [13]. Researchers have explored various methods for estimating rotational speed from vibration signals, including Hilbert–transform–based phase demodulation [14], iterative phase demodulation based on angular resampling [15], and the Teager–Kaiser energy operator [16]. However, these methods still face challenges, such as the presence of multiple harmonic modes in the vibration signal, interactions between rotational frequency harmonics and mechanical structure resonances, and low signal–to–noise ratios, which can lead to errors in instantaneous speed estimation. To address these issues, the multi–order probabilistic approach (MOPA) [17] has been proposed for tracking instantaneous speed. MOPA regards the instantaneous spectrum as a probability density function of instantaneous speed and estimates the speed based on the prior continuity of the function. Recent studies have demonstrated the effectiveness of MOPA in various applications [18–20].

In addition, while order spectrum is effective in identifying time–varying frequency components related to rotational speed in nonstationary conditions [21,22], components unrelated to rotational speed, such as resonances and background noise, are still nonstationary in angular domain. Their energy is not concentrated at a specific order, and result in wide–band feature. Such a result may lead to the misidentification of meaningful order components, hindering effective fault feature extraction. To address this issue, various techniques have been proposed, such as Vold–Kalman filtering [23] and phase–demodulation–based order tracking [24]. However, these methods often rely on prior knowledge of the targeted instantaneous frequency, and cannot distinguish interferences caused by speed–unrelated components from the noises.

In addition to OT, alternative approaches have been proposed for achieving order tracking objectives. One such method, introduced by Blough and Brown [25], involves modifying the Fourier transform kernel to extract order domain information directly from time–domain signals. Borghesani [26] proposed a speed–synchronized discrete Fourier transform, in which the modified Fourier transform kernel is characterized by instantaneous rotational speed, eliminating the need for resampling and interpolation steps. While this approach avoids resampling and interpolation errors associated with COT methods, it still suffers from spectral blurring caused by the nonstationarity of amplitude envelopes. Researchers have also proposed various techniques aimed at reducing spectral aliasing and spectral leakage [27–30]. However, these methods still do not eliminate spectral blurring fundamentally due to the nonstationarity of amplitude envelopes in time–varying conditions.

In this paper, we propose a novel rotating machinery fault diagnosis method, based on adaptive separation of time–varying components and order spectrum analysis. The objectives of the proposed method are as follows:

- Avoid the inconvenience and error of speed measurement.

The proposed method introduces MOPA to directly extract instantaneous rotational speed from the signal without external tachometer.

- Eliminate the interference of irrelevant components and noise.

It effectively separates time–varying components using a combination of time–varying filtering and surrogate test, eliminating interferences from irrelevant components and background noise.

- Avoid spectral energy leakage and achieve fine resolution.

The order spectrum is directly constructed by the average amplitude envelope of each individual order component, thus avoiding spectral energy leakage during Fourier transform and achieving fine resolution. The achieved precise revelation of fault–related order components enables effective fault diagnosis of rotating machinery under nonstationary operating conditions.

The remainder of this article is structured as follows. Section 2 details the proposed fault diagnosis method, including accurate acquisition of instantaneous rotational speed, adaptive separation of time–varying components, elimination of irrelevant components and noise, and precise identification of frequency orders. Section 3 demonstrates the effectiveness of the proposed method through an analysis of time–varying multicomponent simulation signals. Section 4 further validates the effectiveness of the proposed method in diagnosing rolling–element–bearing faults via laboratory experiments. Finally, Section 5 concludes the paper.

2. The Proposed Fault Diagnosis Method

2.1. Estimation of Instantaneous Rotational Speed

Under nonstationary operating conditions, the vibration components of critical components exhibit time–varying characteristics, and their instantaneous frequencies are closely related to rotational speed. Therefore, in the process of extracting time–varying components, instantaneous rotational speed serves as indispensable prior knowledge, providing an important basis for subsequent component identification and ridge extraction. The accurate estimation of instantaneous rotational speed thus becomes a crucial step in the entire signal processing workflow. In practical engineering applications, the acquisition of rotational speed typically relies on specific hardware devices, such as tachometers or encoders. However, the introduction of these devices not only increases additional costs but also poses various inconveniences during actual installation and use. To effectively address this issue, researchers have focused on directly extracting instantaneous rotational speed information from the vibration signal itself, eliminating the need for external measurement instruments.

In this context, methods for extracting rotational speed information by tracking harmonics in the time–frequency representation of vibration signals have emerged. Among them, MOPA has been widely used in various signal processing scenarios due to its high accuracy and excellent robustness in rotational speed identification.

Rotating machinery vibrations are usually dominated by the rotating frequency and its harmonics. The higher the spectral amplitude at a frequency f of a harmonic order H_k, the more likely it is that the rotating frequency will be equal to the frequency divided by the harmonic order f/H_k. Therefore, combining multiple harmonic components can effectively enhance noise resistance performance. Motivated by this fact, MOPA regards the instantaneous spectrum at each moment as a probability density function (PDF) of instantaneous rotational speed and accurately estimates the instantaneous rotational speed based on the prior continuity of the PDF, as shown in Figure 1. The specific procedure is as follows.

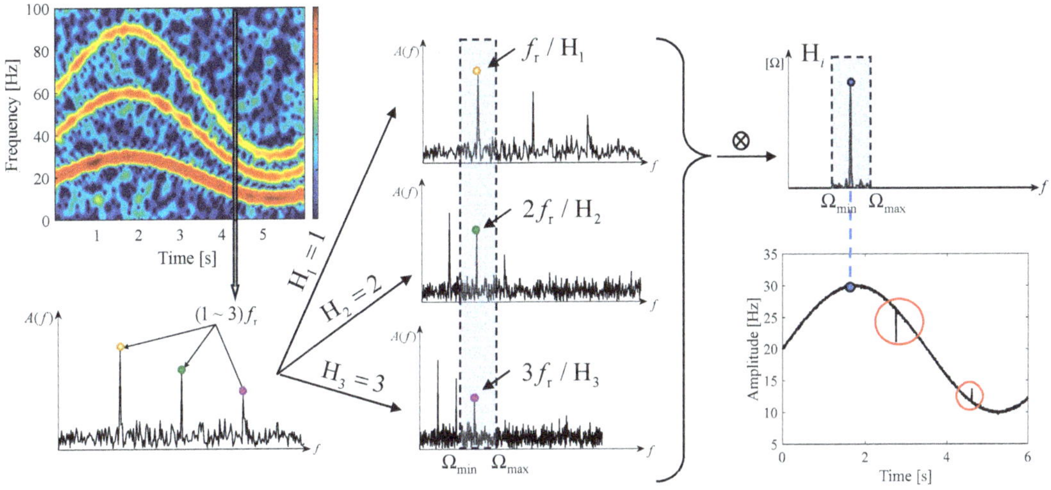

Figure 1. Schematic diagram of instantaneous rotational speed estimation.

Firstly, to mitigate the impact of structural modal responses on the instantaneous spectral amplitude of rotational frequency harmonics, construct a whitened time–frequency representation $W(t, f)$ and regard each column $W(f)$ as the instantaneous spectrum at each time instance.

Then, generate the PDF $[P|O_k]$ by scaling the instantaneous spectrum with all known harmonic orders O_k, maximum speed $S_{\max}$, and minimum speed $S_{\min}$:

$$[P|O_k] = \begin{cases} \frac{1}{\xi_k} W(O_k f), & S_{\min} < f < S_{\max} < f_{\max}/O_k \\ \frac{1}{\xi_k} W(O_k f), & S_{\min} < f < f_{\max}/O_k < S_{\max} \\ \frac{1}{S_{\max}-S_{\min}}, & f_{\max}/H_k < f < S_{\max} \\ 0, & f < S_{\min} \text{ or } f > S_{\max} > f_{\max}/O_k \end{cases}, \tag{1}$$

where $f_{\max}$ is the upper frequency limit and ξ_k is a normalization factor. Multiply PDFs of all orders to yield a global PDF:

$$[P] \propto \prod_{k=1}^{K} [P|O_k]. \tag{2}$$

Next, taking into account the inertia effects of actual mechanical systems, changes in instantaneous rotational speed should be continuous without any step changes. To ensure the continuity of the probability density function, a Gaussian function is introduced as a smoothing process. Perform a priori continuity to smooth the global PDF $[P]$. For any time instance n, the PDF can be estimated by convolving a neighboring PDF $[P_{n+m}]$ with a centered Gaussian [17]:

$$[P_n]_{n+m} = \int_{S_{\min}}^{S_{\max}} [P_n|P_{n+m}][P_{n+m}]\mathrm{d}f \propto \exp\left(-\frac{f^2}{2\sigma_m^2}\right) * [P_{n+m}], \tag{3}$$

where σ_m^2 is a variance and the priori continuity is

$$[P_n|P_{n+m}] = \mathcal{N}(P_{n+m}, \sigma_m) \propto \exp\left[-\frac{(f - P_{n+m})^2}{2\sigma_m^2}\right]. \tag{4}$$

As the variance of the Gaussian function increases, the smoothing effect of convolution also increases accordingly. Multiply PDFs including a priori continuity to yield a smoothed PDF:

$$[P_n]_{\text{total}} \propto \prod_{m=-M}^{M} [P_n]_{n+m}. \tag{5}$$

Finally, the frequencies corresponding to the maximum values at each moment are extracted from the smoothed PDFs. By fitting the relationship between these frequency values and time, an accurate estimation of instantaneous rotational speed is achieved.

2.2. Adaptive Separation of Time–Varying Components

For nonstationary signals, due to the time–varying nature of their sidebands, frequency domain overlap often occurs among various components, making effective separation challenging. As shown in the left figure of Figure 2, there is frequency domain overlap among the components. If filtering is performed directly, it will result in truncation of the time–varying frequencies. To address this issue and ensure the integrity of component extraction, time–varying filtering can be utilized for complete component extraction. The core idea of the time–varying component extraction method based on time–varying filtering is as follows.

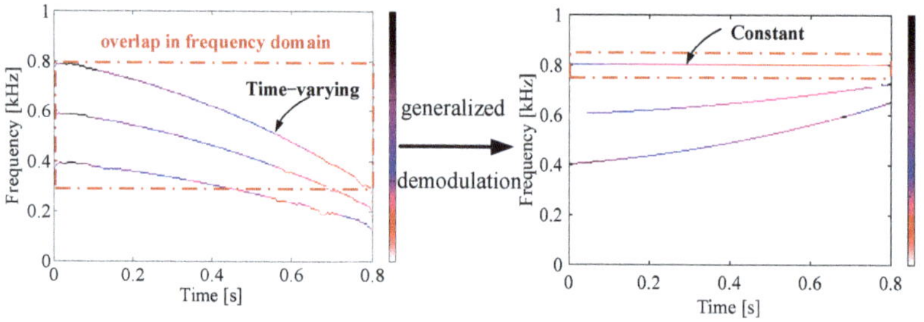

Figure 2. Schematic diagram of time–varying filtering.

According to prior knowledge of instantaneous rotational speed $\omega(t)$, the instantaneous frequency of the kth order harmonic component $y_k(t)$ in the time–frequency domain is calculated as $\omega_k(t) = k\omega(t)$. To achieve complete extraction of components, generalized demodulation is employed to convert time–varying frequency components into constant frequency components, as shown in the right figure of Figure 2. Specifically speaking, the signal $y(t)$ is multiplied with a generalized demodulation function $\exp[-\mathrm{j}w_k(t)]$ to convert the kth order harmonic component $y_k(t)$ into a constant frequency ω_0 component

$$z(t) = y(t)\exp[-\mathrm{j}w_k(t)], \tag{6}$$

where $w_k(t) = \int[\omega_k(t) - \omega_0]\mathrm{d}t$. In this case, the constant frequency components do not overlap with other components in the frequency domain, facilitating their separation through filtering operations.

Then, the component $z_k(t)$ of constant frequency ω_0 is separated by filtering, and the component $z_k(t)$ is multiplied with the inverse generalized demodulation function $\exp[\mathrm{j}w_k(t)]$ to restore the original component $y_k(t)$ by

$$y_k(t) = z_k(t)\exp[\mathrm{j}w_k(t)]. \tag{7}$$

However, the instantaneous frequencies of the rotational frequency harmonic components do not change at the same rate. Specifically, the higher the harmonic order, the greater the rate of change in its instantaneous frequency. This means that during a single generalized demodulation process, only one target component can be converted into a constant frequency component, while other target components may still exhibit time–varying characteristics or even overlap in the frequency domain. To overcome this challenge, iterative application of generalized demodulation is necessary, and demodulation functions must be designed specifically for each harmonic component to achieve complete extraction of all rotational frequency harmonic components. The specific procedure involves designing a generalized demodulation function targeted at a particular harmonic component, followed by applying generalized demodulation to it. Subsequently, the component is separated through filtering and is restored via inverse generalized demodulation. The extracted component is then subtracted from the original signal, and the entire process is repeated for the next harmonic order until all harmonic components have been successfully separated. This approach helps maintain the integrity of the components, effectively resolving the issue of frequency domain overlap.

2.3. Elimination of Pseudo–Components

After processing in the previous section, all time–varying components can be effectively extracted. However, in practical applications, these extracted time–varying components are not limited to those generated by the target component but also include pseudo–components formed by background noise. These pseudo–components can introduce interference and have an impact on the accurate identification of time–varying structures in rotating machinery. To eliminate the interference caused by pseudo–components composed of background noise, the entropy difference between the true components and the pseudo–components formed by noise is considered. By comparing entropy values, it can be determined whether a component is noise or not. Based on this approach, the surrogate test [31,32] has emerged as a solution. The core idea of the surrogate test is to compare the certainty of the target component with its surrogate data. If the target component is more certain than the surrogate data, it is considered a true component; otherwise, it is deemed false. The specific procedure is as follows.

Firstly, construct the surrogates by randomizing the phases of the component $y_k(t)$:

$$s(t) = \frac{1}{2\pi}\int [\int y_k(t)\exp(-j\gamma t)dt]\exp[j\varphi(\gamma)]\exp(j\gamma t)d\gamma, \tag{8}$$

where $\varphi(\gamma)$ is the random phase.

Subsequently, the authenticity of the target component is determined by comparing its statistical data with that of the surrogate data. Estimate the amplitude envelope $e(t)$ and instantaneous frequency $i(t)$ of the surrogate $s(t)$, and then calculate the discriminating statistics by combining spectral entropies of $e(t)$ and $i(t)$:

$$T(\beta_a, \beta_f) = \beta_a L[E(\omega)] + \beta_f L[I(\omega)], \tag{9}$$

where $E(\omega)$ and $I(\omega)$ are the Fourier transform of $e(t)$ and $i(t)$, respectively, and the spectral entropy:

$$L[g(x)] = -\int \frac{|g(x)|^2}{\int |g(x)|_2 dx} \ln \frac{|g(x)|^2}{\int |g(x)|_2 dx} dx. \tag{10}$$

Finally, select the maximum of $T(1,0)$, $T(0,1)$, and $T(1,1)$ as the significance T_s of the sth surrogate. T_0 is the significance of the candidate component $y_k(t)$. Regard the candidate component as true if the number of surrogates with $T_s > T_0$ is equal to or higher than ηN_s. N_s is the number of surrogates and η is the significance level.

2.4. Accurate Identification of Frequency Orders

The traditional approach to order analysis involves resampling the signal in the angle domain, followed by performing Fourier transform on the angle–domain signal to construct an order spectrum. Before resampling in the angle domain, the reference angle $\theta(t_n)$ must be calculated based on the instantaneous rotational speed $f_r(t)$ of the signal $x(t)$:

$$\theta(t_n) = \sum_{i=1}^{n} 2\pi f_r(t_n)\Delta t, \tag{11}$$

where Δt represents the time sampling interval. Then, a correspondence is established between $\theta_m = m\Delta\theta$ $(m = 1, 2, \ldots, M)$ and t_n, where $\Delta\theta$ represents the constant angular sampling interval and M denotes the length of the angle–domain signal. Based on the assumption that the reference axis has a constant angular acceleration, three consecutive sampling points are used to construct an equation in $[t_{p-1}, t_{p+1}]$ that describes the rotation angle of the reference axis:

$$\hat{\theta}(t_n) = b_{p,0} + b_{p,1}t_n + b_{p,2}t_n^2, \tag{12}$$

where the parameters $b_{p,0}$, $b_{p,1}$, and $b_{p,2}$ are determined by (t_{p-1}, θ_{p-1}), (t_p, θ_p), and (t_{p+1}, θ_{p+1}), and their matrix form is

$$\begin{bmatrix} \theta_{p-1} \\ \theta_p \\ \theta_{p+1} \end{bmatrix} = \begin{bmatrix} 1 & t_{p-1} & t_{p-1}^2 \\ 1 & t_p & t_p^2 \\ 1 & t_{p+1} & t_{p+1}^2 \end{bmatrix} \begin{bmatrix} b_{p,0} \\ b_{p,1} \\ b_{p,2} \end{bmatrix}. \tag{13}$$

Afterwards, the transformation relationship between $\hat{t} \in [t_{p-1}, t_{p+1}]$ and $\hat{\theta} \in [\theta_{p-1}, \theta_{p+1}]$ is established:

$$\hat{t} = \frac{1}{2b_{p,2}}\left[\sqrt{4b_{p,2}(m\Delta\theta - b_{p,0}) + b_{p,1}^2} - b_{p,1}\right]. \tag{14}$$

Subsequently, based on the calculated time points, interpolation algorithms are applied to process the signal and determine the amplitude of the vibration signal in the angle domain corresponding to the sampling time points. By performing Fourier transform on these amplitudes, the order spectrum can be obtained.

However, this method has significant limitations. When components with constant instantaneous frequencies undergo angular resampling based on rotational speed, their energy becomes dispersed in the order domain, interfering with the identification of harmonic components. Additionally, errors introduced by the interpolation step and the Fourier transform itself are unavoidable.

To address the aforementioned issues, we explore a novel approach that abandons traditional angular resampling and Fourier transform steps. The aim is to achieve more accurate estimations of frequency orders and amplitudes, thereby enabling precise fault diagnosis.

In actuality, order spectrum aims to pinpoint the average amplitude of each rotating frequency harmonic order in a signal. For a signal of zero mean $x(t) = a(t)\cos[\int \omega(t)\mathrm{d}t]$, suppose that $a(t) \geq 0$ changes slowly compared with $\omega(t)$, which is usually the case for rotating machinery vibration signals because of the mechanical system inertia. In this case, the signal amplitude is equal to the amplitude envelope:

$$a(t) = |x(t) + \mathrm{jHT}[x(t)]| = \left|a(t)\cos[\int \omega(t)\mathrm{d}t] + \mathrm{j}a(t)\sin[\int \omega(t)\mathrm{d}t]\right| = |a(t)| \tag{15}$$

Therefore, it is rational to set the spectral magnitude of each order as the average amplitude envelope of the corresponding mono–component, rather than it being estimated by Fourier transform.

Inspired by the fact that the spectral magnitude of each order is equal to the average amplitude envelope of the corresponding mono–component, we consider using the average amplitude envelope to construct the order spectrum, so as to avoid the spectral blur caused by Fourier spectrum. During the process of determining harmonic components, the method comprehensively traverses all possible orders and utilizes surrogate testing to filter out true harmonic components. Through this processing, true harmonic components $x_k(t)$ are retained, and their theoretical peak positions in the order spectrum, i.e., the corresponding frequency orders $\omega_k(t)$, are precisely located. Subsequently, for the rigorously filtered true time–varying harmonic components, their average amplitude envelopes are further calculated. Specifically, the average amplitude envelopes $e_k(t)$ of the true time–varying harmonic components $y_k(t)$ that pass the surrogate testing are computed:

$$\bar{e}_k = \frac{1}{T}\int e_k(t)\mathrm{d}t, \tag{16}$$

where T is the time length. Based on the essential meaning of the order spectrum, the average amplitude envelope $\bar{e}_k$ is regarded as the spectral magnitude at each order, so as to construct the optimized order spectrum:

$$\mathrm{IOR}(k) = \sum_k \bar{e}_k \delta[\omega - \omega_k(t)]. \tag{17}$$

Due to the concentration of spectral energy on specific frequency orders, the order spectrum constructed using the proposed method exhibits high resolution, enabling precise identification of frequency orders and accurate fault diagnosis.

2.5. Procedure of the Proposed Method

In this section, we propose a novel fault diagnosis method for rotating machinery that combines adaptive separation of time–varying components with precise extraction of order characteristics. This approach introduces a new idea that obviates the need for Fourier transforms and angular resampling steps. Its primary objective is to address the issue of blurred order spectra caused by spectral leakage and resampling errors. By eliminating various types of errors and interference, it ensures accurate acquisition of frequency orders and their corresponding amplitude information, enabling precise representation of the compositional structure of time–varying harmonic components.

In the initial phase of implementing this method, MOPA is employed to directly extract instantaneous rotational speed from vibration signals, eliminating the need for external rotary encoders or tachometers, thereby enhancing the practicality and flexibility of the approach. Subsequently, a time–varying component extraction method based on time–varying filtering is applied to separate each harmonic component individually. During this process, the surrogate test is integrated to identify genuinely existing harmonic components, precisely locate the corresponding order of each component, and eliminate interference from irrelevant components and background noise. Finally, a high–resolution order spectrum is constructed based on the mean amplitude envelope of each harmonic component and its corresponding order. The specific workflow is illustrated in Figure 3.

Step 1: Accurate estimation of instantaneous rotational speed:

(1.1) For the vibration signal $x(t)$, generate its whitened time–frequency representation $A(t,\omega)$. Consider $A(t,\omega)$ as a matrix where the column index represents time instances and the row index denotes frequency positions.

(1.2) Generate the PDF of the instantaneous rotational speed. Each column of the time–frequency representation is regarded as the instantaneous spectrum $A(\omega)$ corresponding to time t. The instantaneous spectrum $A(\omega)$ is scaled by multiple potential orders H_k in the time domain, and the values $A(H_k\omega)$ corresponding to the highest $\Omega_{\max}$ and lowest rotational speeds $\Omega_{\min}$ are considered as the PDF $[\Omega|H_k]$. Multiply the PDFs of all considered orders to generate the global PDF Ω.

(1.3) Apply a priori continuity to the PDF. Around the nth time instance, combine all PDFs within a given time interval to obtain a smoothed PDF $[\Omega_n]_s$, thereby incorporating a priori continuity.

(1.4) Extract the instantaneous rotational speed from the smoothed PDF. At the nth time instance, extract the frequency corresponding to the maximum value of the smoothed

PDF as the instantaneous rotational speed. Fit the instantaneous rotational speed $\omega(n)$ with respect to time t to obtain the global instantaneous rotational speed $\omega(t)$.

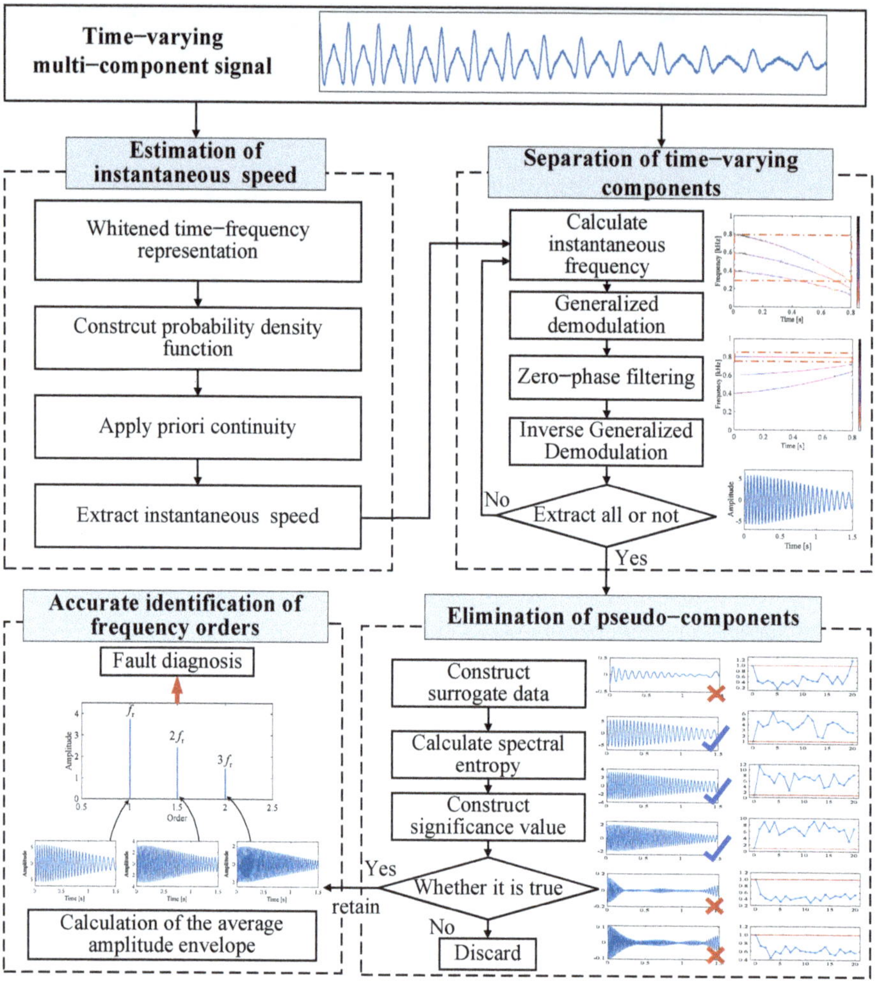

Figure 3. Flowchart of the proposed adaptive high–resolution order spectrum.

Step 2: Time–varying filtering for separation of mono–component:

(2.1) Based on the instantaneous rotational speed $\omega(t)$, the instantaneous frequency of the kth order rotational frequency harmonic component can be calculated using $\omega_k(t) = k\omega(t)$.

(2.2) Multiply the signal $x(t)$ with the generalized demodulation function $\exp[-jv_k(t)]$ to convert the kth order rotational frequency harmonic component $x_k(t)$ into a constant frequency component $u_k(t)$, where $v_k(t) = \int[\omega_k(t) - \omega_0]dt$.

(2.3) Filter the constant frequency component $u_k(t)$ using a zero–phase filter. Set the center frequency to ω_0 and the bandwidth to the minimum frequency spacing between adjacent components in the time–frequency representation.

(2.4) Multiply the mono–component $u_k(t)$ with the inverse generalized demodulation function $\exp[jv_k(t)]$ to reconstruct the original component $x_k(t)$.

(2.5) Check if all target rotational frequency harmonics have been separated. If so, proceed to Step 3; otherwise, repeat Step 2.

Step 3: Surrogate test to remove pseudo–components:

(3.1) For each separated component $x_k(t)$, first perform a Fourier transform, then randomize its phase, and finally conduct an inverse Fourier transform to construct surrogate data.

(3.2) Calculate the amplitude envelope $a(t)$ and instantaneous frequency $f(t)$ of both the target component and the surrogate data. Jointly construct a discrimination statistic using the spectral entropy values of these two sets of data.

(3.3) Determine the authenticity of the separated mono–component $x_k(t)$ by comparing the statistical data of the target component with its surrogate data. If it is true, retain it; otherwise, discard it.

Step 4: Accurate identification of frequency order structure and fault diagnosis:

(4.1) For the real time–varying harmonic component $x_k(t)$, calculate their average amplitude envelope $\bar{a}_k$. Consider the average amplitude envelopes of each order as the corresponding order amplitudes to construct an adaptive high–resolution order spectrum.

(4.2) Identify the fault characteristic frequencies based on the frequency structure displayed in the order spectrum to achieve accurate fault diagnosis.

The proposed method eliminates the need for external measuring instruments, effectively addressing interference from irrelevant components, spectral leakage, and error issues. It achieves precise representation of nonstationary multicomponent signal structure and successfully diagnoses faults.

3. Performance Evaluation by Numerical Simulation Analysis

In this section, the effectiveness of the proposed method is evaluated through the analysis of simulated signals. The simulation includes three amplitude and frequency modulated components to mimic harmonic constituents, along with two gradually enhancing constant–frequency components to simulate potential interferences. This design aims to closely replicate the actual vibration conditions encountered in rotating machinery during acceleration, where the proportional relationship between amplitude and the square of frequency further enhances the realism of the simulation:

$$x(t) = \sum_{k=1}^{3} a_k[c_k f_r(t)]^2 \cos\left\{2\pi c_k \int f_r(t)dt\right\} + Ae^{\beta t}\cos(2\pi f_{n1}t) + Ae^{\beta t}\cos(2\pi f_{n2}t) + n(t) \quad (18)$$

where a_k = 0.012, 0.003, 0.001, c_k = 1, 1.5, 2, A = 0.2, f_{n1} = 10 Hz, f_{n2} = 40 Hz, $\beta = 0.2\pi$, and rotational speed $f_r(t) = -2t^2 + 7t + 20$. $n(t)$ denotes Gaussian white noise characterized by a 2 dB signal–to–noise ratio. The duration of the signal is 1.5 s, and it is sampled at a frequency of 2 kHz.

The waveform and Fourier spectrum of the simulated signal are depicted in Figure 4a,b, respectively. Figure 5a illustrates the short–time Fourier transform spectrogram of the simulated signal, revealing the presence of three time–varying harmonic components and two constant frequency components. These frequency components exhibit significant overlap and even crossover in the frequency domain. Additionally, the interference from background noise and limitations in time–frequency resolution further complicate the identification process, making it challenging to determine the frequency structure of the signal. Given the rotational frequency, the signal is resampled in the angular domain according to $f_r(t)$, as shown in Figure 4c. Subsequently, a Fourier transform is applied to the angular domain resampled signal to obtain the traditional order spectrum, presented in Figure 4d. While three order components can be roughly identified in the figure, their energy is not concentrated at their respective frequency orders but, rather, spreads around them, leading to spectral aliasing between adjacent frequency orders. This is due to the nonstationary nature of the amplitude envelopes in the angular domain, even after equal–angle resampling, resulting in spectral leakage issues in the order spectrum.

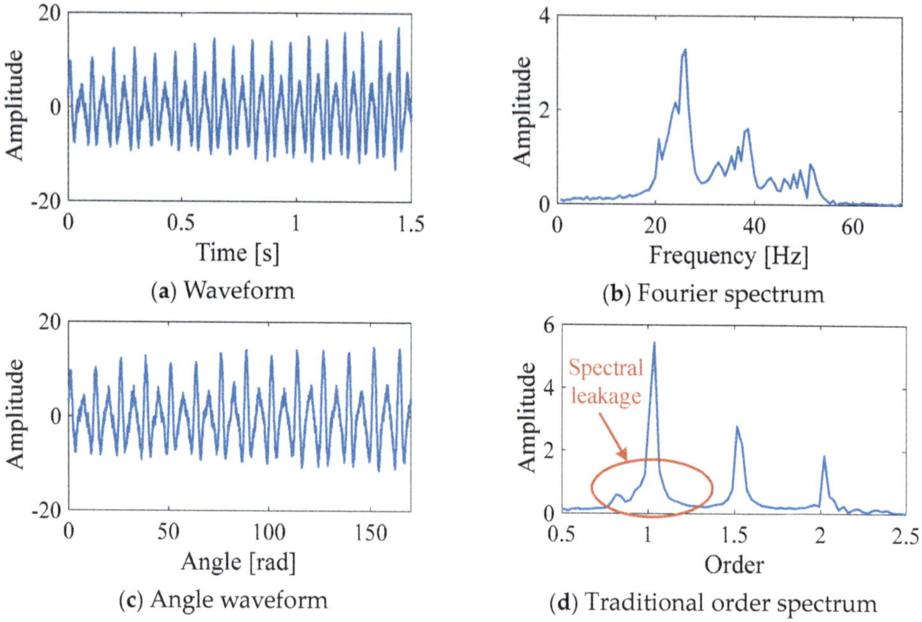

Figure 4. Traditional order analysis results of multi–harmonic component simulation signal.

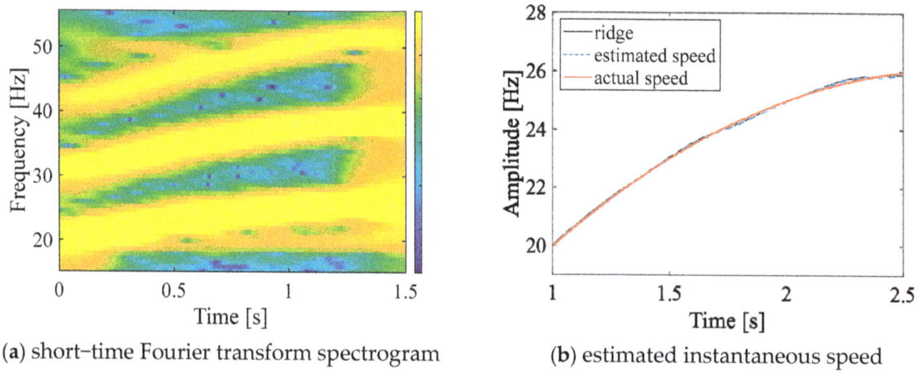

Figure 5. Accurate estimation of instantaneous speed.

To address the above challenges, the proposed method is employed for the identification of frequency components in the signal structure. Initially, the multi–order probabilistic approach is utilized to accurately extract the instantaneous rotational speed from the raw signal, providing a crucial input for subsequent instantaneous frequency calculation. The time–frequency ridge is extracted from the probability density function, and an accurate estimate of the instantaneous rotational speed is obtained through fitting. A comparison of the instantaneous ridge, estimated rotational speed, and actual rotational speed is presented in Figure 5b. The results demonstrate a high degree of agreement between the instantaneous rotational frequency $f_r(t)$ extracted using the multi–order probabilistic approach and the true rotational frequency. This validates the superiority of the proposed method in facilitating effective and efficient estimation of the instantaneous rotational speed without relying on additional rotary encoders or tachometers.

Based on the estimated instantaneous rotational frequency $f_r(t)$, components can be effectively separated through time–varying filtering. The validity of each separated mono–

component is verified through the surrogate test, and the corresponding harmonic orders are determined. Using the obtained rotational frequency $f_r(t)$, an exhaustive calculation of the instantaneous frequencies $kf_r(t)$ for each harmonic order is performed, and they are fully extracted using time–varying filtering. For each component, the generalized demodulation function and inverse demodulation function are denoted as $\exp\{-j2\pi[\int kf_r(t)dt + f_0t]\}$ and $\exp\{j2\pi[\int kf_r(t)dt + f_0t]\}$, respectively, where f_0 is set to one–quarter of the sampling frequency. Through iterative applications of generalized demodulation, zero–phase filtering, and inverse generalized demodulation steps, multiple order components are successfully separated. However, not all of these separated components represent true rotational frequency harmonics, as some may be spurious components arising from background noise. To accurately identify the true components and eliminate noise interference, the surrogate test is conducted on the separated components. As shown in Figure 6, among these components, only those with instantaneous frequencies of $f_r(t)$, $1.5f_r(t)$, and $2f_r(t)$ exhibit lower significance values compared to their surrogate data, indicating that they pass the test and are identified as true components. The other components are recognized as spurious, arising from noise, which aligns perfectly with the actual simulation scenario. Therefore, it can be concluded that the proposed method can adaptively identify and extract true rotational frequency harmonic components while effectively eliminating irrelevant components and background noise interference.

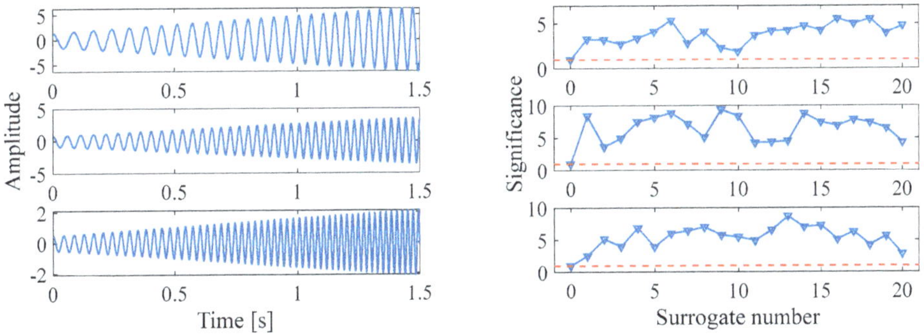

(**a**) order component waveform (**b**) surrogate test of components

Figure 6. Extraction of time–varying component.

After extracting the rotational frequency harmonic components, the mean values of their amplitude envelopes are further calculated. By assigning the mean amplitude envelope values and corresponding orders to each component, an optimized order spectrum is constructed, as illustrated in Figure 7. Compared to the conventional order spectrum in Figure 4b, the optimized version exhibits a higher concentration of energy at respective frequency orders, boasting superior resolution. It clearly reveals the frequency order structure along with the corresponding amplitude information.

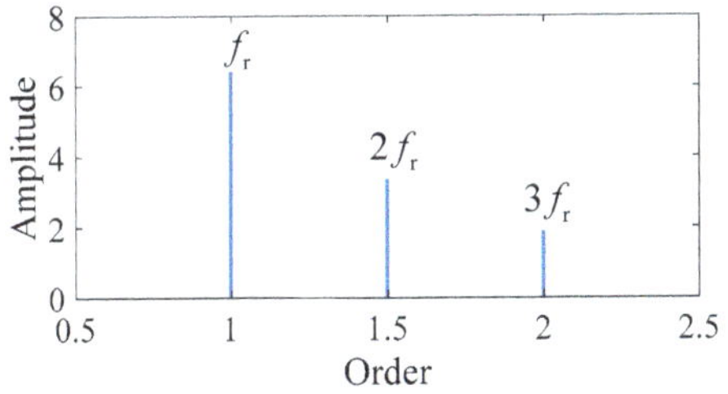

Figure 7. Accurate identification of frequency order structure.

4. Experimental Validations for Rolling Element Bearing Fault Diagnosis

In this section, the proposed method is applied to the amplitude envelopes of bearing vibration signals for the purpose of extracting fault characteristics and facilitating fault diagnosis. This approach serves to validate the effectiveness of the proposed method in accurately identifying and diagnosing faults.

4.1. Experiment Settings

Figure 8 shows a rolling element bearing test rig at the University of Ottawa. Table 1 lists the structural parameters of the bearings. Given the gearbox configuration parameters, the characteristic frequency can be calculated. In the experiment, a defect occurs on outer race of the experimental bearing. An accelerometer is mounted at the top of the experimental bearing, to collect vibration signals at a sampling frequency of 200 kHz for a duration of 7 s.

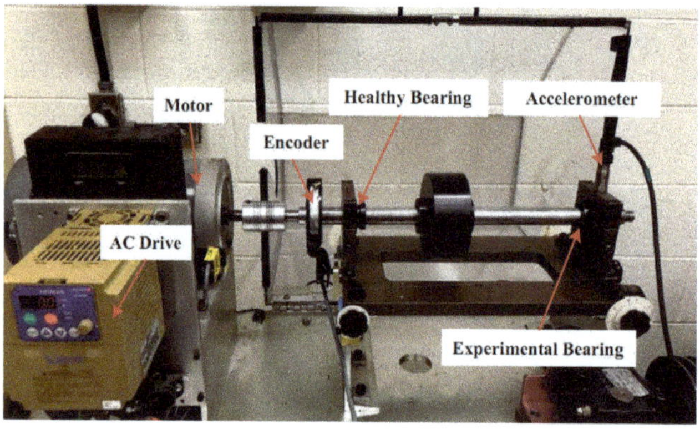

Figure 8. Rolling element bearing test rig configuration.

Table 1. Bearing configuration.

Bearing Parameters			Characteristic Frequency	
Pitch Diameter	**Ball Diameter**	**Number of Balls**	**BPFI**	**BPFO**
38.52 mm	7.94 mm	9	$5.43f_r$	$3.57f_r$

4.2. Signal Analysis

Figure 9a,b present the waveform and Fourier spectrum, respectively. Upon examination of the short–time Fourier transform spectrogram in Figure 10a, the nonstationary nature of the signal, along with noise interference components, becomes evident.

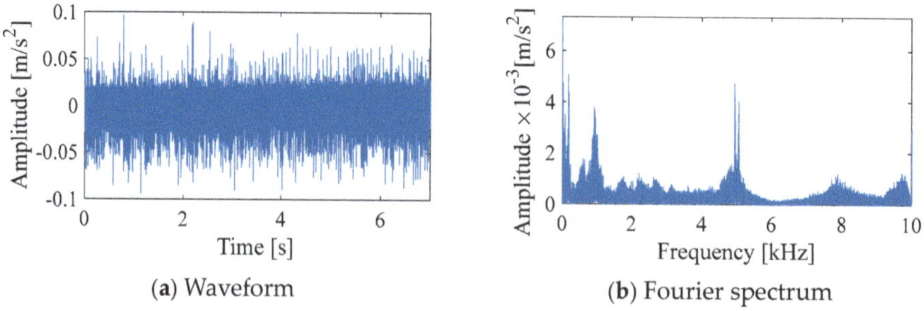

Figure 9. Outer race fault signal of rolling element bearing.

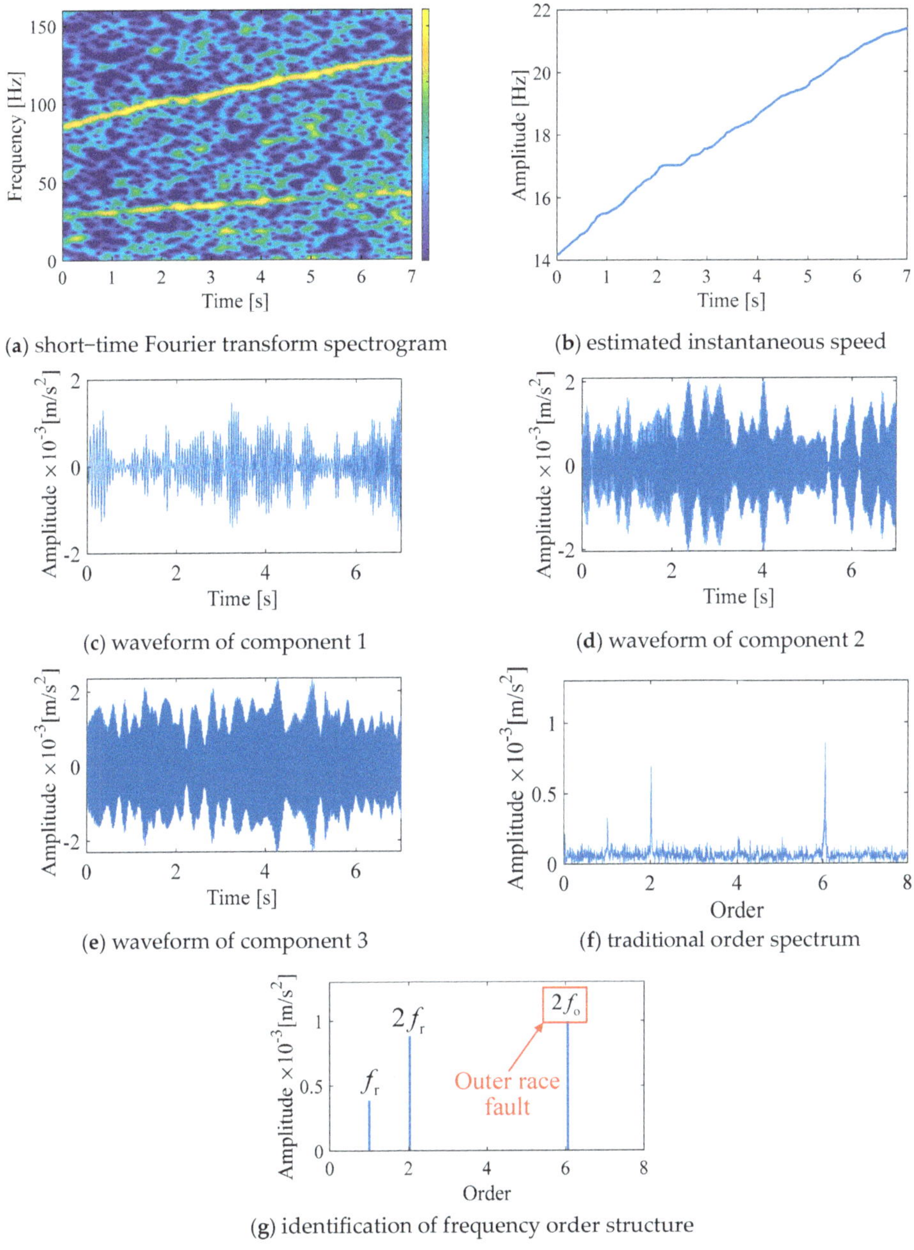

Figure 10. Fault diagnosis of rolling element bearing signal.

To accurately identify these components, we employed the proposed method. Initially, MOPA was utilized to precisely extract the instantaneous rotational speed information of the main shaft, which is directly connected to the motor, from the intricate raw signal. Subsequently, extraction of the time–frequency ridge lines from the probability density function was conducted, followed by an accurate fitting process to derive the precise instantaneous speed curve, as illustrated in Figure 10b.

Leveraging this instantaneous speed information, we further conducted a comprehensive traversal calculation of the instantaneous frequencies corresponding to various harmonic components. Through an iterative application of sophisticated steps including generalized demodulation, zero–phase filtering, and inverse generalized demodulation, we successfully achieved complete separation of the multi–order components. The surrogate test confirmed the authenticity of the components presented in Figure 10c–e. Subsequently, we calculated the mean amplitude envelopes of these components and combined them with their corresponding order information to construct an optimized order spectrum. Compared to the traditional order spectrogram shown in Figure 10f, the proposed spectrogram depicted in Figure 10g not only reveals the frequency order details with remarkable clarity but also effectively eliminates noise and irrelevant component interference, ensuring purity and accuracy in the analysis results. Within the order features, we distinctly identified the presence of the shaft rotation frequency and its harmonics $(1,2)f_r$, along with the outer race fault frequency f_o. This discovery aligns perfectly with the actual fault condition of the bearing outer race, providing a robust foundation for fault diagnosis.

Based on the analysis results of simulation signals and experiment signals, Rényi entropy calculations were performed for both the traditional order spectrum and the proposed order spectrum. The comparison results are presented in Figure 11. Rényi entropy represents the degree of energy concentration in the spectrum, where a smaller entropy value indicates a higher degree of concentration. As shown in the figure, the proposed order spectrum exhibits lower entropy values in both simulation and actual signal applications. This significant result demonstrates the excellent performance of the proposed method in revealing frequency structures, with high resolution, strong readability, and effective energy concentration.

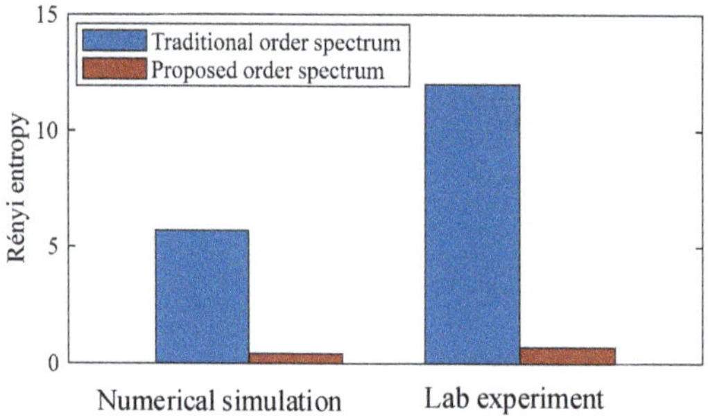

Figure 11. Comparison of Rényi entropy.

The comparison between the proposed method and the traditional order analysis method is shown in Table 2. The proposed method effectively avoids energy leakage caused by irrelevant components and nonstationarity, and prevents errors introduced by resampling. Compared with the traditional order spectrum, it exhibits more concentrated energy and higher resolution.

Table 2. Comparison of our study with existing work.

	Traditional Order Spectrum	Proposed Order Spectrum
Is rotational speed data required?	Yes	No
Does energy leakage exist due to irrelevant components?	Yes	No
Does energy leakage exist due to nonstationarity?	Yes	No
Does error exist due to resampling?	Yes	No
Spectral energy concentration	Low	High

5. Conclusions

Order spectrum analysis is widely applied in rotating machinery fault diagnosis under time–varying operating conditions. However, it faces the challenges of inconvenient instantaneous speed measurement, interference from irrelevant components, angular resampling errors, and spectral leakage. To address these issues, a novel method based on adaptive separation of time–varying components and order spectrum construction is proposed. Compared with available methods, no tachometer is required, the interferences from speed–unrelated components are eliminated, and fine spectral resolution is achieved. Results of the numerical simulation prove the correctness of the constructed order spectrum based on adaptive component identification and time–varying amplitude envelope. The outer race fault on a real–world rolling element bearing is successfully diagnosed under time–varying speed conditions, by identifying twice the fault frequency on the constructed order spectrum. In summary, the proposed method is validated to be capable of characterizing time–varying multicomponent signals with high accuracy and little interference, thus supporting rotating machinery fault diagnosis under time–varying speed conditions. However, it should be noted that the proposed method has its limitations. The process involved is relatively complex, requiring multiple steps and computations. Additionally, the method demands a significant amount of computational power, which may pose challenges in real–time applications or when dealing with large datasets. Despite these limitations, the method still offers valuable insights into rotating machinery fault diagnosis under time–varying speed conditions.

Author Contributions: Conceptualization, X.Y.; methodology, X.C.; software, M.D.; validation, Y.Y.; formal analysis, M.D.; investigation, Y.Y.; resources, Z.F.; data curation, Z.F.; writing—original draft preparation, X.Y.; writing—review and editing, X.C.; visualization, Z.F. All authors have read and agreed to the published version of the manuscript.

Funding: This research received no external funding.

Data Availability Statement: All data generated or analyzed during the study are included in this published article.

Conflicts of Interest: The authors declare no conflict of interest.

References

1. Cheng, Y.; Chen, B.Y.; Zhang, W.H. Adaptive Multipoint Optimal Minimum Entropy Deconvolution Adjusted and Application to Fault Diagnosis of Rolling Element Bearings. *IEEE Sens. J.* **2019**, *19*, 12153–12164. [CrossRef]
2. Rafiq, H.J.; Rashed, G.I.; Shafik, M.B. Application of multivariate signal analysis in vibration–based condition monitoring of wind turbine gearbox. *Int. Trans. Electr. Energy Syst.* **2021**, *31*, e12762. [CrossRef]
3. Qin, A.S.; Hu, Q.; Zhang, Q.H.; Lv, Y.R.; Sun, G.X. Application of sensitive dimensionless parameters and PSO–SVM for fault classification in rotating machinery. *Assem. Autom.* **2019**, *40*, 175–187. [CrossRef]
4. Xu, X.L.; Liu, X.L. Fault Diagnosis Method for Wind Turbine Gearbox Based on Image Characteristics Extraction and Actual Value Negative Selection Algorithm. *Int. J. Pattern. Recognit. Artif. Intell.* **2020**, *34*, 2054034. [CrossRef]
5. Peng, G.; Xu, D.; Zhou, J.; Yang, Q.; Shen, W. A novel curve pattern recognition framework for hot–rolling slab camber. *IEEE Trans. Ind. Inform.* **2022**, *19*, 1270–1278. [CrossRef]
6. Peng, G.; Cheng, Y.; Zhang, Y.; Shao, J.; Wang, H.; Shen, W. Industrial big data–driven mechanical performance prediction for hot–rolling steel using lower upper bound estimation method. *J. Manuf. Syst.* **2022**, *65*, 104–114. [CrossRef]
7. Fyfe, K.R.; Munck, E.D.S. Analysis of Computed Order Tracking. *Mech. Syst. Signal Process.* **1997**, *11*, 187–205. [CrossRef]
8. Lu, S.; Yan, R.; Liu, Y.; Wang, Q. Tacholess Speed Estimation in Order Tracking: A Review With Application to Rotating Machine Fault Diagnosis. *IEEE Trans. Instrum. Meas.* **2019**, *68*, 2315–2332. [CrossRef]
9. Xu, L.; Ding, K.; He, G.; Li, W.; Li, J. A novel tacholess order tracking method for gearbox vibration signal based on extremums search of gear mesh harmonic. *Mech. Syst. Signal Process.* **2023**, *189*, 110070. [CrossRef]
10. Potter, R.; Gribler, M. Computed Order Tracking Obsoletes Older Methods. In Proceedings of the SAE Noise and Vibration Conference, West Palm Beach, FL, USA, 8–10 February 1989; pp. 63–67.
11. Li, B.; Zhang, X. A New Strategy of Instantaneous Angular Speed Extraction and Its Application to Multistage Gearbox Fault Diagnosis. *J. Sound. Vib.* **2017**, *396*, 340–355. [CrossRef]
12. Rémond, D.; Antoni, J.; Randall, R. Editorial for the special issue on Instantaneous Angular Speed (IAS) processing and angular applications. *Mech. Syst. Signal Process.* **2014**, *44*, 1–4. [CrossRef]

13. Peeters, C.; Leclère, Q.; Antoni, J.; Lindahl, P.; Donnal, J.; Leeb, S.; Helsen, J. Review and comparison of tacholess instantaneous speed estimation methods on experimental vibration data. *Mech. Syst. Signal Process.* **2019**, *129*, 407–436. [CrossRef]
14. Zhao, M.; Lin, J.; Xu, X.; Lei, Y. Tacholess envelope order analysis and its application to fault detection of rolling element bearings with varying speeds. *Sensors* **2013**, *13*, 10856–10875. [CrossRef]
15. Hu, Y.; Cui, F.; Tu, X.; Li, F. Bayesian estimation of instantaneous speed for rotating machinery fault diagnosis. *IEEE Trans. Ind. Electron.* **2021**, *68*, 8842–8852. [CrossRef]
16. Gu, L.; Tian, Q.; Ma, Z. Extraction of the instantaneous speed fluctuation based on normal time–frequency transform for hydraulic system. *Proc. Inst. Mech. Engineers. Part C J. Mech. Eng. Sci.* **2020**, *234*, 1196–1211. [CrossRef]
17. Leclère, Q.; André, H.; Antoni, J. A multi–order probabilistic approach for instantaneous angular speed tracking debriefing of the CMMNO' 14 diagnosis contest. *Mech. Syst. Signal Process.* **2016**, *81*, 375–386. [CrossRef]
18. Peeters, C.; Leclere, Q.; Antoni, J.; Guillaume, P.; Helsen, J. Vibration–based angular speed estimation for multi–stage wind turbine gearboxes. *J. Phys. Conf. Ser.* **2017**, *842*, 012053. [CrossRef]
19. Schmidt, S.; Heyns, P.S.; De Villiers, J.P. A tacholess order tracking methodology based on a probabilistic approach to incorporate angular acceleration information into the maxima tracking process. *Mech. Syst. Signal Process.* **2018**, *100*, 630–646. [CrossRef]
20. Gryllias, K.; Andre, H.; Leclere, Q.; Antoni, J. Condition monitoring of rotating machinery under varying operating conditions based on cyclononstationary indicators and a multi–order probabilistic approach for instantaneous angular speed tracking. *IFAC–PapersOnLine* **2017**, *50*, 4708–4713.
21. Wang, K.; Heyns, P.S. The combined use of order tracking techniques for enhanced fourier analysis of order components. *Mech. Syst. Signal Process.* **2011**, *25*, 803–811. [CrossRef]
22. Wang, K.; Guo, D.; Heyns, P.S. The application of order tracking for vibration analysis of a varying speed rotor with a propagating transverse crack. *Eng. Fail. Anal.* **2012**, *21*, 91–101. [CrossRef]
23. Zhao, M.; Lin, J.; Wang, X.; Lei, Y.; Cao, J. A tacho–less order tracking technique for large speed variations. *Mech. Syst. Signal Process.* **2013**, *40*, 76–90. [CrossRef]
24. Coats, M.D.; Randall, R.B. Single and multi–stage phase demodulation based order–tracking. *Mech. Syst. Signal Process.* **2014**, *44*, 86–117. [CrossRef]
25. Blough, J.R.; Brown, D.L.; Vold, H. The time variant discrete Fourier transform as an order tracking method. *SAE Trans.* **1997**, *106*, 3037–3045.
26. Borghesani, P.; Pennacchi, P.; Chatterton, S.; Ricci, R. The velocity synchronous discrete Fourier transform for order tracking in the field of rotating machinery. *Mech. Syst. Signal Process.* **2014**, *44*, 118–133. [CrossRef]
27. Xu, S.; Zhang, Y.; Pham, D. Antileakage Fourier transform for seismic data regularization. *Geophysics* **2005**, *70*, V87–V95. [CrossRef]
28. Xu, S.; Zhang, Y.; Lambaré, G. Antileakage Fourier transform for seismic data regularization in higher dimensions. *Geophysics* **2010**, *75*, WB113–WB120. [CrossRef]
29. Schonewille, M.; Klaedtke, A.; Vigner, A. Anti–alias anti–leakage Fourier transform. In Proceedings of the SEG International Exposition and Annual Meeting, Houston, TX, USA, 25–30 October 2009; pp. 3249–3253.
30. Zhang, F.; Geng, Z.; Yuan, W. The algorithm of interpolating windowed FFT for harmonic analysis of electric power system. *IEEE Trans. Power Deliv.* **2001**, *16*, 160–164. [CrossRef]
31. Schreiber, T.; Schmitz, A. Surrogate time series. *Phys. D Nonlinear Phenom.* **2000**, *142*, 346–382. [CrossRef]
32. Theiler, J.; Eubank, S.; Longtin, A.; Galdrikian, B.; Farmer, J.D. Testing for nonlinearity in time series: The method of surrogate data. *Phys. D Nonlinear Phenom.* **1992**, *58*, 77–94. [CrossRef]

processes

MDPI

Article

Modeling and Switched Control of Modular Reconfigurable Flight Array for Faulty Redundancy

Bin Ren [1,2], Chunxi Yang [2,3,*], Xiufeng Zhang [2,3] and Wenyuan Mao [1]

1 Faculty of Chemical Engineering, Kunming University of Science and Technology, No. 727 Jingming South Road, Chenggong District, Kunming 650032, China; r1038039846@163.com (B.R.); maowenyuanyn@163.com (W.M.)
2 Yunnan International Joint Laboratory of Intelligent Control and Application of Advanced Equipment, Kunming University of Science and Technology, No. 727 Jingming South Road, Chenggong District, Kunming 650032, China; 20220056@kust.edu.cn
3 Faculty of Mechanical and Electrical Engineering, Kunming University of Science and Technology, No. 727 Jingming South Road, Chenggong District, Kunming 650032, China
* Correspondence: ycx@kmust.edu.cn

Abstract: The modular reconfigurable flight array (MRFA) is composed of multiple identical flight unit modules, which has several advantages such as structural variability, strong versatility, and low cost. Due to the redundant properties of MRFA, it keeps stable by adopting a suitable control law when it suffers actuator fault or actively stops some actuators. To address the attitude stability issue of the modular flight array when actuators actively stop or encounter failures during the flight process, a modeling method based on a switched system is proposed at first, and an arbitrary switched controller design method based on the segmented Lyapunov functions and the average dwell time is also given. By introducing the actuator efficiency matrix, the dynamic switched model of the modular flight array is described. Then, a group of arbitrary switched linear feedback gains is designed to ensure the exponential stability of the flight array if the switched process satisfies the constraint of the average dwell time. Simulation and experiment results indicate that when there is an accident in the actuator states, the switched controllers can achieve precise tracking of the desired trajectory, thus confirming the effectiveness of the proposed modeling method and controller.

Keywords: modular reconfigurable flight array; switched system; segmented Lyapunov function; average dwell time

1. Introduction

In recent years, a variety of unmanned aerial vehicles (UAVs) have been proposed to be adapted to complex dynamic environments or to undertake more challenging missions [1]. Many concepts such as UAV swarms that collaborate with multiple unmanned aerial vehicles [2], omnidirectional heterogeneous UAV [3], coaxial drones [4,5], multi-modal drone with three different flight modes (rotor wing, tailsitter, cruise) [6], and MRFA [7–10] have also emerged. The multi-linked aerial robot, named DRAGON, described in reference [7], incorporates a two-degree-of-freedom (DoF) force vectoring apparatus in each link. This aircraft connects individual flight modules through a flexible swinging motion framework, enabling the aircraft to possess the capability to flexibly change its configuration during flight. Reference [8] proposes a novel modular design-based reconfigurable transport drone that can adapt to the shape, size, and weight of payloads, thus avoiding the lack of flexibility in traditional drones, which require updating specific cargo dimensions to customize different drone fleets. References [9,10] introduce another novel modular aircraft for distributed flight array systems. The flight system is capable of aerial configuration reconstruction, cooperative transportation, and self-reconfiguring fault tolerance in the event of system failures. Among those different types of UAVs, the MRFA has certain advantages for its variable structure and higher environmental adaptability. It is composed of several standard and common flight modular. Different MRFA configurations can be rapidly constructed by assembling multiple universal modules to fit for various mission requirements and environmental constraints. Consequently, MRFA has promising applications on complex environments and multiple mission scenes.

When an MRFA possesses no less than four universal flight modules, stable flight can be achieved in the event of actuator deactivation or actuator failure by full use of the redundancy of actuators. This feature enhances the resilience and versatility of the MRFA, and then improve its fault-tolerant capability.

In addressing the fault-tolerant control challenges of rotorcraft, comprehensive and systematic research has been conducted by scholars [11,12]. In [11], an adaptive control strategy is proposed to estimate the fault parameter of the actuator, with which the stabilization problem of the quadrotor with actuator fault can be effectively solved. Reference [12] compares the static hover of two different configurations of hexacopters (coaxial Y-shaped and regular hexagonal) when one actuator fails. Following the geometric interpretation, it demonstrates that the hexacopter with a hexagonal configuration is unable to maintain static hover when one of its propellers fails, whereas the Y-shaped hexacopter can maintain static hover even when any one propeller fails. The control allocation technique, which is widely used in UAV is applied in [13], wherein the control signals of UAV actuators are directly reassigned to the remaining properly functioning actuators. This approach achieves fault redundancy without altering the control laws of the UAV. In [14], a gain scheduling mechanism based on PID control is devised, involving the design of multiple PID controllers to address the fault-tolerant control issues of a quadrotor under normal operation and in the event of different actuator failures. Through gain scheduling, the UAV undergoes flight mode switching from normal flight to the occurrence of actuator failures, ultimately endowing the UAV with fault tolerance. However, due to the disparate control gains associated with the UAV in different states, the stability problem of the UAV control system at controller switching time can not be guarantee from theoretical prospects. This drawback may result in a degradation in UAV stability performance or even attitude instability. In [15], a robust optimization approach is used to deal with uncertain power outputs, which can be used to solve the problem of abrupt energy output in the event of actuator failure in multi-rotor UAVs.

Considering the inherent property of actuator redundancy in MRFA and the practical demand for fault-tolerant control of UAVs, an attitude control method of MRFA based on the switched control approach is proposed. This approach ensures the stability of MRFA during modal transitions triggered by actuator deactivation or actuator failure. This ensures exponential stability under any switched law, realizing stable flight during mode transitions. The aim of this paper is to investigate control strategies for redundant actuator systems. In MRFA systems, traditional control methods may not effectively address actuator failures. To tackle this issue, a design scheme based on dwell time arbitrary switching systems is proposed in this paper. According to pre-defined switching rules, the controllers are automatically switched in the event of actuator failures to maintain the stability and flight performance of the MRFA. The paper is organized as follows: Section 2 provides a brief description of the mechanical design and optimization of the MRFA, Section 3 elaborates on the mathematical model of the MRFA, and Section 4 introduces the MRFA switched controller and its design methodology. The effectiveness of the designed switched controller is verified through numerical simulations and experiments in Section 5, accompanied by details on the hardware setup of the experimental platform and its structure. Finally, conclusions are drawn in Section 6.

2. Design of the Mechanical System

MRFA is assembled by arbitrarily combining multiple structurally identical flight modules. Each flight module comprises components of carbon fiber panels, brushless motors, propellers, communication modules, and mechanical connectors. The connection between multiple flight modules is achieved through bolt-fastened connector components (Figure 1). It allows for the adjustment of the quantity and spatial positioning of modules based on mission requirements, ensuring that a wider array of complex and dynamic tasks can be accomplished. Through the arbitrary combination of flight modules, various configurations of the aircraft can be assembled, significantly expanding the functionality and application scenarios beyond traditional aircraft. To facilitate dynamic modeling and control design, we assume that all flight modules are uniform, including shape, mass, inertia, and actuators.

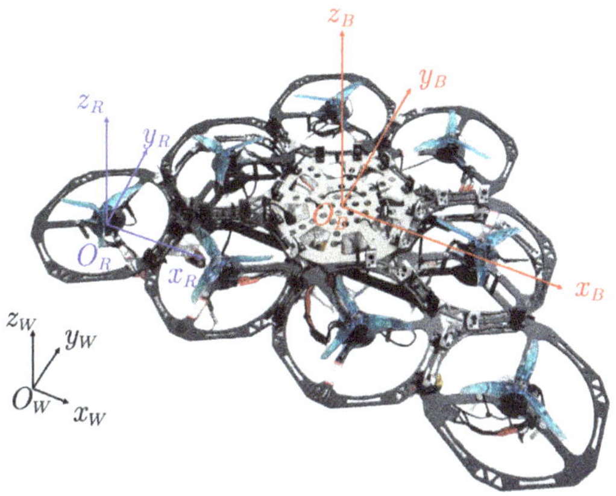

Figure 1. Coordinate schematic of modular reconfigurable flight array (MRFA). The earth coordinate frames, module coordinate frames, and flight array coordinate frames are represented by the black, red, and blue axes, respectively.

3. Modeling of MRFA

3.1. Coordinate Frames

Three different coordinate systems are employed for describing the positional relationships between flight modules and the MRFA. The three coordinate systems shown in Figure 1 are defined as follows.

The earth coordinate frame W: or inertial coordinate system, is characterized by adherence to the right-hand rule, with its three axes following the convention of the East–North–Up (ENU) direction. The position of the MRFA in the earth coordinate system is denoted by $\mathbf{p} = [x, y, z]^T$. The attitude of the MRFA is denoted by $\Theta = [\phi, \theta, \psi]^T$, where ϕ, θ, ψ represent the roll, pitch, and yaw angles, respectively.

The flight module coordinate frame R_i: representing the position of each individual flight module. R_i corresponds to the coordinate system of the i-th flight module, where the subscript $i \in \{1, 2, \cdots, N\}$. The origin of the coordinate system coincides with the center of mass of the flight module. The z-axis aligns with the thrust direction generated by the propellers. For model simplification, it is assumed that the x-axis of all flight modules points in the same direction.

The structure coordinate frame B: The origin of the coordinate system is positioned at the center of mass of the MRFA, adhering to the right-hand rule. The x-axis aligns with the forward direction of the MRFA, and the z-axis is oriented vertically upward from the horizontal plane of MRFA.

3.2. Dynamic Model of the MRFA

The dynamic model of the MRFA is employed to describe the relationship between the forces and torques acting on the UAV array and its motion. The modeling approach in [16] is adopted, assuming equal mass for each flight module. Based on the relationships between lift, torque, motor speed, and control channels, a generalized and universal dynamic model for the irregularly structured MRFA is established, as presented below.

The kinematic equations can be expressed as:

$$\begin{bmatrix} \ddot{x} \\ \ddot{y} \\ \ddot{z} \end{bmatrix} = \frac{1}{Nm}\begin{bmatrix} (c\phi s\theta c\psi + s\phi s\psi)T \\ (c\phi s\theta s\psi - s\phi c\psi)T \\ (c\phi c\theta)T \end{bmatrix}, \tag{1}$$

where N represents the number of modules in the MRFA, m denotes the mass of a single module, $s\phi = \sin(\phi), c\phi = \cos(\phi)$, T represents the resultant force exerted on the MRFA in the structure coordinate system.

Based on the inertia matrix of a single flight module $\mathbf{I} = \mathrm{diag}(I_{xx}, I_{yy}, I_{zz})$, using the parallel axis theorem, once the MRFA is determined, the relationship between MRFA and the rotational inertia can be established.

$$\mathbf{I_B} = N\mathbf{I} + m\begin{bmatrix} \sum_{i=1}^{N} y_{Bi}^2 & -\sum_{i=1}^{N} x_{Bi}y_{Bi} & 0 \\ -\sum_{i=1}^{N} y_{Bi}x_{Bi} & \sum_{i=1}^{N} x_{Bi}^2 & 0 \\ 0 & 0 & \sum_{i=1}^{N} x_{Bi}^2 + y_{Bi}^2 \end{bmatrix} \tag{2}$$

where x_{Bi} and y_{Bi} denote the positions of the geometric centers of each flight module in the structure coordinate system. Equation (2) can be rearranged as:

$$\mathbf{I_B} = \begin{bmatrix} I_{xx}^B & -I_{xy}^B & 0 \\ -I_{yx}^B & I_{yy}^B & 0 \\ 0 & 0 & I_{zz}^B \end{bmatrix}. \tag{3}$$

Its inverse matrix is given by:

$$\mathbf{I_B}^{-1} = \begin{bmatrix} I_{11} & I_{12} & 0 \\ I_{21} & I_{22} & 0 \\ 0 & 0 & I_{33} \end{bmatrix}, \tag{4}$$

where

$$\begin{cases} I_{11} = I_{yy}^B / \left(I_{xx}^B I_{yy}^B - I_{xy}^B I_{yx}^B\right) \\ I_{12} = I_{xy}^B / \left(I_{xx}^B I_{yy}^B - I_{xy}^B I_{yx}^B\right) \\ I_{21} = I_{yx}^B / \left(I_{xx}^B I_{yy}^B - I_{xy}^B I_{yx}^B\right) \\ I_{22} = I_{xx}^B / \left(I_{xx}^B I_{yy}^B - I_{xy}^B I_{yx}^B\right) \\ I_{33} = 1 / I_{zz}^B \end{cases}.$$

The rotational motion of the MRFA can be given as

$$\begin{bmatrix} \ddot{\phi} \\ \ddot{\theta} \\ \ddot{\psi} \end{bmatrix} = \mathbf{I_B}^{-1}\left(\begin{bmatrix} \tau_x \\ \tau_y \\ \tau_z \end{bmatrix} - \begin{bmatrix} \dot{\phi}\dot{\psi}\left(I_{zz}^B - I_{yy}^B\right) + \dot{\phi}\dot{\psi}I_{yx}^B \\ \dot{\phi}\dot{\psi}\left(I_{xx}^B - I_{zz}^B\right) - \dot{\theta}\dot{\psi}I_{xy}^B \\ \dot{\phi}\dot{\theta}\left(I_{yy}^B - I_{xx}^B\right) + \dot{\theta}^2 I_{xy}^B - \dot{\phi}^2 I_{yx}^B \end{bmatrix}\right), \tag{5}$$

where $[\tau_x, \tau_y, \tau_z]^T$ represents the rotational torques exerted on the MRFA due to rotation about the x_B, y_B , and z_B axes of the structure coordinate system.

$$\begin{cases} T = \sum\limits_{i=1}^{N} (f_i - m_i g), \\ \tau_x = \sum\limits_{i=1}^{N} y_i (f_i - m_i g), \\ \tau_y = \sum\limits_{i=1}^{N} (-x_i)(f_i - m_i g), \\ \tau_z = \sum\limits_{i=1}^{N} (-1)^k M_i, \end{cases} \tag{6}$$

where k represents the rotation direction of the propellers in the MRFA, with $k = 1$ when the blade rotates clockwise and $k = 2$ when the blade rotates counterclockwise.

To further simplify the model, it is assumed that the MRFA satisfies the small angle assumption [17], and nonlinear terms such as aerodynamic damping are neglected. Therefore, the simplified model of the MRFA is represented by

$$\begin{bmatrix} \ddot{x} \\ \ddot{y} \\ \ddot{z} \end{bmatrix} = \frac{1}{Nm} \begin{bmatrix} \theta T \\ -\phi T \\ T \end{bmatrix}, \tag{7}$$

$$\begin{bmatrix} \ddot{\phi} \\ \ddot{\theta} \\ \ddot{\psi} \end{bmatrix} = \mathbf{I_B}^{-1} \begin{bmatrix} \tau_x \\ \tau_y \\ \tau_z \end{bmatrix}. \tag{8}$$

To achieve the attitude control of the MRFA, it is necessary to establish the error-dynamic equations based on the mathematical model of the MRFA. For each switching subsystem, we design a feedback gain matrix to ensure that the states of each subsystem quickly track the reference signal.

Define $z_d, \phi_d, \theta_d, \psi_d$ as the reference signals. The tracking error states are $\varepsilon_1 = z_d - z$, $\varepsilon_2 = \dot{z}_d - \dot{z}$, $\varepsilon_3 = \phi_d - \phi$, $\varepsilon_4 = \dot{\phi}_d - \dot{\phi}$, $\varepsilon_5 = \theta_d - \theta$, $\varepsilon_6 = \dot{\theta}_d - \dot{\theta}$, $\varepsilon_7 = \psi_d - \psi$, $\varepsilon_8 = \dot{\psi}_d - \dot{\psi}$. Substituting Equations (7) and (8), the error-dynamic equations of the MRFA can be expressed as follows:

$$\begin{cases} \dot{\varepsilon}_1 = \varepsilon_2, \\ \dot{\varepsilon}_2 = \ddot{z}_d - \frac{1}{Nm} T, \\ \dot{\varepsilon}_3 = \varepsilon_4, \\ \dot{\varepsilon}_4 = \ddot{\phi}_d - I_{11}\tau_x - I_{12}\tau_y, \\ \dot{\varepsilon}_5 = \varepsilon_6, \\ \dot{\varepsilon}_6 = \ddot{\theta}_d - I_{21}\tau_x - I_{22}\tau_y, \\ \dot{\varepsilon}_7 = \varepsilon_8, \\ \dot{\varepsilon}_8 = \ddot{\psi}_d - I_{33}\tau_z. \end{cases} \tag{9}$$

The error-dynamic equations are represented in state–space form as follows:

$$\dot{\epsilon} = \mathbf{A}\epsilon + \mathbf{B}\mathbf{u_v} + \mathbf{r}, \tag{10}$$

where

$$\mathbf{A} = \begin{bmatrix} 0 & 1 & 0 & 0 & 0 & 0 & 0 & 0 \\ 0 & 0 & 0 & 0 & 0 & 0 & 0 & 0 \\ 0 & 0 & 0 & 1 & 0 & 0 & 0 & 0 \\ 0 & 0 & 0 & 0 & 0 & 0 & 0 & 0 \\ 0 & 0 & 0 & 0 & 0 & 1 & 0 & 0 \\ 0 & 0 & 0 & 0 & 0 & 0 & 0 & 0 \\ 0 & 0 & 0 & 0 & 0 & 0 & 0 & 1 \\ 0 & 0 & 0 & 0 & 0 & 0 & 0 & 0 \end{bmatrix}, \mathbf{B} = \begin{bmatrix} 0 & 0 & 0 & 0 \\ -1/Nm & 0 & 0 & 0 \\ 0 & 0 & 0 & 0 \\ 0 & -I_{11} & -I_{12} & 0 \\ 0 & 0 & 0 & 0 \\ 0 & -I_{21} & -I_{22} & 0 \\ 0 & 0 & 0 & 0 \\ 0 & 0 & 0 & -I_{33} \end{bmatrix},$$

$$\epsilon = [\varepsilon_1, \varepsilon_2, \varepsilon_3, \varepsilon_4, \varepsilon_5, \varepsilon_6, \varepsilon_7, \varepsilon_8]^T,$$
$$\mathbf{u_v} = [T, \tau_x, \tau_y, \tau_z]^T,$$
$$\mathbf{r} = [0\ \ddot{z}_d\ 0\ \ddot{\phi}_d\ 0\ \ddot{\theta}_d\ 0\ \ddot{\psi}_d]^T.$$

3.3. Control Allocation

The MRFA generates lift and torque by driving the propeller blades to rotate through motors, enabling motion in various directions. As there are no actuators directly generating thrust and torque in the system, the thrust produced by the motors is considered as the input vector for the actuators, and the input to sustain the motion of the flight array is treated as a virtual control input vector. Figure 2 illustrates the schematic structure of control allocation in the control system.

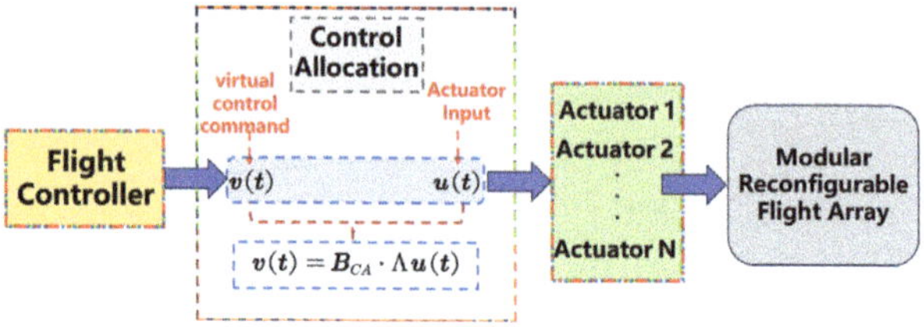

Figure 2. Control allocation.

Control allocation reasonably and uniquely assigns the virtual control commands corresponding to the control laws to each actuator, in order to obtain the desired control performance. The relationship between virtual control input and the actual input to the actuators is expressed by the following equation:

$$\mathbf{u_v} = \mathbf{B_{CA}} \mathbf{u_m}, \tag{11}$$

where $\mathbf{u_m} = [u_1, u_2, \cdots, u_N]^T$ represents the actual inputs to the actuators of the MRFA, $\mathbf{u_v}$ represents the virtual control input, and $\mathbf{B_{CA}}$ is the control allocation matrix, reflecting the mapping relationship between virtual control input and actual input.

For quadrotor, the pulse width modulation output signal u from the electronic speed controller is approximately proportional to the square of the speed of the brushless motor [18]. The lift f_i and torque M_i of the flight unit module are proportional to the square of the rotor speed [19]. Details of the relationship can be seen in Equation (12).

$$\begin{cases} u_i = p\omega_i^2 \\ f_i = L_f\omega_i^2 \\ M_i = L_m\omega_i^2 \end{cases}, \tag{12}$$

where p is the proportional coefficient, ω_i represents the speed of the brushless motor, L_f is the lift coefficient, and L_m is the torque coefficient.

The characteristic of the arbitrary interconnection and variable structure among flight modules in the MRFA allows for various configurations. When substituting the control allocation matrix $\mathbf{B_{CA}}$ into Equation (11), the following equation is obtained.

$$\begin{bmatrix} T_z \\ \tau_x \\ \tau_y \\ \tau_z \end{bmatrix} = \mathbf{L} \begin{bmatrix} 1 & 1 & \cdots & 1 \\ y_1 & y_2 & \cdots & y_N \\ -x_1 & -x_2 & \cdots & -x_N \\ (-1)^k & (-1)^k & \cdots & (-1)^k \end{bmatrix} \begin{bmatrix} u_1 \\ u_2 \\ \vdots \\ u_N \end{bmatrix}, \tag{13}$$

where $\mathbf{L} = \mathrm{diag}\left(L_f/p, L_f/p, L_f/p, L_m/p\right)$.

In accordance with Equation (13), the values of the control allocation matrix are solely dependent on the distribution of actuators within the MRFA and the rotation direction of the propellers. Based on the positional information of each module and the rotation direction of the propellers, the control allocation matrix for the corresponding MRFA can be determined (Figure 3). The red arrow indicates counterclockwise rotation of the propeller, while the blue arrow indicates clockwise rotation of the propeller. The Moore–Penrose pseudoinverse ultimately is employed to allocate the virtual control commands to each actuator. One has

$$\begin{aligned} \mathbf{u_m} &= \mathbf{B_{CA}^T}\left(\mathbf{B_{CA}}\mathbf{B_{CA}^T}\right)^{-1}\mathbf{u_v} \\ &= \mathbf{B_{CA}^{\dagger}}\mathbf{u_v}, \end{aligned} \tag{14}$$

where $\mathbf{B_{CA}^{\dagger}}$ represents the right pseudo-inverse of the control allocation matrix $\mathbf{B_{CA}}$.

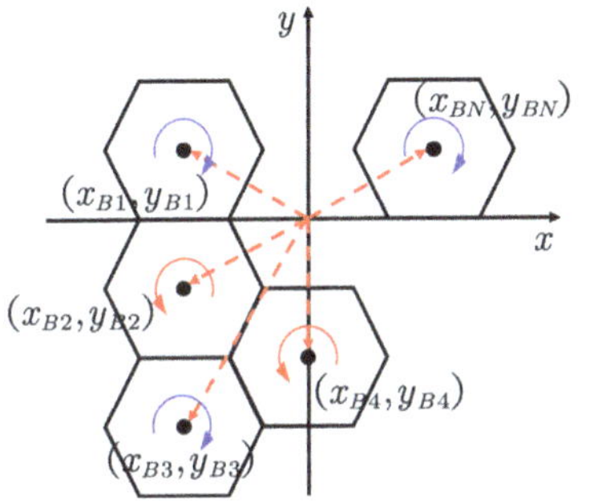

Figure 3. The distribution of modular position.

Remark 1. *The mixer matrix of the MRFA is a matrix used to convert the control inputs of the aircraft into speed commands for each actuator. From Equation* (14)*, it can be derived that the rotor speeds of the actuators can be obtained through the control inputs of the flight array transformed by matrix* $\mathbf{B_{CA}^{\dagger}}$*, thus inferring the equivalence between the mixing matrix and matrix* $\mathbf{B_{CA}^{\dagger}}$.

4. Arbitrary Switched Controller Design

To ensure robust control performance when no less than one actuator suffers an accident and has to switch to alternative actuator combinations, a switching controller is designed to accommodate the arbitrary switching law. To characterize the controllability of the MRFA in the presence of faults or failures in the power system, a diagonal matrix $\mathbf{\Lambda}$ is introduced to depict the states of each rotor.

$$\mathbf{\Lambda} = diag(\lambda_1, \lambda_2, \cdots, \lambda_N), \lambda_l \in [0, 1], \tag{15}$$

where $\mathbf{\Lambda}$ denotes the actuator state matrix. Here, $\lambda_l = 0$ indicates a complete failure of the l-th motor; $0 < \lambda_l < 1$ signifies a partial failure of the actuator; $\lambda_l = 1$ represents the full health of the actuator. Consequently, the error-dynamic equation of the MRFA, encompassing the actuator states, can be derived:

$$\dot{\epsilon} = \mathbf{A}\epsilon + \mathbf{B}\mathbf{B_{\Gamma}} \cdot \mathbf{u_m} + \mathbf{r}, \tag{16}$$

where $\mathbf{B_{\Gamma}} = \mathbf{B_{CA}}\mathbf{\Lambda}$ denotes the control allocation matrix with actuator states.

When a failure of actuator occurs, the actuator state matrix $\mathbf{\Lambda}$ undergoes changes. A linear switched system with actuator fault can be described as Equation (16).

$$\dot{\epsilon} = \mathbf{A}\epsilon + \mathbf{B}_{\sigma(t)}\mathbf{u_m} + \mathbf{r}, \tag{17}$$

where the switching law $\sigma(t)$ is a right-continuous constant function with time t as a variable. In the time sequence $\{t_0 < t_1 < \cdots < t_i < t_{i+1} < \cdots\}$, when $t \in [t_i, t_{i+1})$, $\sigma(t) = p$ represents the activation of the p-th system, corresponding to the selection of the system matrix $(\mathbf{A}, \mathbf{B}_p)$.

Remark 2. *When the number of flight modules in the MRFA is more than 4 ($N > 4$), the MRFA, as an actuator redundant system, has a subset of actuators $\mathbf{M}_i$, where $i = 1, 2, \cdots, m$, which is mainly used to enhance and improve the performance of the actuators. If a failure occurs in these actuators, it will not affect the stability of the system. Another subset of actuators $\overline{\mathbf{M}}_i$ can affect the stability of the system if a failure occurs, making the system lack fault tolerance. In an MRFA, when the control allocation matrix $\mathbf{B}_{\Gamma_i} = \mathbf{B}_{\mathbf{CA}}\mathbf{\Lambda}_i$ containing the state of the actuators is less than full rank, the corresponding set of actuators belong to $\overline{\mathbf{M}}_i$. For ease of study, we consider the system switching among controllable combinations of actuators. That is, the actuator combination of MRFA belongs to $\mathbf{M}_i$.*

According to the description mentioned above, MRFA is easy to facilitate redundant control. When certain actuators proactively cease operation or encounter failures, a fault-tolerant switched controller is designed based on segmented Lyapunov methods [20]. This method can keep stable flight during mode transitions in the MRFA.

Lemma 1. *A positive number τ_a is considered as the average dwell time if there exists a switching signal $\sigma(t)$ and the number of switches $N_\sigma(t, \tau)$ during the time interval (τ, t), such that a non-negative integer $N_0 \geq 0$ is satisfied* [21].

$$N_\sigma(t, \tau) \leq N_0 + \frac{t - \tau}{\tau_a} \tag{18}$$

Theorem 1. *For the MRFA satisfying the switched model in* (17), *if for each switched subsystem, there exists a state feedback controller*

$$\mathbf{u}_{mi} = -\mathbf{K}_i \epsilon - \left(\mathbf{B} \cdot \mathbf{B}_{\Gamma_i}\right)^{+} \mathbf{r}, \tag{19}$$

where the feedback gain of the controller is $\mathbf{K}_i = \alpha_i \left(\mathbf{B}\mathbf{B}_{\Gamma_i}\right)^T \cdot \mathbf{P}_i$,$i \in \Omega$,$\mathbf{P}_i$ is a positive definite matrix for subsystem i, positive numbers λ_0,α_i and $\mu \geq 1$ are chosen such that the following matrix inequalities hold:

$$\mathbf{A}_i^T \mathbf{P}_i + \mathbf{P}_i \mathbf{A}_i - 2\alpha_i \mathbf{P}_i^T \mathbf{B}\mathbf{B}_{\Gamma_i} \left(\mathbf{B}\mathbf{B}_{\Gamma_i}\right)^T \mathbf{P}_i + 2\lambda_0 \mathbf{P}_i < 0, \tag{20}$$

$$\mathbf{P}_i \leq \mu \mathbf{P}_j,\ \forall i, j \in \{1, 2, \cdots, m\},\ i \neq j. \tag{21}$$

And the average dwell time satisfies

$$\tau_a \geq \tau_a^* = \frac{\ln \mu}{2\lambda},\ \lambda \in (0, \lambda_0). \tag{22}$$

The system can be exponentially stabilized under arbitrary switched law.

Proof of Theorem 1. For each subsystem of the switched system in MRFA, the following piecewise Lyapunov function is defined:

$$V_{\sigma(t)}(\epsilon(t)) = \epsilon^T \mathbf{P}_{\sigma(t)} \epsilon, \tag{23}$$

where the positive definite matrix $\mathbf{P}_{\oe(t)}$ satisfies the matrix inequalities (20) and (21). When the MRFA switches to the i-th subsystem, the time derivative of $V_i(\epsilon(t))$ is given by:

$$\dot{V}_i(\epsilon(t)) = \dot{\epsilon}^T \mathbf{P}_i \epsilon + \epsilon^T \mathbf{P}_i \dot{\epsilon}. \tag{24}$$

Substituting the error state space (Equation (16)) into the above equation, we obtain:

$$\begin{aligned}\dot{V}_i(\epsilon(t)) =& (\mathbf{A}\epsilon + \mathbf{B}\mathbf{B}_{\Gamma_i}\mathbf{u_m} + \mathbf{r})^T \mathbf{P}_i \epsilon + \epsilon^T \mathbf{P}_i (\mathbf{A}\epsilon + \mathbf{B}\mathbf{B}_{\Gamma_i}\mathbf{u_m} + \mathbf{r}) \\ =& \epsilon^T \left(\mathbf{A}^T \mathbf{P}_i + \mathbf{P}_i \mathbf{A}\right)\epsilon + \mathbf{u}_\mathbf{m}^T \mathbf{B}_{\Gamma_i}^T \mathbf{B}^T \mathbf{P}_i \epsilon + \epsilon^T \mathbf{P}_i \mathbf{B}\mathbf{B}_{\Gamma_i}\mathbf{u_m} + \mathbf{r}^T \mathbf{P}_i \epsilon + \epsilon^T \mathbf{P}_i \mathbf{r}.\end{aligned} \tag{25}$$

Rearranging the equation by substituting the state feedback controller (19), we obtain:

$$\dot{V}_i(\epsilon(t)) = \epsilon^T \left(\mathbf{A}^T \mathbf{P}_i + \mathbf{P}_i \mathbf{A}\right)\epsilon - \epsilon^T \mathbf{K}_i^T \mathbf{B}_{\Gamma_i}^T \mathbf{B}^T \mathbf{P}_i \epsilon - \epsilon^T \mathbf{P}_i \mathbf{B}\mathbf{B}_{\Gamma_i}\mathbf{K}_i \epsilon. \tag{26}$$

Then, we have

$$\begin{aligned}\dot{V}_i(\epsilon(t)) =& \epsilon^T \left(\mathbf{A}^T \mathbf{P}_i + \mathbf{P}_i \mathbf{A}\right)\epsilon - 2\epsilon^T \left(\alpha_i \mathbf{K}_i^T \mathbf{B}_{\Gamma_i}^T \mathbf{B}^T \mathbf{P}_i\right)\epsilon \\ =& \epsilon^\mathbf{T} \left(\mathbf{A}^T \mathbf{P}_i + \mathbf{P}_i \mathbf{A} - 2\alpha_i \mathbf{K}_i^T \mathbf{B}_{\Gamma_i}^T \mathbf{B}^T \mathbf{P}_i\right)\epsilon.\end{aligned} \tag{27}$$

To analyze Equation (20), we have

$$\dot{V}_i(\epsilon(t)) = \epsilon^T \left(\mathbf{A}^T \mathbf{P}_i + \mathbf{P}_i \mathbf{A} - 2\alpha_i \mathbf{K}_i^T \mathbf{B}_{\Gamma_i}^T \mathbf{B}^T \mathbf{P}_i\right)\epsilon < -2\lambda_0 \epsilon^T \mathbf{P}_i \epsilon < 0, \tag{28}$$

then

$$\dot{V}_i < -2\lambda_0 V_i. \tag{29}$$

Assuming that the switching signal of each subsystem in the MRFA satisfies $0 = t_0 < t_1 < \cdots < t_k = t_{N_{\sigma(t,0)}} < t$, the result from Equations (20) and (29) can be obtained as follows:

$$\begin{aligned}& V(t) < V(t_k)e^{-2\lambda_0(t-t_k)} \le \mu V(t_k^-)e^{-2\lambda_0(t-t_k)} \le \\ & \mu\left[V(t_{k-1})e^{-2\lambda_0(t_k-t_{k-1})}\right]e^{-2\lambda_0(t-t_k)} = \\ & \mu V(t_{k-1})e^{-2\lambda_0(t_k-t_{k-1})} \le \\ & \mu^2 V\left(t_{k-1}^-\right)e^{-2\lambda_0(t-t_{k-1})} \le \\ & \mu^2\left[V(t_{k-2})e^{-2\lambda_0(t_{k-1}-t_{k-2})}\right]e^{-2\lambda_0(t-t_{k-1})} = \\ & \mu^2 V(t_{k-2})e^{-2\lambda_0(t-t_{k-2})} \le \\ & \mu^3 V\left(t_{k-2}^-\right)e^{-2\lambda_0(t-t_{k-2})} \le \cdots \le \\ & \mu^k V(t_0)e^{-2\lambda_0 t} = \mu^{N_{\sigma(t,0)}} V(0)e^{-2\lambda_0 t} = V(0)e^{N_{\sigma(t,0)}\ln\mu - 2\lambda_0 t}.\end{aligned} \tag{30}$$

If τ_a satisfies Equation (22), combined with (18), it can be obtained:

$$N_{\sigma(t,0)} \le \frac{t}{\tau_a^*},\ \tau_a^* = \frac{\ln\mu}{2\lambda}. \tag{31}$$

Then,

$$V(t) < e^{-2(\lambda_0-\lambda)t}V(0). \tag{32}$$

Therefore, for each subsystem of the MRFA, given the satisfaction of the average dwell time with Equation (30), it can be inferred that upon exiting the previous subsystem, the Lyapunov function value of the current subsystem at the current moment is less than the Lyapunov function value of the last activated subsystem. Hence, the MRFA switched system is exponentially stable under the condition of satisfying the average dwell time (22). □

To obtain the switched controller, let $\mathbf{X}_i = \mathbf{P}_i^{-1}$ and $\mathbf{X}_j = \mathbf{P}_j^{-1}$. By left-multiplying and right-multiplying both sides of Equation (20) with $\mathbf{X}_i$ and both sides of Equation (21) with $\mathbf{X}_j$, we obtain:

$$\begin{cases} \mathbf{X}_j \mathbf{P}_i \mathbf{X}_j \leq \mu \mathbf{X}_j, \\ \mathbf{A}_i \mathbf{X}_i + \mathbf{X}_i \mathbf{A}_i^T - 2\alpha_i \mathbf{B}\mathbf{B}_{\Gamma_i} (\mathbf{B}\mathbf{B}_{\Gamma_i})^{\mathrm{T}} + 2\lambda_0 \mathbf{X}_i < 0. \end{cases}$$

By using the Schur complement, the above equation is equivalent to:

$$\begin{bmatrix} -\mu \mathbf{X}_j & \mathbf{X}_j \\ \mathbf{X}_j & -\mathbf{X}_i \end{bmatrix} < 0, \tag{33}$$

$$(\mathbf{A}_i + \lambda_0 \mathbf{E}) \mathbf{X}_i + \mathbf{X}_i (\mathbf{A}_i + \lambda_0 \mathbf{E})^T - 2\alpha_i \mathbf{B}\mathbf{B}_{\Gamma_i} (\mathbf{B}\mathbf{B}_{\Gamma_i})^T < 0. \tag{34}$$

5. Experiments

5.1. Simulation

Considering the MRFA with eight modules shown in Figure 4, the tracking performance of the proposed switched control algorithm for reference trajectories is evaluated through numerical simulation and experimental testing under scenarios with no actuator failure, failure of actuator 5, and failure of actuators 1 and 4.

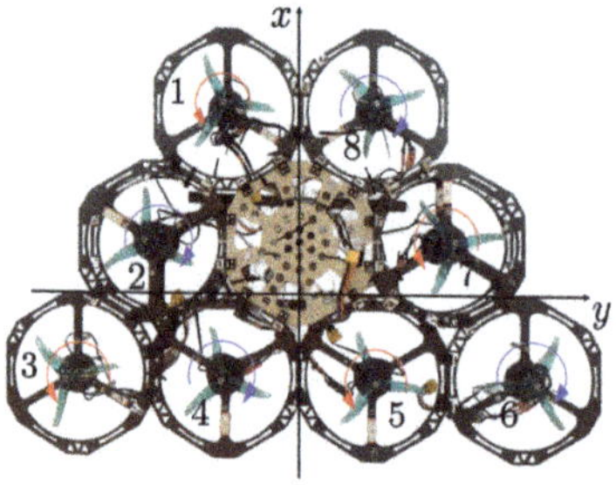

Figure 4. Modular reconfigurable flight array (MRFA) with eight modules.

The relevant parameters of the flight module (as shown in Table 1) can be used to compute the physical parameters of the MRFA, as depicted in Figure 4. Subsequently, the values for **A** and **B** in the expression for the error tracking state space (16) can be determined.

$$\mathbf{A} = \begin{bmatrix} 0 & 1 & 0 & 0 & 0 & 0 & 0 & 0 \\ 0 & 0 & 0 & 0 & 0 & 0 & 0 & 0 \\ 0 & 0 & 0 & 1 & 0 & 0 & 0 & 0 \\ 0 & 0 & 0 & 0 & 0 & 0 & 0 & 0 \\ 0 & 0 & 0 & 0 & 0 & 1 & 0 & 0 \\ 0 & 0 & 0 & 0 & 0 & 0 & 0 & 0 \\ 0 & 0 & 0 & 0 & 0 & 0 & 0 & 1 \\ 0 & 0 & 0 & 0 & 0 & 0 & 0 & 0 \end{bmatrix}, \mathbf{B} = \begin{bmatrix} 0 & 0 & 0 & 0 \\ -0.382 & 0 & 0 & 0 \\ 0 & 0 & 0 & 0 \\ 0 & -14.12 & 0 & 0 \\ 0 & 0 & 0 & 0 \\ 0 & 0 & -17.00 & 0 \\ 0 & 0 & 0 & 0 \\ 0 & 0 & 0 & -7.52 \end{bmatrix} \tag{35}$$

Table 1. Parameters of flight module.

Parameter	Value	Description
m	327.2 g	Weight of flight module
g	9.8 m/s	Acceleration of gravity
l	110 mm	The length of the carbon fiber panel
I_{xx}	1132 kg·mm^2	Moment of inertia about the x-axis
I_{yy}	1132 kg·mm^2	Moment of inertia about the y-axis
I_{zz}	2409 kg·mm^2	Moment of inertia about the z-axis

The position information of each module is computed based on the edge length information of the MRFA unit module. The lift coefficient $L_f = 1.496 \times 10^{-6}$ Ns^2/rad^2, the torque coefficient $L_m = 1.446 \times 10^{-8}$ Nms^2/rad^2, and the proportionality coefficient $p = 1.042 \times 10^{-7}$ [17]. With this information, the MRFA control allocation matrix $\mathbf{B_{CA}}$ can be determined:

$$\mathbf{B_{CA}} = \mathbf{B}_{\Gamma_1} = \begin{bmatrix} 14.37 & 14.37 & 14.37 & 14.37 & 14.37 & 14.37 & 14.37 & 14.37 \\ 1.3689 & 2.7378 & 4.1035 & 1.3689 & -1.3689 & -4.1035 & -2.7378 & -1.3689 \\ 3.1614 & 0.7903 & -2.3710 & -2.3710 & -2.3710 & -2.3710 & 0.7903 & 3.1614 \\ 0.1388 & -0.1388 & 0.1388 & -0.1388 & 0.1388 & -0.1388 & 0.1388 & -0.1388 \end{bmatrix} \quad (36)$$

Remark 3. *Due to the fact that L_f and L_m are functions of the propeller's radius, geometry, and air density, the chosen dimensions of the propeller in this experiment are the same as those in reference* [17]. *However, due to differences in air density and propeller geometric shape during experiments, the output of the actuators can be adjusted by introducing positive constant matrices β and ζ. The corrected MRFA actuator output can be expressed as*

$$\mathbf{u_m}_i = -\beta * \mathbf{K}_i\epsilon - \zeta\left[\left(\mathbf{B} \cdot \mathbf{B}_{\Gamma_i}\right)^{\dagger}\mathbf{r}\right],$$

where the symbol $*$ denotes the Hadamard product. The actuator output consists of two terms: the first term is related to the state of the flight array, and the second term is related to the desired attitude trajectory of the MRFA. It can be used to suppress the oscillation of the actuator output. By adjusting the constant matrices β and ζ, the proportion of these two terms can be adjusted.

Sequentially we compute the control allocation matrices $\mathbf{B}_{\Gamma_i}$, $i \in \{1,2,3\}$, where $\mathbf{B}_{\Gamma_1}$, $\mathbf{B}_{\Gamma_2}$ and $\mathbf{B}_{\Gamma_3}$ represent the control allocation matrices when there is no actuator failure, actuator 5 fails, and actuators 1 and 4 fail, respectively.

$$\mathbf{B}_{\Gamma_2} = \begin{bmatrix} 14.37 & 14.37 & 14.37 & 14.37 & 0 & 14.37 & 14.37 & 14.37 \\ 1.3689 & 2.7378 & 4.1035 & 1.3689 & 0 & -4.1035 & -2.7378 & -1.3689 \\ 3.1614 & 0.7903 & -2.3710 & -2.3710 & 0 & -2.3710 & 0.7903 & 3.1614 \\ 0.1388 & -0.1388 & 0.1388 & -0.1388 & 0 & -0.1388 & 0.1388 & -0.1388 \end{bmatrix}$$

$$\mathbf{B}_{\Gamma_3} = \begin{bmatrix} 0 & 14.37 & 14.37 & 0 & 14.37 & 14.37 & 14.37 & 14.37 \\ 0 & 2.7378 & 4.1035 & 0 & -1.3689 & -4.1035 & -2.7378 & -1.3689 \\ 0 & 0.7903 & -2.3710 & 0 & -2.3710 & -2.3710 & 0.7903 & 3.1614 \\ 0 & -0.1388 & 0.1388 & 0 & 0.1388 & -0.1388 & 0.1388 & -0.1388 \end{bmatrix}$$

The parameters are set as $\alpha_1 = 0.1$, $\alpha_2 = 0.5$, $\alpha_3 = 0.7$, $\lambda_0 = 0.4$, $\mu = 1.5$, and an LMI solver is utilized to solve Equations (33) and (34) to obtain matrices $\mathbf{X}_1, \mathbf{X}_2$ and $\mathbf{X}_3$. From these, we derive the positive definite matrices $\mathbf{P}_1, \mathbf{P}_2$ and $\mathbf{P}_3$ corresponding to scenarios with no actuator failure, actuator 5 failure, and actuators 1 and 4 failure, respectively.

Remark 4. *The process of solving the LMIs mentioned above is performed offline on a ground station, utilizing the Python programming language for implementation and storing the results.*

If an onboard computer is installed on the reconfigurable flight array, LMIs can be solved online, and the controller gains can be determined accordingly. Finally, the results can be transmitted to the flight controller via USB communication, which will be implemented in future work.

$$\mathbf{P}_1 = \begin{bmatrix} 0.0123 & 0.0154 & 0.0000 & 0.0000 & 0.0001 & 0.0002 & 0.0000 & 0.0000 \\ 0.0154 & 0.0364 & 0.0000 & 0.0000 & 0.0002 & 0.0004 & -0.0001 & -0.0001 \\ 0.0000 & 0.0000 & 0.0003 & 0.0003 & 0.0000 & 0.0000 & -0.0012 & -0.0015 \\ 0.0000 & 0.0000 & 0.0003 & 0.0008 & 0.0000 & 0.0000 & -0.0015 & -0.0037 \\ 0.0001 & 0.0002 & 0.0000 & 0.0000 & 0.0002 & 0.0003 & 0.0000 & 0.0000 \\ 0.0002 & 0.0004 & 0.0000 & 0.0000 & 0.0003 & 0.0007 & 0.0000 & 0.0000 \\ 0.0000 & -0.0001 & -0.0012 & -0.0015 & 0.0000 & 0.0000 & 0.3446 & 0.4303 \\ 0.0000 & -0.0001 & -0.0015 & -0.0037 & 0.0000 & 0.0000 & 0.4303 & 1.0189 \end{bmatrix}$$

$$\mathbf{P}_2 = \begin{bmatrix} 0.0107 & 0.0134 & -0.0000 & -0.0000 & 0.0001 & 0.0002 & 0.0001 & 0.0002 \\ 0.0134 & 0.0294 & 0.0000 & 0.0000 & 0.0002 & 0.0003 & -0.0003 & -0.0003 \\ 0.0000 & 0.0000 & 0.0002 & 0.0003 & 0.0000 & 0.0000 & -0.0010 & -0.0012 \\ 0.0000 & 0.0000 & 0.0003 & 0.0006 & 0.0000 & 0.0000 & -0.0012 & -0.0026 \\ 0.0001 & 0.0002 & 0.0000 & 0.0000 & 0.0002 & 0.0003 & 0.0000 & 0.0000 \\ 0.0002 & 0.0003 & 0.0000 & 0.0000 & 0.0003 & 0.0006 & 0.0000 & 0.0000 \\ 0.0001 & -0.0003 & -0.0010 & -0.0012 & 0.0000 & 0.0000 & 0.2823 & 0.3356 \\ 0.0002 & -0.0003 & -0.0012 & -0.0026 & 0.0000 & 0.0000 & 0.3356 & 0.7408 \end{bmatrix}$$

$$\mathbf{P}_3 = \begin{bmatrix} 0.0107 & 0.0134 & 0.0000 & 0.0000 & 0.0001 & 0.0002 & 0.0000 & 0.0002 \\ 0.0134 & 0.0294 & 0.0000 & 0.0000 & 0.0002 & 0.0003 & -0.0002 & -0.0002 \\ 0.0000 & 0.0000 & 0.0002 & 0.0003 & 0.0000 & 0.0000 & -0.0010 & -0.0012 \\ 0.0000 & 0.0000 & 0.0003 & 0.0006 & 0.0000 & 0.0000 & -0.0012 & -0.0026 \\ 0.0001 & 0.0002 & 0.0000 & 0.0000 & 0.0002 & 0.0003 & -0.0001 & -0.0001 \\ 0.0002 & 0.0003 & 0.0000 & 0.0000 & 0.0003 & 0.0006 & -0.0003 & -0.0005 \\ 0.0000 & -0.0002 & -0.0010 & -0.0012 & -0.0001 & -0.0003 & 0.2796 & 0.3284 \\ 0.0002 & -0.0002 & -0.0012 & -0.0026 & -0.0001 & -0.0005 & 0.3284 & 0.7224 \end{bmatrix}$$

Let $\lambda = 0.5$, from Equation (22), we obtain $\tau_a^* = 0.405$. When the dwell time $\tau_a = 6 > \tau_a^*$, the switching signal $\sigma(t)$ is illustrated in Figure 5.

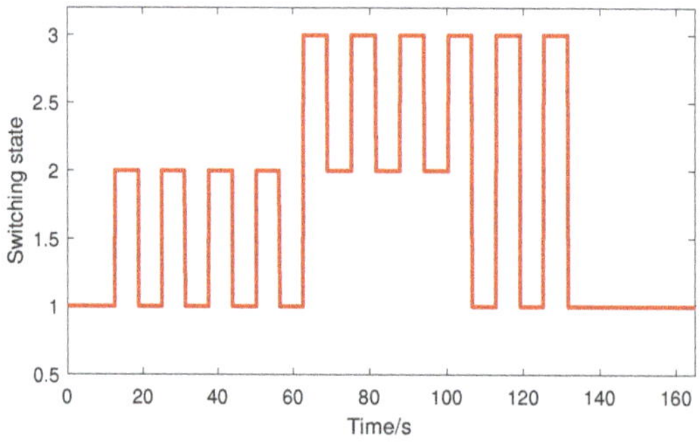

Figure 5. The switching laws of modular reconfigurable flight array (MRFA).

The MRFA has three states. When $\sigma(t) = 1$, it indicates that the MRFA is in a fault-free state. For $\sigma(t) = 2$, the fifth actuator is faulty. When $\sigma(t) = 3$, the first and fourth actuators are faulty. Calculating the controller feedback gains $\mathbf{K}_i = \alpha_i(\mathbf{B}\mathbf{B}_{\Gamma_i})^T \cdot \mathbf{P}_i$ yields:

$$\mathbf{K}_1 = \begin{bmatrix} -0.0094 & -0.0223 & -0.0005 & -0.0012 & -0.0017 & -0.0040 & -0.0418 & -0.0990 \\ -0.0087 & -0.0206 & -0.0015 & -0.0034 & -0.0005 & -0.0012 & 0.0510 & 0.1206 \\ -0.0077 & -0.0183 & -0.0018 & -0.0042 & 0.0011 & 0.0026 & -0.0361 & -0.0853 \\ -0.0077 & -0.0183 & -0.0008 & -0.0019 & 0.0011 & 0.0026 & 0.0478 & 0.1133 \\ -0.0077 & -0.0183 & 0.0008 & 0.0019 & 0.0011 & 0.0026 & -0.0480 & -0.1135 \\ -0.0077 & -0.0183 & 0.0018 & 0.0042 & 0.0011 & 0.0026 & 0.0359 & 0.0851 \\ -0.0087 & -0.0206 & 0.0015 & 0.0034 & -0.0005 & -0.0012 & -0.0508 & -0.1204 \\ -0.0094 & -0.0223 & 0.0005 & 0.0011 & -0.0017 & -0.0040 & 0.0421 & 0.0996 \end{bmatrix}$$

$$\mathbf{K}_2 = \begin{bmatrix} -0.0411 & -0.0897 & -0.0022 & -0.0048 & -0.0073 & -0.0160 & -0.1633 & -0.3613 \\ -0.0377 & -0.0832 & -0.0062 & -0.0137 & -0.0021 & -0.0047 & 0.1982 & 0.4372 \\ -0.0337 & -0.0737 & -0.0078 & -0.0171 & 0.0047 & 0.0103 & -0.1410 & -0.3111 \\ -0.0335 & -0.0740 & -0.0034 & -0.0075 & 0.0047 & 0.0104 & 0.1870 & 0.4126 \\ 0 & 0 & 0 & 0 & 0 & 0 & 0 & 0 \\ -0.0335 & -0.0740 & 0.0078 & 0.0171 & 0.0047 & 0.0104 & 0.1424 & 0.3131 \\ -0.0380 & -0.0828 & 0.0062 & 0.0137 & -0.0021 & -0.0047 & -0.1968 & -0.4357 \\ -0.0408 & -0.0900 & 0.0022 & 0.0048 & -0.0073 & -0.0160 & 0.1648 & 0.3623 \end{bmatrix}$$

$$\mathbf{K}_3 = \begin{bmatrix} 0 & 0 & 0 & 0 & 0 & 0 & 0 & 0 \\ -0.0528 & -0.1163 & -0.0087 & -0.0192 & -0.0031 & -0.0070 & 0.2753 & 0.6039 \\ -0.0471 & -0.1031 & -0.0109 & -0.0239 & 0.0066 & 0.0147 & -0.1995 & -0.4367 \\ 0 & 0 & 0 & 0 & 0 & 0 & 0 & 0 \\ -0.0471 & -0.1029 & 0.0048 & 0.0106 & 0.0067 & 0.0148 & -0.2631 & -0.5769 \\ -0.0469 & -0.1032 & 0.0109 & 0.0240 & 0.0065 & 0.0141 & 0.1852 & 0.4090 \\ -0.0530 & -0.1158 & 0.0087 & 0.0191 & -0.0029 & -0.0061 & -0.2683 & -0.5923 \\ -0.0572 & -0.1259 & 0.0030 & 0.0066 & -0.0102 & -0.0227 & 0.2356 & 0.5134 \end{bmatrix}$$

The computed feedback gains $\mathbf{K}_i$, $i = 1,2,3$ are substituted into Equation (19), resulting in arbitrary switched controllers for the MRFA with eight modules, as depicted in Figure 4, under scenarios without actuator faults, with a fault in the fifth actuator, and with faults in the first and fourth actuators.

Trajectory tracking of the MRFA's attitude angles (roll, pitch, and yaw) yields the simulation results depicted in Figure 6.

From Figure 6, it can be observed that the error curves for the MRFA with the configuration shown in Figure 4 are convergent both without controller switching and under the switching law depicted in Figure 5. However, the error curve for the switched control based on the law in Figure 5 is smoother and converges faster. In the case of actuator failure, the error curve without switched control exhibits significant overshooting, and the convergence time of the error curve is longer compared to the error curve with the addition of switched control, as illustrated by the blue dashed line in Figure 6. In the event of actuator failure, the switched control based on dwell time needs to redistribute the power loss caused by the failed actuator to the remaining functional actuators, thereby reducing the sudden change in tracking error.

Therefore, the simulation validates the effectiveness of the arbitrarily designed switched controller for the MRFA.

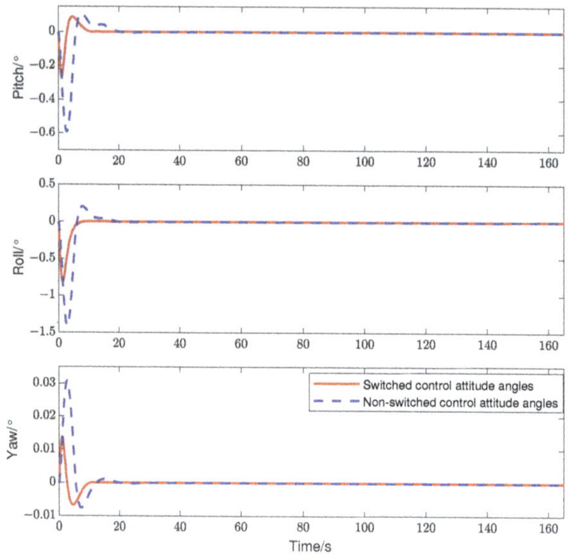

Figure 6. Comparison of attitude tracking curves between switched controller and non-switched controller.

5.2. *Experiment*

In this subsection, the attitude stability of the MRFA is experimentally investigated using a three-degrees-of-freedom testbed. The attitude control frequency of the MRFA is set to 200 Hz, and the IMU sensor update frequency is 1000 Hz. The experimental platform is shown in Figure 7.

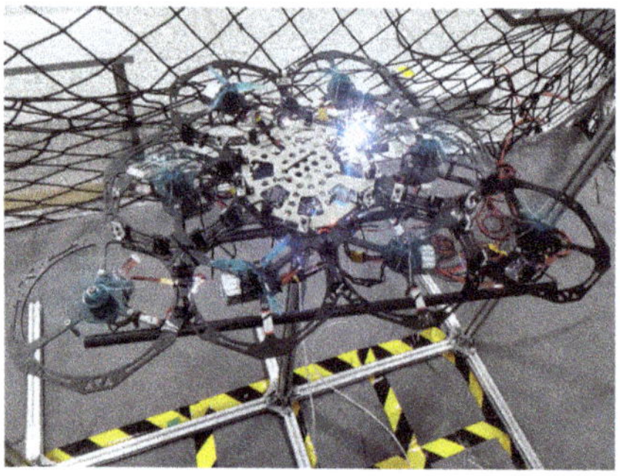

Figure 7. Three-degrees-of-freedom testbed.

The MRFA utilizes an Ano flight controller developed by an Anonymous technology startup as the controller. The flight controller is equipped with an STM32F407VGT6 microcontroller, with a maximum frequency of up to 168 MHz, and employs an ICM20602 IMU sensor produced by the Japanese electronic component manufacturer TDK to measure the attitude information of the flight array. The actuators of the flight array are Tmotor F80Pro brushless DC motors, manufactured by the Tmotor brushless motor manufacturer, and powered by a 4S lithium battery.

The MRFA is connected to the three-degrees-of-freedom test bed using custom connectors, restricting the translational motion of the MRFA. To validate the effectiveness of the switched controller, both the switched controller and the cascaded Proportional-Integral-Derivative (PID) controller were employed for attitude trajectory tracking control of the MRFA. To ensure comparability between the two controllers, PID parameters were selected to match the tracking curves of both controllers in the absence of actuator faults. The PID parameters are listed in Table 2.

Table 2. Proportional-Integral-Derivative parameters of modular reconfigurable flight array (MRFA).

		Proportional	Integral	Derivative
Pitch	Angular Velocity Loop	6000	0	60
	Angle Loop	12,000	100	200
Roll	Angular Velocity Loop	9500	0	120
	Angle Loop	10,000	50	60
Yaw	Angular Velocity Loop	9000	0	500
	Angle Loop	2250	0	0

When conducting attitude trajectory tracking with the switched controller, adjustments were made to track the reference trajectory, considering the simplified linear model of the MRFA model and the fact that the lift coefficient L_f and torque coefficient L_m of the experimental platform's propellers differed from the coefficients in [17]. This was achieved by introducing the adjustment coefficient matrices β and ζ.

$$\beta = \begin{bmatrix} 1 & 1 & 34 & 34 & 25 & 25 & 12 & 12 \\ 1 & 1 & 27 & 27 & 36 & 36 & 8 & 8 \\ 1 & 1 & 46 & 46 & 13 & 13 & 11 & 11 \\ 1 & 1 & 21 & 21 & 12 & 12 & 11 & 11 \\ 1 & 1 & 18 & 18 & 20 & 20 & 11 & 11 \\ 1 & 1 & 36 & 36 & 15 & 15 & 11 & 11 \\ 1 & 1 & 69 & 69 & 16 & 16 & 8 & 8 \\ 1 & 1 & 28 & 28 & 22 & 22 & 12 & 12 \end{bmatrix}$$

$$\zeta = \text{diag}(500, 367, 648, 489, 260, 214, 448, 425)$$

According to the switching law shown in Figure 5, attitude trajectory tracking control was performed on the MRFA for each case, as illustrated in Figure 8.

The red curve in Figure 8 represents the reference attitude trajectory of the MRFA, the blue dashed line represents the attitude tracking curve under the arbitrary switching controller, and the black dashed line represents the attitude tracking curve without using the switched controller. Under the switching law in Figure 5, for the MRFA with no actuator failure during the time intervals $0 \text{ s} \leq t \leq 15 \text{ s}$ and $132 \text{ s} < t \leq 165 \text{ s}$, the tracking curves under the two different control strategies overlap substantially. During the transitions of the MRFA to states $\sigma(t) = 1$ and $\sigma(t) = 2$, where actuator 4 continuously switches between healthy and completely failed states, the attitude tracking curves under the switched controller and without the switched controller can both follow the reference curve. However, the tracking curve under the switched controller is closer to the reference curve, because the switched controller can dynamically adjust the control strategy based on the current system state to adapt to different flight conditions. When some actuators in the MRFA fail or malfunction, the switched controller can promptly adjust the control strategy, redistribute power, allowing the system to maintain a stable flight state and get closer to the reference trajectory. When the MRFA transitions to state $\sigma(t) = 3$, where both actuators 1 and 5 simultaneously suffer complete failure, the lost power of the MRFA cannot be compensated by the PID controller within a short time. This results in a situation where the remaining actuators are insufficient to precisely track the reference attitude curve, leading to a sudden increase. In contrast, for systems utilizing the switched controller, when an actuator failure occurs, the control gains of each actuator can be switched to redistribute the outputs of each actuator, as illustrated in Figure 9. This enables the rapid convergence of tracking errors in the MRFA. As a result, the system can quickly adapt to the new power distribution and ensure rapid convergence of the MRFA's tracking error.

Therefore, employing a switching controller can enhance the fault tolerance of the MRFA against actuator failures, ensuring system stability and performance.

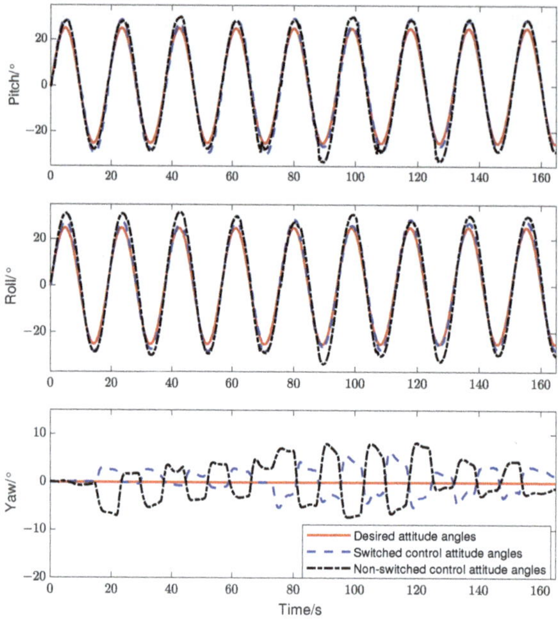

Figure 8. Comparison of attitude tracking curves at arbitrary switched controller and non-switched controller.

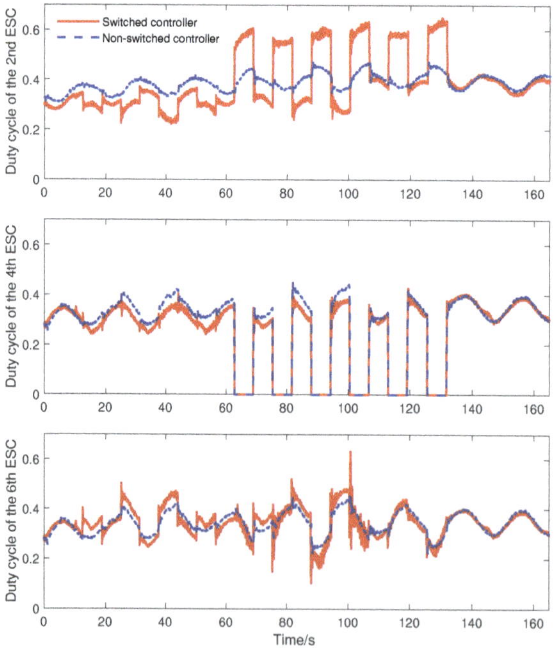

Figure 9. The partial actuator output curve of modular reconfigurable flight array (MRFA).

When employing different control strategies, the outputs of the actuators also undergo changes. Under the switching law depicted in Figure 5, the partial actuator output curves of the MRFA during pitch angle trajectory tracking are illustrated in Figure 9. When a state transition occurs and an actuator fails, there is an instantaneous reduction in power, leading to a sudden change in tracking error. Therefore, the actuator outputs under the two control strategies used in this paper exhibit abrupt changes. The blue dashed line corresponds to the case without using the switched controller is smoother compared to the curve with the switched controller. This is due to the fact that the arbitrary switched control redistributes the power loss caused by actuator failure due to the remaining properly functioning actuators.

From Figure 9, it can be observed that the behavior of actuators with different indices varies during the occurrence of actuator failures in the MRFA. In the configuration of the MRFA shown in Figure 4, between 15 s < t $\leq$ 63 s, the MRFA switches back and forth between states $\sigma(t) = 1$ and $\sigma(t) = 2$. When the MRFA tracks the reference pitch angle curve, actuator 2 is located in the positive half-axis of the body coordinate system's x-axis, and actuators 4 and 6 are in the negative half-axis of the x-axis. Since the four actuators in the negative half-axis of x generate smaller torques along the y-axis to ensure precise pitch angle tracking after the failure of actuator 5, the output of actuator 2 needs to decrease, while the outputs of actuators 4 and 6 need to increase. During the time interval 63 s < t $\leq$ 132 s, when $\sigma(t) = 3$, actuators 1 and 4 fail. Therefore, the output of actuator 4 is 0 when $\sigma(t) = 3$. In the process of switching back and forth between states $\sigma(t) = 2$ and $\sigma(t) = 3$, the MRFA is always in a state where the actuators cannot work completely. As shown in Figure 9, for any switched control strategy, the actuator undergoes the maximum mutation amplitude at this point, facilitating the redistribution of power to achieve precise tracking of the reference attitude.

6. Conclusions

If there is a failure in the actuators of the MRFA, the instantaneous reduction in power may cause a sudden change in tracking error. In addressing the issue of partial actuator failures in the MRFA, a modeling approach based on a switched system is proposed. Furthermore, a robust arbitrary switched controller design method is introduced. When the MRFA suffers partial actuator failures, it is necessary to design a state feedback controller for each subsystem that satisfies arbitrary switching. This controller enables the MRFA to keep better stable performance if the actuators have to switch arbitrarily among different combinations of actuators. The proposed method ensures that the MRFA can mitigate or eliminate disturbances caused by actuator failures through switching, thereby achieving rapid tracking of the desired trajectory. We conclude our results as two points.

1. Through simulation and experimental validation, we have found that compared to the cascade PID controller, the proposed control strategy based on dwell time can achieve more accurate attitude trajectory tracking in the presence of multiple actuator failures.

2. Under the two control strategies used in this work, the outputs of the actuators will experience sudden changes. The main differences between two methods is that the output generated by the cascade PID controller is smoother than that of the switched controller. It means that the cascade PID controller does not compensate power loss by actuator fault while the switched controller does. Therefore, the method based on the dwell time switching controller can achieve more accurate attitude trajectory tracking in the presence of multiple actuator failures.

In practical applications, actuator faults in the MRFA may significantly deteriorate system performance, so timely identification and handling of actuator fault are crucial. Therefore, our future work may introduce fault detection of the actuator into the control strategy to enhance the robustness and reliability of the MRFA system against actuator faults.

Author Contributions: Conceptualization, C.Y.; software, B.R.; writing—original draft, B.R.; writing—review and editing, C.Y., W.M. and X.Z.; supervision, C.Y. All authors have read and agreed to the published version of the manuscript.

Funding: This work is supported by the National Nature Science Foundation under Grant 62063011, 62303199, Yunnan Major Scientific and Technological Projects under Grant 202202AG050002, 202302AD 080005, Yunnan Fundamental Research Projects (Grant No. 202301AT070401) and partially supported by the Scientific Research Project of Yunnan Provincial Department of Education (Grant No. 2023Y0418).

Data Availability Statement: All data generated or analyzed during the study are included in this published article.

Conflicts of Interest: The authors declare no conflicts of interest.

References

1. Saeed, A.S.; Younes, A.B.; Cai, C.; Cai, G. A survey of hybrid unmanned aerial vehicles. *Prog. Aerosp. Sci.* **2018**, *98*, 91–105. [CrossRef]
2. Zhou, X.; Wen, X.; Wang, Z.; Gao, Y.; Li, H.; Wang, Q.; Yang, T.; Lu, H.; Cao, Y.; Xu, C.; et al. Swarm of micro flying robots in the wild. *Sci. Robot.* **2022**, *7*, Eabm5954. [CrossRef] [PubMed]
3. Park, S.; Lee, J.; Ahn, J.; Kim, M.; Her, J.; Yang, G.H.; Lee, D. Odar: Aerial manipulation platform enabling omnidirectional wrench generation. *IEEE/ASME Trans. Mechatron.* **2018**, *23*, 1907–1918. [CrossRef]
4. Schafroth, D.; Bermes, C.; Bouabdallah, S.; Siegwart, R. Modeling, system identification and robust control of a coaxial micro helicopter. *Control Eng. Pract.* **2010**, *18*, 700–711. [CrossRef]
5. Pastor, D.; Izraelevitz, J.; Nadan, P.; Bouman, A.; Burdick, J.; Kennedy, B. Design of a ballistically-launched foldable multirotor. In Proceedings of the 2019 IEEE/RSJ International Conference on Intelligent Robots and Systems (IROS), Macau, China, 3–8 November 2019; pp. 5212–5218.
6. Sufiyan, D.; Win, L.S.T.; Win, S.K.H.; Pheh, Y.H.; Soh, G.S.; Foong, S. An Efficient Multimodal Nature Inspired Unmanned Aerial Vehicle Capable of Agile Maneuvers. *Adv. Intell. Syst.* **2023**, *5*, 2200242. [CrossRef]
7. Zhao, M.; Okada, K.; Inaba, M. Enhanced modeling and control for multilinked aerial robot with two dof force vectoring apparatus. *IEEE Robot. Autom. Lett.* **2020**, *6*, 135–142. [CrossRef]
8. Mu, B.; Chirarattananon, P. Universal Flying Objects (UFOs): Modular Multirotor System for Flight of Rigid Objects. *IEEE Trans. Robot.* **2019**, *36*, 458–471. [CrossRef]
9. Oung, R.; D'Andrea, R. The distributed flight array: Design, implementation, and analysis of a modular vertical take-off and landing vehicle. *Int. J. Robot. Res.* **2014**, *33*, 375–400. [CrossRef]
10. Saldana, D.; Gabrich, B.; Li, G.; Yim, M.; Kumar, V. Modquad: The flying modular structure that self-assembles in midair. In Proceedings of the 2018 IEEE International Conference on Robotics and Automation (ICRA), Brisbane, QLD, Australia, 21–25 May 2018; pp. 691–698.
11. Zhao, G.L.; Gao, R.S.; Chen, J.N. Adaptive prescribed performance control of quadrotor with unknown actuator fault. *Control Decis.* **2021**, *36*, 2103–2112.
12. Baskaya, E.; Hamandi, M.; Bronz, M.; Franchi, A. A novel robust hexarotor capable of static hovering in presence of propeller failure. *IEEE Robot. Autom. Lett.* **2021**, *6*, 4001–4008. [CrossRef]
13. Merheb, A.; Nourra, H.; Bateman, F. Active fault tolerant control of octorotor uav using dynamic control allocation. In Proceedings of the International Conference on Intelligent Unmanned Systems, Montreal, QC, Canada, 29 September–1 October 2014.
14. Nguyen, D.T.; Saussie, D.; Saydy, L. Design and experimental validation of robust self-scheduled fault-tolerant control laws for a multicopter UAV. *IEEE/ASME Trans. Mechatronics* **2020**, *26*, 2548–2557. [CrossRef]
15. Wei, C.; Wang, Y.; Shen, Z.; Xiao, D.; Bai, X.; Chen, H. AUQ-ADMM Algorithm-Based Peer-to-Peer Trading Strategy in Large-Scale Interconnected Microgrid Systems Considering Carbon Trading. *IEEE Syst. J.* **2023**, *17*, 6248–6259. [CrossRef]
16. Yang, J.; Yang, C.; Zhang, X.; Na, J. Fixed-time sliding mode control with varying exponent coefficient for modular reconfigurable flight arrays. *IEEE/CAA J. Autom. Sin.* **2024**, *11*, 514–528. [CrossRef]
17. Wang, Q.; Wang, T.; Dong, C.Y.; Jiang, W.L. Chained smooth switching control for morphing aircraft. *Control Theory Appl.* **2015**, *32*, 949–954.
18. Ma, D.; Xia, Y.; Shen, G.; Jiang, H.; Hao, C. Practical fixed-time disturbance rejection control for quadrotor attitude tracking. *IEEE Trans. Ind. Electron.* **2020**, *68*, 7274–7283. [CrossRef]
19. Liao, W.Z.; Zong, Q.; Ma, Y.L. Modeling and finite-time control for quad-rotor mini unmanned aerial vehicles. *Control Theory Appl.* **2015**, *32*, 1343–1350.
20. Hong, X.F. Fault-Tolerant Control of Switched Systems. Master's Thesis, Xiamen University, Xiamen, China, 2007.
21. Hespanha, J.P.; Morse, A.S. Stability of switched systems with average dwell-time. In Proceedings of the 38th IEEE Conference on Decision and Control, Phoenix, AZ, USA, 7–10 December 1999; pp. 2655–2660.

processes MDPI

Article
Measurement Method of Bar Unmanned Warehouse Area Based on Binocular Vision

Shuzong Yan [1], Dong Xu [1,*], He Yan [2], Ziqiang Wang [1], Hainan He [1], Xiaochen Wang [1] and Quan Yang [1]

1 National Engineering Research Center of Flat Rolling Equipment, University of Science and Technology Beijing, Beijing 100083, China; ustb_yanzong@163.com (S.Y.); m17600598175@163.com (Z.W.); csuhhn@163.com (H.H.); xcwangustb@163.com (X.W.); yangquan@nercar.ustb.edu.cn (Q.Y.)
2 Shougang Research Institute of Technology, Beijing 100043, China; yanhe2024@163.com
* Correspondence: xudong@ustb.edu.cn

Abstract: With the development of Industry 4.0 and the implementation of the 14th Five-Year Plan, intelligent manufacturing has become a significant trend in the steel industry, which can propel the steel industry toward a more intelligent, efficient, and sustainable direction. At present, the operation mode of unmanned warehouse area for slabs and coils has become relatively mature, while the positioning accuracy requirement of bars is getting more stringent because they are stacked in the warehouse area according to the stacking position and transferred by disk crane. Meanwhile, the traditional laser ranging and line scanning method cannot meet the demand for precise positioning of the whole bundle of bars. To deal with the problems above, this paper applies machine vision technology to the unmanned warehouse area of bars, proposing a binocular vision-based measurement method. On the one hand, a 3D reconstruction model with sub-pixel interpolation is established to improve the accuracy of 3D reconstruction in the warehouse area. On the other hand, a feature point matching algorithm based on motion trend constraint is established by means of multi-sensor data fusion, thus improving the accuracy of feature point matching. Finally, a high-precision unmanned 3D reconstruction of the bar stock area is completed.

Keywords: bar unmanned warehouse area; binocular vision; multisensory fusion; 3D reconstruction

1. Introduction

With the development of information technology such as the Internet of Things and artificial intelligence, the steel industry is transforming into a smart manufacturing model with the smart factory as the carrier as well as the intelligence of key manufacturing as the core and is based on end-to-end data flow and supported by NetCom interconnection technology. At present, unmanned warehouse areas have become a very representative and important technology for smart factories. As the main lifting execution unit in the unmanned warehouse area, the unmanned crane accurately identifies the position of the crane and the location of materials in the warehouse through the crane and material automatic positioning system. Then, the material can be accurately lifted and moved under unmanned conditions. Therefore, the intelligent automatic positioning system for cranes and materials is the basis for realizing unmanned cranes and warehouse areas, which directly affects the operational accuracy of unmanned warehouse areas.

The positioning of the crane in the bar unmanned warehouse area mainly relies on the Gray bus and the data feedback in real-time to the dispatching system through the PLC, which is able to pinpoint the crane during the movement. In terms of material positioning in the warehouse area, unlike the steel coil warehouse area, where each steel coil can be located precisely through the fixed saddle, the bars in the warehouse area are tightly stacked on a large scale. Therefore, the precise positioning of each bundle of bar is vital during the lifting process. The traditional bar warehouse area uses laser ranging and line scanning to locate the material, but this method can only obtain 2D contour information of the material in the warehouse area, and the lack of information in one dimension along the width of the stack leads to incomplete measurement of the stack contour information. As the steel materials in the bar unmanned warehouse area are stored in large quantities and close piles, the above material positioning method often leads to accidents such as the material falling due to the lack of suction force caused by the inaccurate disk position, the disk crane being destroyed due to the exceeding number of bundles, the stack collapsing during the palletizing and so on. Therefore, in order to solve the above problems, it is necessary to use a more accurate method to realize the precise positioning of the material in the bar unmanned warehouse area.

Therefore, in order to solve the above problems, this paper introduces machine vision technology into the bar unmanned warehouse area and puts forward a binocular vision detection method. On the basis of the subpixel interpolation stereo matching algorithm, the IMU and camera information fusion method are used to improve the matching accuracy of visual feature points and, finally, complete the 3D reconstruction of the bar unmanned warehouse area.

2. Literature Review

In recent years, with the development of machine vision technology, binocular vision-based inspection technology and applications have been widely used in various fields [1–5]. Simultaneous Localization and Mapping (SLAM) [6] based on binocular vision has undergone rapid development, too, which has attracted extensive attention from researchers.

Early researchers divided SLAM into visual SLAM and LIDAR SLAM according to the different types of sensors used in SLAM technology, such as cameras and LIDAR [7]. Davision [8] constructed the first monocular vision SLAM system framework, MonoSLAM, using sparse feature point-based vision front-end tracking and using the current frame pose of the monocular camera and observed feature point information as back-end optimization variables. Subsequently, Klein [9] proposed PTAM, which uses two separate tasks for parallel processing of split tracking and mapping while introducing the keyframe mechanism to reduce the system computation, but it has problems such as small detection scenes and easy loss of front-end target tracking. In order to solve the problem of SLAM being insensitive to features, feature extraction and matching are used to ensure the accuracy of pose estimation in SLAM tracking. Mur-Artal [10,11] proposed the ORB-SLAM and ORB-SLAM2 systems, which ensure the consistency of movement trajectories and constructed maps by tracking targets and constructing maps. Then, the LSD-SLAM system proposed by Engle [12] bit-pose estimation in the visual front-end by constructing image gradient-based wayfinding points. In order to overcome the problem of front-end feature point tracking failure caused by environmental factors during camera movement.

In recent years, Visual SLAM algorithms have witnessed significant advancements and demonstrated stable and robust operation across various scenarios. However, challenges arise when low-quality images result from fast camera movements and varying light conditions, limiting the effectiveness of current visual sensors [13]. To address this, researchers have explored composite SLAM approaches such as VINS (IMU + visual) and RTAB-MAP (LIDAR + visual). IMU sensors offer superior angular velocity measurement accuracy and local position measurement accuracy compared to odometers. The IMU can capture clear images of dynamic objects during fast camera movements, and the camera can correct cumulative errors generated by the IMU during slow speeds [14]. The synergistic combination of the two significantly enhances SLAM performance.

Furthermore, due to the affordability and ease of use of vision and IMU sensors, an increasing number of scholars are directing their attention towards them [15]. Presently, visual-inertial fusion methods can be categorized into tight coupling and loose coupling based on whether image feature information is incorporated into the state vector [16]. In loose coupling, the IMU and camera independently estimate their motion and then fuse their pose estimates. In contrast, tight coupling involves initially fusing IMU and camera states, then jointly constructing motion and observation equations, and finally performing state estimation [17]. OKVIS was proposed by [18,19] Leutenegger, which uses the IMU measurement error and the visual reprojection error to construct the objective function, and uses the graph optimization method to solve the objective and finally obtain the state volume to be estimated, and the experimental results show that the system has higher accuracy compared with the filter-based method.

With the development of SLAM technology, SLAM based on binocular vision has gradually been widely used in various fields. Li [20] used multi-channel fusion SLAM technology for road detection and positioning. Pai [21] used SLAM technology to build a 2D map and complete path planning. Shao [22] applied semantic SLAM to autonomous indoor parking and achieved great results. Yan [23] presented a state-of-the-art real-time localization and mapping system by SLAM to achieve precise pose estimation and dense (3D) point cloud mapping in complex greenhouses.

Compared with the bar unmanned warehouse area, the above SLAM application research environment is better, which means the graphic texture and feature points are clear and easy to detect, so it has achieved good application results. In the bar unmanned warehouse area, the bar as the detection target is characterized by too much texture and a large number of repetitive textures, which leads to a large number of false matches in feature point detection. Therefore, in order to solve this problem, this paper, on the basis of the above research, adopts the feature point matching algorithm based on motion trend constraint to complete the inter-frame feature point matching and uses the point cloud splicing algorithm to finally get the accurate 3D position information of the bar unmanned warehouse area, which provides the basis for the crane lifting. The field application shows that this system can realize the synchronous positioning and map construction of unmanned bar warehouse areas and provide an effective location basis for crane lifting, which has good application prospects.

3. 3D Dimensional Measurement of Stack Based on Binocular Vision

3.1. Binocular Camera Joint Calibration

The joint calibration of binocular cameras is based on the completion of calibration of both the left and right cameras separately. The process of binocular camera calibration is as follows. Firstly, assume that there is a feature point P on the checkerboard whose coordinate under the world coordinate system is P_w. The coordinate would be recorded by the left and right cameras as $P_L = [X_L, Y_L, Z_L]^T$, $P_R = [X_R, Y_R, Z_R]^T$, respectively, and the expression of the point P in the imaging model of the left and right cameras can be found in Equation (1) as follows:

$$\begin{cases} P_L = R_L P_W + T_L \\ P_R = R_R P_W + T_R \end{cases} \tag{1}$$

By establishing a binocular stereo vision reference coordinate system with the left camera coordinate system as the origin, the expression can be obtained as:

$$P_R = R P_L + T \tag{2}$$

Combining Equations (1) and (2) yields the coordinate system conversion relationship between the binocular cameras as:

$$\begin{cases} R = R_R R_L^{-1} \\ T = T_R - R_R R_L^{-1} T_L \end{cases} \tag{3}$$

3.2. Stereo Matching Algorithm Based on Sub-Pixel Interpolation

During the 3D reconstruction process, the pixel-level parallax is incapable of adapting to the space that the depth changes continuously, especially in the scene with a large inclination or a curved surface, and therefore cannot be applied to the 3D dimensional measurement of bar stacks. Therefore, the accuracy of the stereo matching algorithm needs to be improved from pixel level to sub-pixel level to make the parallax of the measurement surface in a natural smooth transition. The sub-pixel accuracy parallax value can be obtained by curve fitting, which utilizes the neighboring parallax values and the cost values of the best parallax value, measured by WTA in the cost space, as supplementary information to calculate, as shown in Figure 1.

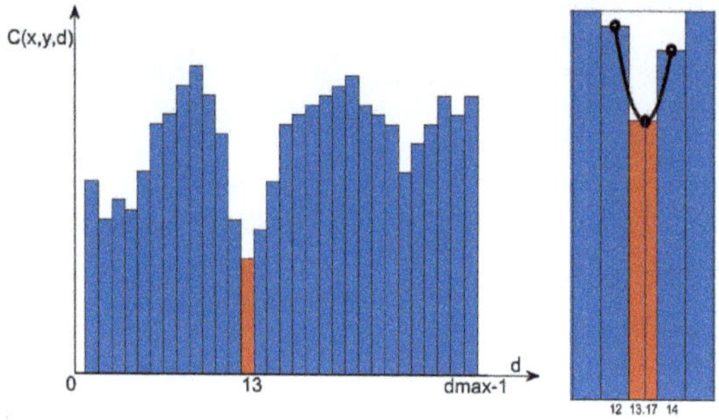

Figure 1. Parallax sub-pixel difference.

For any pixel point (x, y), a quadratic curve can be fitted by extracting three sets of data corresponding to it, which are $(d_{\min-1}, C(x, y, d_{\min-}))$, $(d_{\min}, C(x, y, d_{\min}))$, $(d_{\min+1}, C(x, y, d_{\min+}))$. The extreme point of the fitted curve reveals the best sub-pixel difference for that pixel point, and the expression is

$$d^* = d_{\min} - \frac{C(x, y, d_{\min+}) - C(x, y, d_{\min-})}{2(C(x, y, d_{\min+}) + C(x, y, d_{\min-}) - 2C(x, y, d_{\min}))} \quad (4)$$

where $d_{\min}$ is the best parallax obtained by WTA in the cost space; $C(x, y, d_{\min})$ is the best parallax corresponding to the pixel point (x, y); $(d_{\min-1}, C(x, y, d_{\min-}))$, $(d_{\min+1}, C(x, y, d_{\min+}))$ are the corresponding parallax at the same pixel of the best parallax; d^* is the sub-pixel precision parallax calculated through the parallax post-processing process.

In order to obtain more accurate subpixel-level parallax values, this paper uses a high interpolation method based on the cost space, as shown in Figure 2.

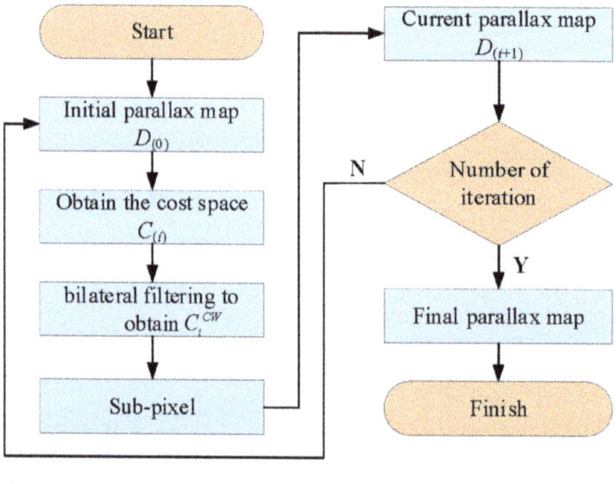

Figure 2. Flow chart of the stereo matching algorithm.

It reconstructs the cost space C_0 based on the initial parallax map D_0, then utilizes bilateral filtering on the cost space to smooth the image as well as to keep its edge features for the purpose of dealing with the abrupt changes in the depth change region, to obtain the cost space C_i^{CW}, and extracts the data needed for Equation (4) from the newly generated cost space, applies subpixel interpolation to updates the new depth estimates, and at last feeds them back into the process until the final parallax map comes out.

The expression for constructing the cost space based on the initial parallax map is:

$$C_{(i)}(x,y,d) = \min(\eta * L, (d - D_{(i)}(x,y))^2) \quad (5)$$

where L is the search for parallax levels range; d is the candidate parallax; $D_{(i)}(x,y)$ is the parallax corresponding to the L level; η is the scaling coefficient while dividing the parallax levels.

As shown in Figure 3, the sub-pixel parallax 3D reconstruction result is verified according to the actual collected images in the warehouse area. Due to the stratification caused by the discontinuous depth of the pixel-level parallax 3D reconstruction, it could not be applied to the region where the depth changes continuously. The 3D reconstructed image becomes obviously smooth after the sub-pixel parallax post-processing, which, to a certain extent, solves the inapplicability of binocular vision measurement to the region with continuously changing depth of large field of view.

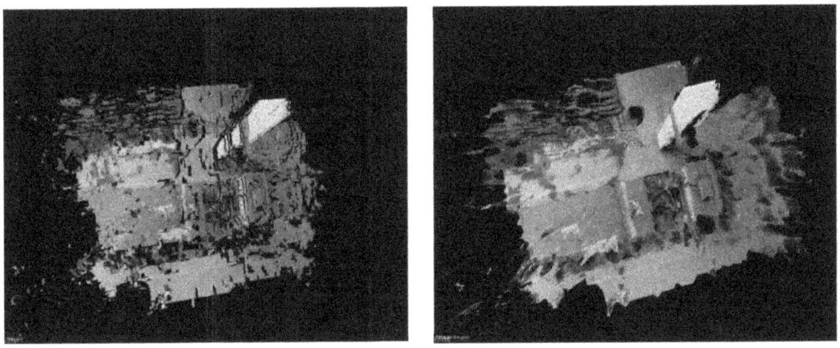

(**a**) Pixel precision 3D reconstruction (**b**) Sub-pixel precision 3D reconstruction

Figure 3. Comparison of pixel and sub-pixel precision 3D reconstruction.

The method in this paper is also compared with SGBM [24] and CSCA [25] methods, as shown in Table 1, to verify that the measurement obtained by this method has a significant accuracy advantage under the long-range measurement conditions.

Table 1. 3D dimensional measurement results.

Measurement Range (mm)	Target Size (mm)	SGBM (mm)	Error (%)	CSCA (mm)	Error (%)	Proposed Algorithm (mm)	Error (%)
1800	90	89.413	0.65	91.324	1.47	91.703	1.89
2500	90	88.161	2.04	91.815	2.02	91.133	1.26
3000	90	91.858	2.06	89.978	4.01	93.608	1.34
3600	90	93.330	3.70	88.679	1.47	89.596	0.45
4500	90	95.212	5.79	88.420	1.76	88.912	1.21
5000	90	92.472	2.75	95.360	5.96	91.747	1.94
5600	90	92.228	2.48	87.991	2.23	88.991	1.12
8500	90	108.265	20.29	95.995	6.66	92.543	2.83

4. Vision-Based Multi-Bit Pose Measurement Technology for Bar Warehouse Area

Due to the large measurement range in the metallurgical warehouse area, the partial 3D information measured from a fixed viewpoint cannot meet the demand for 3D point cloud reconstruction of the stack surface in the bar warehouse area of a large field of view. Therefore, the motion structure is used, which is to extract and match the feature points between adjacent images to solve the spatial transformation relationship and to add the newly collected point cloud information to the original one to obtain the final field point cloud information under a larger detection range. Take the feature detection of adjacent two frames in the image sequence, and the matching to obtain correlation information as the prerequisite and foundation for the realization of visual front-end tracking, then the number of correlation information obtained and the rate of correct matching would affect the accuracy of detecting the carrier pose as well as motion state. With the research and development of computer vision, some algorithms of feature point detection that can adapt to different environments and can perform well under various working conditions have been gradually derived, such as SIFT [26], SURF [27], ORB [28], etc.

4.1. Bar Warehouse Area Feature Point Detection

It is because the texture on the surface of the bars has a complex structure and repeats too much, and there are labels bound on the side of the bars that distinguish their specifications and material numbers, which makes the feature measurement and matching process more difficult. In order to verify and compare the measurement effect of several feature point measurement algorithms in terms of bar images, the images of the side and top surface of the stack in the bar warehouse area were collected and measured, and the measurement result of each algorithm is shown in Figure 4.

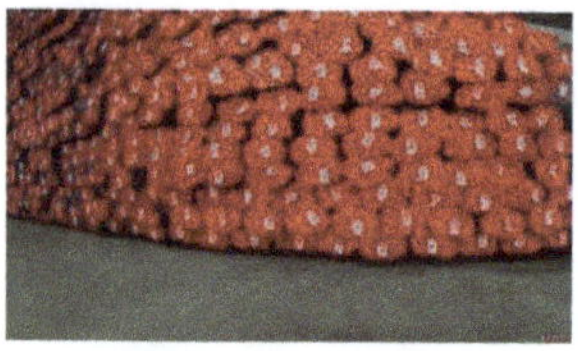

(**a**) SIFT measurement result

(**b**) SURF measurement result

(**c**) ORB measurement result

(**d**) SIFT measurement result

(**e**) SURF measurement result

(**f**) ORB measurement result

Figure 4. Bar stack feature point measurement comparison results.

Among the measurement results of the above three feature point measurement algorithms on the side surface of the bar stack, the SIFT obtains the largest number of feature points. In terms of the measurement on the top surface, the number of feature points measured is obviously sparse due to the long measurement distance (about 10 m). Thus, the three feature measurements obtain a comparable number of feature points, while the ORB spends the shortest time. The results of these feature measurement algorithms are shown in Table 2.

Table 2. Comparison of feature point measurement results.

	Side Surface of Bar Stack		Top Surface of Bar Stack	
Feature Point Type	**Number of Measurements**	**Time (ms)**	**Number of Measurements**	**Time (ms)**
SIFT	18972	7840	107	1980
SURF	8240	6070	36	2160
ORB	500	1580	41	1580

4.2. Feature Point Matching Algorithm Based on Motion Trend Constraint

In order to complete the overall 3D reconstruction of the warehouse area, it is necessary to match the feature points of the two adjacent frames collected by the moving camera. During the initial matching process, the matching point pairs are usually filtered by setting the Hamming distance. Assume that the binary descriptors of feature point A and feature point B are $D(A)$ and $D(B)$, respectively, then the Hamming distance between them is defined as:

$$H(A,B) = \sum_{i=1}^{N} \delta(D(A)_i \neq D(B)_i) \tag{6}$$

where δ( ) is an indicator function, which is 1 when the conditions in the brackets are true (otherwise, it is 0). Equation (6) means that for each binary bit, if A and B have different values at that position, increase the Hamming distance by 1. In feature point matching, the feature point pair with a smaller Hamming distance is usually selected as the matching point because a smaller Hamming distance indicates that the two binary descriptors are more similar.

However, lots of mismatches are generated due to a large number of similar or repeated textures in the images. As shown in Figure 5, a great number of false matches can be seen in the images, which are generated through the three feature detection algorithms, respectively.

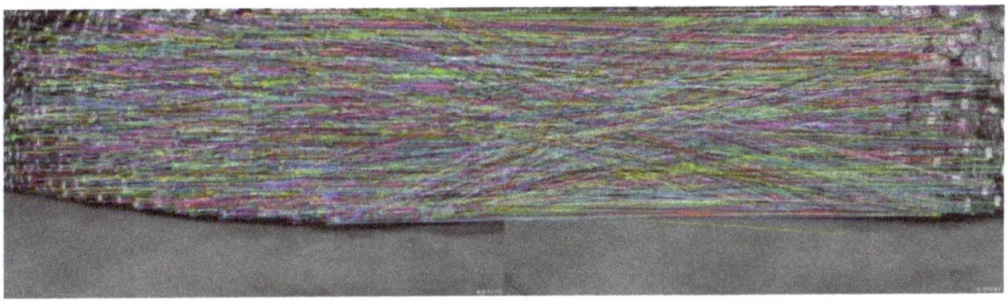

(**a**) SIFT matching results

(**b**) SURF matching results

(**c**) ORB matching results

Figure 5. Initial feature matching point pair images.

Any mismatch would lead to calculation errors in the position and posture of the adjacent images. Thus, in order to eliminate the mismatch and to take the long measurement distance in the metallurgical warehouse area as well as the large amount of repetitive texture information in the bar stack into consideration, a feature point mismatch elimination strategy combining RANSAC algorithm and motion constraints is designed.

The RANSAC algorithm, in an iterative manner, is applied to estimate the optimal homography matrix **H** from a set of observation data, including outliers, and then to preliminarily reject the mismatched pairs of points from the matched images. Its expression E_s for constructing the optimal solution function is:

$$E_s = \operatorname{argmin}\left(\sum_{i=1}^{n}\left(\left(x_i' - \frac{h_{11}x_i + h_{12}y_i + h_{13}}{h_{31}x_i + h_{32}y_i + h_{33}}\right)^2 + \left(y_i' - \frac{h_{21}x_i + h_{22}y_i + h_{23}}{h_{31}x_i + h_{32}y_i + h_{33}}\right)^2\right)\right) \tag{7}$$

For images of bar stacks that have repetitive textures, the RANSAC algorithm can only reject a small number of mismatched feature points. Due to the strong correlation between feature point matching and vision system pose transformation, the inertial measurement module (IMU) is used to solve the value of vision system pose transformation in real-time. However, the error would increase non-linearly over long periods of operation because of the zero point offset and the accumulated error. In this paper, the binocular camera and IMU are combined to form an inertial vision navigation system, and then the inertial information and visual information could be mutually complementary to reduce the limitations of single sensor detection and, therefore, the reliability of the whole system could be improved. The system adopts a strap-down carrying mode and realizes the inertial navigation function with the help of a computer. The structure of the inertial vision navigation system is shown in Figure 6.

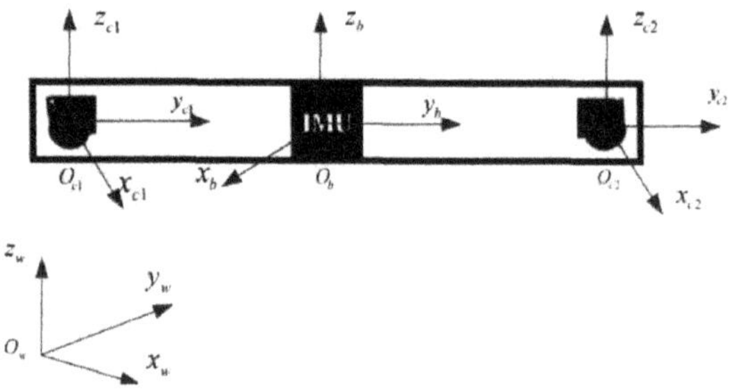

Figure 6. Inertial vision navigation architecture.

The inertial measurement module mainly consists of an accelerometer and gyroscope. It obtains the structural motion trend through the two modules, supplements the camera dynamic information, and uses the structural motion trend as a constraint to filter the feature points of the two frames of images. As shown in Figure 7, it is the feature point matching process based on motion trend constraints.

The specific steps are as follows:

1. System Calibration

During the measurement process, the IMU sensor might generate zero offset, Gaussian white noise, and random noise due to factors such as internal mechanical and temperature changes. Therefore, the optimal value should be obtained after calibration compensation and back-end optimization.

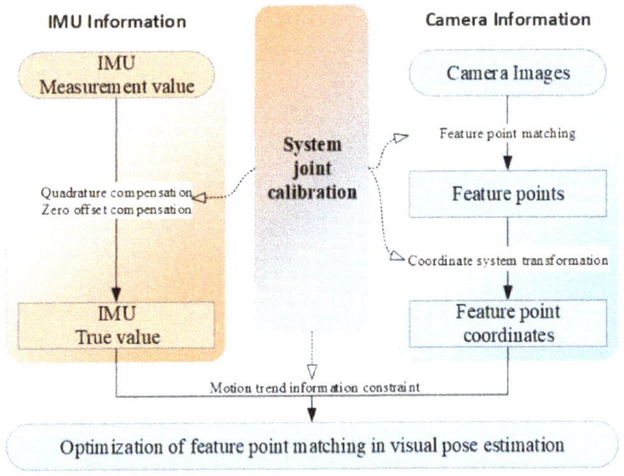

Figure 7. Feature point matching process based on motion trend constraints.

First, the calibration coefficient is obtained through joint calibration of the camera and IMU. This joint calibration uses the caliber [29] method to obtain the system parameters. This method calculates the calibration data by minimizing the objective function (Equation (8)).

$$\sum_{n=1}^{N}\sum_{m=1}^{M} \| J_{nm} - \widehat{J}(\mathbf{T}_{c,i}, \mathbf{T}_{g,i}, g_{\omega}, d, b^{a}(t), b^{g}(t), \eta^{a}(t), \eta^{g}(t)) \| \quad (8)$$

where N is the number of calibration images; M is the number of landmarks; $\mathbf{T}_{c,i}$ is the transformation between the camera and the IMU; $\mathbf{T}_{g,i}$ is the pose of the IMU; g_w is the gravity direction; d is the offset between camera time and IMU time; $\widetilde{a}(t)$, $\widetilde{w}(t)$ are measurements of the accelerometer and gyroscope; $a(t)$, $w(t)$ are real values of the accelerometer and gyroscope; $b^a(t)$, $b^g(t)$ are zero offsets of the accelerometer and gyroscope; $\eta^a(t)$, $\eta^g(t)$ are gaussian white noises.

2. Establish an IMU measurement model.

The joint calibration coefficients are obtained through the joint calibration of the camera and IMU, and the measurement model of the inertial model is established according to the calibration coefficients as follows.

$$\widetilde{a}(t) = a(t) + b^{a}(t) + \eta^{a}(t) \quad (9)$$

$$\widetilde{w}(t) = w(t) + b^{g}(t) + \eta^{g}(t) \quad (10)$$

3. Establish IMU kinematics model.

After obtaining the measurements from the accelerometer and gyroscope of the inertial module, the IMU kinematic model could be used to calculate the spatial position transformation in the interval $[t, t + \Delta t]$, and the calculation model can be expressed as:

$$\begin{cases} R(t+\Delta t) = R(t)Exp((\widetilde{w}(t) - b^{g}(t) - \eta^{gd}(t))\Delta t) \\ v(t+\Delta t) = v(t) + g\Delta t + R(t)(\widetilde{a}(t) - b^{a}(t) - \eta^{ad}(t))\Delta t \\ p(t+\Delta t) = p(t) + v(t)\Delta t + \frac{1}{2}g\Delta t^{2} + \frac{1}{2}R(t)(\widetilde{a}(t) - b^{a}(t) - \eta^{ad}(t))\Delta t^{2} \end{cases} \quad (11)$$

where R is the relative rotation angle, V is velocity, and P is the position.

4. Eliminate mismatched feature points using the inertia model.

Equation (11) represents the spatial pose change of the system from moment t to $t + \Delta t$. Therefore, after the system calibration is completed, the relative rotation angle R, velocity V, and position P of the IMU carrier at the moment $t + \Delta t$ can be calculated based on the moment t. R, V, and P is the motion constraint. The matched feature points are verified based on the calculated pose as follows:

$$F(t + \Delta t) = f(R, V, P)F(t) \tag{12}$$

where f ( ) is the state transition function. The feature point pairs that do not conform to Equation (12) are mismatched feature points that are different from the pose changes, and they are eliminated to retain the correctly matched feature points.

Finally, the precise matching of adjacent image feature points is completed through the above steps. As shown in Figures 8 and 9, it is the matching result of feature points between adjacent images that are completely consistent under several algorithms.

(**a**) SIFT matching results

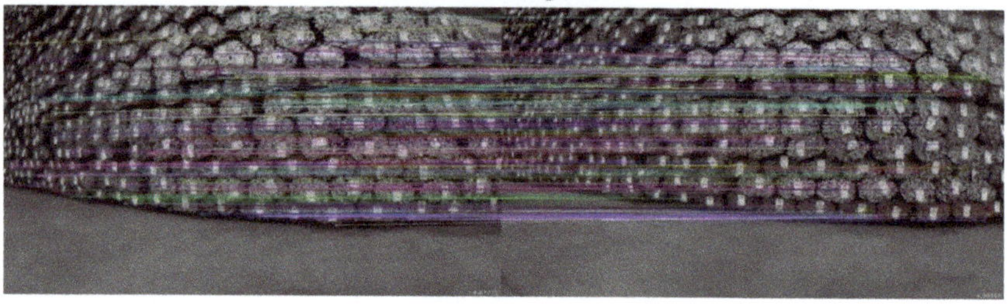

(**b**) SURF matching results

(**c**) ORB matching results

Figure 8. Matching point pair images of bar stack side surface after filtering.

(**a**) SIFT matching results

(**b**) SURF matching results

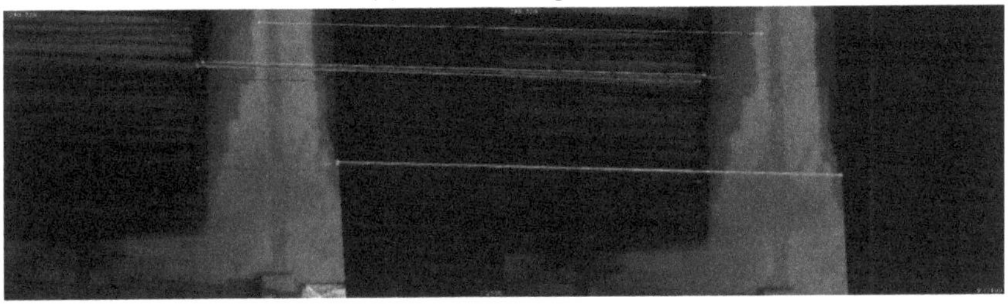

(**c**) ORB matching results

Figure 9. Matching point pair images of bar stack top surface after filtering.

According to the random consistent check results of several feature point detection methods under both side and top acquisition angles of the bar pile, it can be seen that all three feature detection methods fit the conformance. However, there are differences in the final results of the three feature detection algorithms, as shown in Table 3.

Table 3. Comparison of feature point matching results after filtering.

	Side Surface of Bar Stack			Top Surface of Bar Stack		
Feature Point Type	Number of Initial Matching Point Pairs	Number of Matching Point Pairs after Filtering	Time (ms)	Number of Initial Matching Point Pairs	Number of Matching Point Pairs after Filtering	Time (ms)
SIFT	2823	837	12,050	107	4	1930
SURF	534	208	6630	69	17	2170
ORB	59	28	2620	41	20	1750

It is found that the three measurement methods could obtain enough matching point pairs on the side of the bar stack and are visually consistent, but in terms of shooting at the top of the stack, the SIFT method cannot obtain a sufficient number of correct matching pairs and thus cannot meet the demand for long-distance detection of the bar stack, while the ORB has the advantages on both the measurement speed and the number of matching point pairs obtained from the top, compared with the SURF. Therefore, this paper chooses the ORB feature point measurement methods hereinafter, as they are the most stable and accurate.

4.3. Bar Library Multi-Position Point Cloud Stitching Algorithm

In order to get the complete 3D position data of the warehouse area, the point cloud of each frame image needs to be stitched together. The ICP problem is constructed on the basis of the congruent matching point pairs between frames and the two sets of corresponding 3D points obtained by the spatial depth information calculated by the binocular vision detection system to solve the image stitching problem.

Construct the source image point cloud data $P = \{p_1, p_2 \dots p_n\}$ and the target image point cloud data $Q = \{q_1, q_2 \dots q_n\}$, and the points in the two sets of data correspond to each other. To obtain an optimal Euclidean transformation R, t, which makes $\forall i, p_i = Rq_i + t$. By constructing the error term of a single point e_i and the ensemble error term of the whole point cloud E_i, the least squares problem is constructed and then solved by using the SVD decomposition. The error term is defined as:

$$e_i = p_i - (Rq_i + t) \tag{13}$$

$$E_i = \min_{R,t} \frac{1}{2} \sum_{i=1}^{n} \|p_i - (Rq_i + t)\|_2^2 \tag{14}$$

Define the center of mass p', q' of the set of points P and Q as:

$$p' = \frac{1}{n} \sum_{i=1}^{n} (p_i) \tag{15}$$

$$q' = \frac{1}{n} \sum_{i=1}^{n} (q_i) \tag{16}$$

The error term is integrated and decomposed into three parts, as shown in Equation (17), where the first term is only related to the rotation matrix R, and after it is solved, t can be solved by making the second term zero.

$$\begin{aligned} E_i &= \tfrac{1}{2} \sum_{i=1}^{n} \|p_i - Rq_i - t - p' + Rq' + p'Rq'\|^2 \\ &= \tfrac{1}{2} \sum_{i=1}^{n} \|(p_i - p' - R(q - q')) + (p' - Rq' - t)\|^2 \\ &= \tfrac{1}{2} \sum_{i=1}^{n} (\|(p_i - p' - R(q - q'))\|^2 + \|p' - Rq' - t\|^2 \\ &\quad + 2(p_i - p' - R(q - q'))^T (p' - Rq' - t)) \end{aligned} \tag{17}$$

Combining Equations (15)–(17) yields the rotational error term in decentered coordinates p'' and q'' as

$$E_R = \frac{1}{2} \sum_{i=1}^{n} \|p'' - Rq''\|^2 = -tr\left(R \sum_{i=1}^{n} q'' p''^T \right) \tag{18}$$

Constructive the matrix W and conduct SVD decomposition to it. Additionally, solve the spatial rotation matrix R, combining which with the Equation (13). Then, the displacement vector t could be solved.

$$W = \sum_{i=1}^{n} q'' p''^T = U \sum V^T \tag{19}$$

$$R = UV^T \tag{20}$$

By the methods above, the images of the bar warehouse area are collected and point-cloud fused. The acquisition and point cloud fusion effect of the visual odometer image of multi-position is shown in Figure 10.

(**a**) Visual odometer source image

(**b**) Visual odometer target image

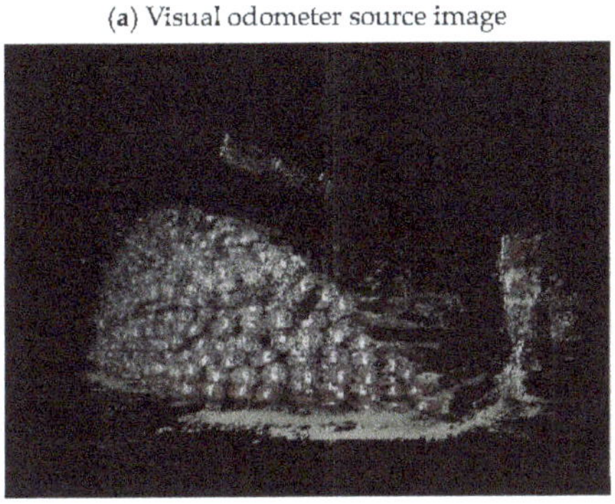

(**c**) 3D reconstruction image after fusion

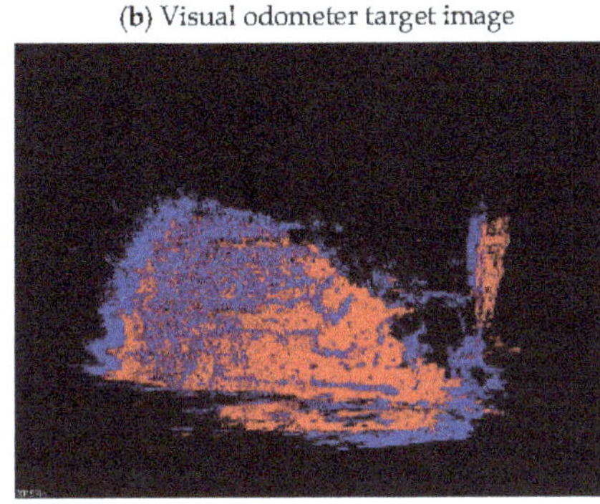

(**d**) Post-differential point cloud after fusion

Figure 10. Point cloud registration based on multi-position visual pose estimation.

After the visual odometry solves the spatial transformation relationship of the camera at different locations, the 3D information of the target image could be added to the source image in an incremental way. As shown in Figure 10, two images acquired at different locations are stitched together by the point cloud and displayed in a differential color image. The two common imaging areas overlap partially in the stitched point cloud image, while the separate imaging areas of both the target image and the original image also remain. Therefore, a more complete scene and material pile point cloud information could be obtained.

5. Fields Test and Application

The measurement device was applied in the unmanned bar warehouse area in a factory to verify the measurement accuracy by means of 3D reconstruction of the bar stack. A binocular camera was set up on an unmanned crane, and continuous sequences of images of the warehouse area were collected with the movement of the crane, organizing into a sequence of 100 images from the images of the whole 35 m bar stack. Then, by fitting the visual coordinate system to the warehouse area coordinate system, the spatial information measurement results that are consistent with the crane laser scanning coordinate system were formed. Consequently, a complete 3D point cloud image of the stack surface was plotted (Figure 11).

Figure 11. Point cloud image of bar stack in unmanned warehouse area.

In order to verify the reconstruction accuracy of the vision system, the 2D contour data of the stack was selected, which was collected by the vision system at the same position as the crane laser scanning. Additionally, the median of the redundant points of the splicing position was calculated, and the cubic polynomial interpolation method was used to no data points in the data post-processing stage to form the bar stack contour data with the same length of the crane laser scanning, and finally, the two sets of data were filtered. The comparison of the 2D contour of the stack measured by the vision system and the crane laser scanning system is shown in Figure 12.

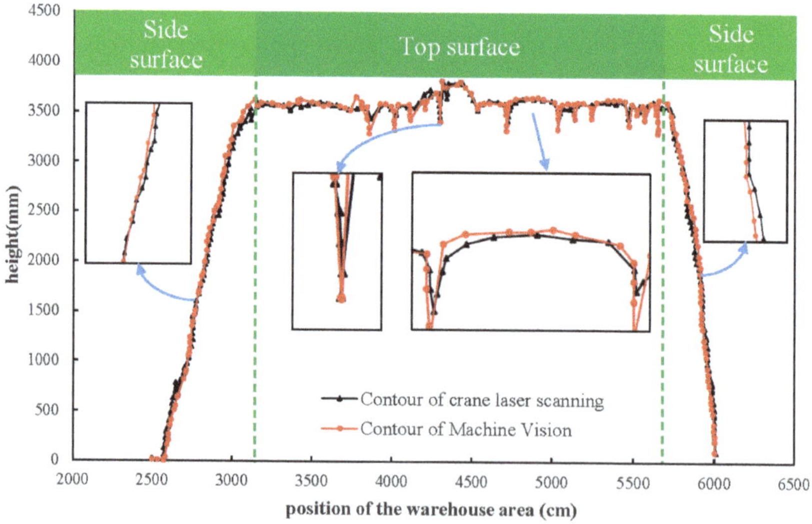

Figure 12. Bar stack contours from Binocular vision and crane laser scanning.

Under the premise of using the crane laser scanning results as standard data, it can be found from Figure 12 that the binocular vision-based bar stack contour measurement method used in this paper can obtain bar stack contours that are relatively consistent, and the visual scanning results are in gully areas also perform well. Compared with the laser scanning results, the mean absolute error of the visual scanning result is 9.82 mm, the mean relative error is 0.26%, the maximum absolute error is 17.73 mm, and the maximum relative error is 0.47%. The scene of the bar warehouse area is large, and the above data shows that the reconstruction of the bar warehouse area based on binocular vision has high accuracy.

The reconstruction of warehouse areas based on binocular vision primarily addresses the dimensional deficiencies caused by the current laser scanning, which leads to lifting issues. It also establishes a data foundation for subsequent unmanned warehouse operations. Therefore, this paper compares the proposed method with the traditional laser scanning approach. With the premise of using crane laser scanning results as the reference data, the results are presented in Table 4.

Table 4. Comparison results between laser scanning and binocular vision.

	MAE (mm)	MRE	Maximum Absolute Error (mm)	Maximum Relative Error	Bar Dimension	Method	Unit Distance (mm)	Frequency (Hz)
Laser scanning	0	0	0	0	2D	Single Point Scan	10	2
Binocular vision	9.82	0.26%	17.73	0.47%	3D	Face Scan	9.56	30

From Table 4, it can be observed that the use of binocular vision in the unmanned warehouse for bar materials ensures synchronized positioning and map construction while maintaining the data accuracy of the original laser scanning method. Simultaneously, it enhances data dimensions, obtaining the 3D profile of stack positions, thereby improving the accuracy of overhead crane hoisting. Furthermore, the adoption of high-frequency surface scanning facilitates the capture of a greater number of data points in the warehouse map, meeting on-site application requirements and demonstrating promising practicality and value.

6. Conclusions

(1) In order to overcome the inadequacy of the existing contour detection methods of the stack in the bar warehouse area, a 3D dimension detection method of stack based on binocular vision and IMU is proposed. It obtains the 3D information of the stack position by constructing a three-dimensional point cloud image of the stack with a binocular camera to ensure the accurate positioning of the crane during lifting.

(2) A stereo matching algorithm based on subpixel interpolation is proposed to solve the interpolation using the neighboring parallax value and the corresponding surrogate value of the best parallax value, which is obtained by WTA in the cost space, as supplementary information, so as to improve the accuracy of 3D reconstruction. After verifying using the acquisition images of the warehouse area, it can be found that the images have been greatly improved compared with those before subpixel processing, which, to a certain extent, solves the adaptation problem of binocular vision detection applied in the continuous transformation region with a large field of view depth.

(3) In order to solve the feature point mismatching problem caused by the consistent texture in the warehouse area, a feature point matching algorithm based on the motion trend constraint is proposed; then, by solving the system positional change with IMU, the mismatched feature points are eliminated, and finally, the fused image of the warehouse area point cloud could be obtained.

(4) The field application results show that the binocular vision-based bar stacking measurement method proposed in this paper could reconstruct the three-dimensional image of the warehouse area and obtain the three-dimensional contour of the stack. Compared with the traditional detection method, it has a higher data dimension to improve the lifting

accuracy, and it can meet the field demand in terms of processing speed as well as detection accuracy, which has good application value and promotion prospects.

(5) The system proposed in this paper changes the current situation of lack of laser scanning dimensions in the bar warehouse area, making the bar warehouse area more intelligent, helping to improve the work efficiency of steel enterprises, and providing a solid data foundation for automatic hoisting in bar unmanned warehouse areas.

(6) This paper realized a 3-D reconstruction of the bar unmanned warehouse area based on binocular vision and obtained certain research conclusions and results. However, further questions need to be solved in future work. For the problem of the impact of complex environmental factors in the bar warehouse area on system robustness, more in-depth research is needed, and sensors need to give full play to their respective advantages to achieve multi-information fusion 3-D reconstruction of the reservoir area.

Author Contributions: S.Y.: Conceptualization, Methodology, Formal Analysis, Writing—Original Draft. D.X.: Project administration, Funding acquisition, Writing—Review & Editing. H.Y.: Methodology, Software. Z.W.: Data curation. H.H.: Visualization. X.W.: Writing—Review & Editing. Q.Y.: Resources, Supervision. All authors have read and agreed to the published version of the manuscript.

Funding: This work was supported by the National Natural Science Foundation of China (Grant No. 62273032). The authors would like to acknowledge the financial support from the Royal Society with grant IEC\NSFC\211254.

Data Availability Statement: Data are contained within the article.

Conflicts of Interest: The authors declare no competing interests.

References

1. Fang, J.; Xu, S.; Yang, Y.; Wang, Y. Localization and measurement method of continuous casting slab model based on binocular vision. *Microw. Opt. Technol. Lett.* **2020**, *62*, 53–59. [CrossRef]
2. Niu, M.; Song, K.; Huang, L.; Wang, Q.; Yan, Y.; Meng, Q. Unsupervised saliency detection of rail surface defects using stereoscopic images. *IEEE Trans. Ind. Inform.* **2020**, *17*, 2271–2281. [CrossRef]
3. Zhao, S.; Kang, F.; Li, J. Displacement monitoring for slope stability evaluation based on binocular vision systems. *Optik* **2018**, *171*, 658–671. [CrossRef]
4. Shi, B.; Liu, Z.; Zhang, G. Online stereo vision measurement based on correction of sensor structural parameters. *Opt. Express* **2021**, *29*, 37987–38000. [CrossRef]
5. Liu, L.; Cui, J.; Huan, Y.; Zou, Z.; Hu, X.; Zheng, L. A Design of Smart Unmanned Vending Machine for New Retail Based on Binocular Camera and Machine Vision. *IEEE Consum. Electron. Mag.* **2021**, *11*, 21–31. [CrossRef]
6. Smith, R.C.; Cheeseman, P. On the representation and estimation of spatial uncertainty. *Int. J. Robot. Res.* **1986**, *5*, 56–68. [CrossRef]
7. Bresson, G.; Alsayed, Z.; Yu, L.; Glaser, S. Simultaneous localization and mapping: A survey of current trends in autonomous driving. *IEEE Trans. Intell. Veh.* **2017**, *2*, 194–220. [CrossRef]
8. Davison, A.J.; Reid, I.D.; Molton, N.D.; Stasse, O. MonoSLAM: Real-Time Single Camera SLAM. *IEEE Trans. Pattern Anal. Mach. Intell.* **2007**, *29*, 1052–1067. [CrossRef] [PubMed]
9. Klein, G.; Murray, D. Parallel tracking and mapping for small AR workspaces. In Proceedings of the 2007 6th IEEE and ACM International Symposium on Mixed and Augmented Reality, Washington, DC, USA, 13–16 November 2007; IEEE Computer Society: Washington, DC, USA, 2007; pp. 1–10.
10. Mur-Artal, R.; Montiel, J.M.M.; Tardos, J.D. ORB-SLAM: A Versatile and Accurate Monocular SLAM System. *IEEE Trans. Robot* **2015**, *31*, 1147–1163. [CrossRef]
11. Mur-Artal, R.; Tardós, J.D. Orb-SLAM2: An open-source slam system for monocular, stereo, and RGB-D cameras. *IEEE Trans Robot.* **2017**, *33*, 1255–1262. [CrossRef]
12. Engel, J.; Schöps, T.; Cremers, D. LSD-SLAM: Large-scale direct monocular SLAM. In Proceedings of the European Conference on Computer Vision, Zurich, Switzerland, 6–12 September 2014; Springer: Cham, Switzerland, 2014; pp. 834–849.
13. Liu, Y.; Zhao, C.; Ren, M. An Enhanced Hybrid Visual–Inertial Odometry System for Indoor Mobile Robot. *Sensors* **2022**, *22*, 2930. [CrossRef]
14. Xie, H.; Chen, W.; Wang, J. Hierarchical forest based fast online loop closure for low-latency consistent visual-inertial SLAM. *Robot. Auton. Syst.* **2022**, *151*, 104035. [CrossRef]
15. Lee, W.; Eckenhoff, K.; Yang, Y.; Geneva, P.; Huang, G. Visual-inertial-wheel odometry with online calibration. In Proceedings of the 2020 IEEE/RSJ International Conference on Intelligent Robots and Systems (IROS), Las Vegas, NV, USA, 24 October 2020–24 January 2021.

16. Cheng, J.; Zhang, L.; Chen, Q. An Improved Initialization Method for Monocular Visual-Inertial SLAM. *Electronics* **2021**, *10*, 3063. [CrossRef]
17. Jung, J.H.; Cha, J.; Chung, J.Y.; Kim, T.I.; Seo, M.H.; Park, S.Y.; Yeo, J.Y.; Park, C.G. Monocular visual-inertial-wheel odometry using low-grade IMU in urban areas. *IEEE Trans. Intell. Transp. Syst.* **2020**, *23*, 925–938. [CrossRef]
18. Leutenegger, S.; Furgale, P.; Rabaud, V.; Chli, M.; Konolige, K.; Siegwart, R. Keyframe-based visual-inertial slam using nonlinear optimization. In Proceedings of the Robotis Science and Systems (RSS), Berlin, Germany, 24 June–28 June 2013.
19. Leutenegger, S.; Lynen, S.; Bosse, M.; Siegwart, R.; Furgale, P. Keyframe-based visual–inertial odometry using nonlinear optimization. *Int. J. Robot. Res.* **2014**, *34*, 314–334. [CrossRef]
20. Li, J.; Shi, X.X.; Cheng, Z.P.; Wang, J.Z. Road detection and location based on multi-channel fusion and polar constraint. *J. Beijing Inst. Technol.* **2020**, *40*, 867–872.
21. Pai, N.S.; Huang, W.Z.; Chen, P.Y.; Chen, S.A. Optimization and Path Planning of Simultaneous Localization and Mapping Construction Based on Binocular Stereo Vision. *Sens. Mater.* **2022**, *34*, 1091–1104. [CrossRef]
22. Shao, X.; Zhang, L.; Zhang, T.; Shen, Y.; Zhou, Y. MOFIS SLAM: A Multi-Object Semantic SLAM System with Front-View, Inertial, and Surround-View Sensors for Indoor Parking. *IEEE Trans. Circuits Syst. Video Technol.* **2021**, *32*, 4788–4803. [CrossRef]
23. Yan, Y.; Zhang, B.; Zhou, J.; Zhang, Y.; Liu, X.A. Real-Time Localization and Mapping Utilizing Multi-Sensor Fusion and Visual–IMU–Wheel Odometry for Agricultural Robots in Unstructured, Dynamic and GPS-Denied Greenhouse Environments. *Agronomy* **2022**, *12*, 1740. [CrossRef]
24. Yang, Q. A non-local cost aggregation method for stereo matching. In Proceedings of the 2012 IEEE Conference on Computer Vision and Pattern Recognition, Providence, RI, USA, 16–21 June 2012; pp. 1402–1409.
25. Peng, X.; Han, J.; Tang, Y.; Shang, Y.; Yu, Y. Anti-noise stereo matching algorithm based on improved Census transform and outlier elimination. *Acta Opt. Sin.* **2017**, *37*, 223–231.
26. Lowe, G. Distinctive image features from scale-invariant keypoints. *Int. J. Comput. Vis.* **2004**, *60*, 91–110. [CrossRef]
27. Bay, H.; Ess, A.; Tuytelaars, T.; Van Gool, L. Speeded-up robust features. *Comput. Vis. Image Underst.* **2008**, *110*, 346–359. [CrossRef]
28. Rosten, E.; Drummond, T. Machine learning for high-speed corner detection. In Proceedings of the Computer Vision–ECCV 2006: 9th European Conference on Computer Vision, Graz, Austria, 7–13 May 2006; pp. 430–443.
29. Furgale, P.; Rehder, J.; Siegwart, R. Unified temporal and spatial calibration for multi-sensor systems. In Proceedings of the 2013 IEEE/RSJ International Conference on Intelligent Robots and Systems, Tokyo, Japan, 3–7 November 2013.

processes

Article

Online Partition-Cooling System of Hot-Rolled Electrical Steel for Thermal Roll Profile and Its Industrial Application

Qiuna Wang [1,2,*], Jiquan Sun [1], Jiaxuan Yang [3], Haishen Wang [2], Lijie Dong [2], Yanlong Jiao [2], Jieming Li [2], Zhenyang Zhi [3] and Lipo Yang [3,*]

[1] Institute of Engineering Technology, University of Science and Technology Beijing, Beijing 100083, China
[2] Hot Rolling Department of Qian'an Iron and Steel Corporation, Shougang Company Limited, Tangshan 064404, China
[3] National Engineering Research Center for Equipment and Technology of Cold Strip Rolling, Yanshan University, Qinhuangdao 066004, China; yangjiaxuan1996@126.com (J.Y.)
* Correspondence: wangqiuna2160@sgqg.com (Q.W.); yanglp@ysu.edu.cn (L.Y.)

Abstract: The shape and convexity are crucial quality assessment indicators for hot-rolled electrical steel strips. Besides bending rolls, shifting rolls, and the original roll profile, the thermal roll profile also plays a significant role in controlling the shape and convexity during the hot-rolling process. However, it is always overlooked due to its dynamic uncertainty. To solve this problem, it is necessary to achieve online cooling-status control for the local thermal expansion of rolls. Based on the existing structure of a mill, a pair of special partition-cooling beams with an intelligent cooling system was designed. For high efficiency and practicality, a new online predictive model was established for the dynamic temperature field of the hot-rolling process. An equivalent treatment was applied to the boundary condition corresponding to the practical cooling water flow. In addition, by establishing the corresponding target distribution curve for the partitioned water flow cooling, online water-flow-partitioning control of the thermal roll profile was achieved. In the practical application process, a large number of onsite results exhibited that the predicted error was within 5% compared to the experimental results. The temperature difference between the upper and lower rolls was within 5 °C, and the temperature difference on both sides of the rolls was controlled within 0.7 °C. The hit rate of convexity (C40) increased by 33%. It was demonstrated that the partition-cooling processes of hot rolling are effective for the local shape and special convexity. They are able to serve as a better control method in the hot-rolling process.

Keywords: thermal roll profile; roll temperature field; partition cooling; electrical steel; hot-rolled strip

1. Introduction

As the most widely utilized material, electrical steel is used in diverse sectors, such as the aerospace, bridge construction, and petrochemical industries [1,2]. Strips hold the largest share in the production structure of the steel variety, so their production capacity and product quality are always used as key indicators of the country's industry development level [3]. As a pivotal product, the demands for the quality of hot-rolled strips are continually escalating [4]. During the hot-rolling process, there is always a bottleneck in regulating some crucial challenges, such as the convexity, the wedge, and the transverse thickness difference of the strip [5]. At the same time, to achieve efficient and energy-saving production, the intelligence control of the hot-rolling process is imperative [2].

In terms of shape control, the partition-cooling method has been successfully applied in the cold-rolling and hot-rolling processes of aluminum alloys, achieving effective control over the strip shape. The main purpose of partition-cooling control in cold rolling is to adjust the high-order shape and local shape defects of the strips [6]. During the cold-rolling process, the roll temperature is influenced by various factors, such as the rolling speed, the rolling force, and deformation, which makes it a highly coupled and nonlinear process [7,8]. Different from the cold-rolling process, the convexity of hot-rolled strips is a more crucial indicator for challenging aspects in shape control. By adjusting the water flow on the work roll, the thermal roll profile is achieved within a significant range to accomplish co-operative control of the shape and convexity. Guo et al. simulated partition cooling in aluminum rolling to control the local thermal expansion of the work rolls and established a closed-loop control model for partition cooling by using an adaptive neural network trained with on-site data [9,10]. Bohacek et al. improved the cooling capacity of the gap by the nozzles [11]. Wang et al. adjusted the local temperature and thermal expansion of the roll edges by applying a thermal spraying method to the edges of the hot-rolling strips [12]. All in all, due to the high-temperature conditions of hot rolling, it is challenging to achieve the precise control of roll thermal expansion at different positions. Obviously, the dynamic temperature field model of hot rolling is crucial for implementing thermal profile control. The temperature field model primarily involves the finite element method, finite difference method, and analytical regression method [13]. Gavalas et al. attained the steady boundary conditions of a work roll three-dimensional model [14]. Han et al. established a roll temperature field model for the deformation zone and analyzed influencing factors, such as the deformation heat, friction heat, and strip temperature [15]. Wu et al. similarly employed a fully coupled finite element model for the temperature, phase transformation, and strain [16]. Hu et al. analyzed the thermal fatigue life of work rolls through a simplified finite element temperature field model combined with material property parameters [17]. Li et al. established a roll temperature field model to evaluate the practical application effects of the cooling water on the rolling speed and roll temperature distribution [18]. With the development of intelligent algorithms, some intelligent algorithms and big data methods are playing increasingly important roles in modeling the temperature field and thermal expansion of mills. Bao et al. established a data-driven dynamic neural network model for temperature forecasting in heated strips [19]. Li et al. established a comprehensive forecasting model for hot-rolled strips based on a multi-granularity cascaded forest framework through real-time data collection and analysis, and they achieved accurate predictions under the conditions of limited samples [20]. Meng et al. established a predictive model for the thermal roll profile and roll wear based on the informer network algorithm [21]. Overall, with the development of the hot-rolling process, online control of the shape and convexity is playing an increasingly important role. Firstly, accurate online prediction or actual measurement of the lateral temperature change is required. Secondly, it is necessary to perform high-precision control of the dynamic heat transfer coefficient. On this basis, partition cooling also needs to be co-ordinated and regulated by methods such as roll tilting, roll bending, and roll shifting. The current research focus of hot rolling is on roll gap control, as well as on the difficulty of regulating the shape and convexity of hot-rolled strips. Especially for silicon steel, because the local convexity is more pronounced, the high-precision, real-time control of the convexity is necessary to reduce the edge drop and achieve the best shape state.

For the analysis of the partition-cooling rules on the thermal roll profile, a practical finite difference method with dynamic boundary conditions was established to improve the shape and convexity of hot-rolled electrical steel. Combined with the actual production process, the analysis results were obtained by the predictive capabilities of the model and the application effectiveness of the partition-cooling device.

2. New Partition-Cooling Device of Hot Rolling and Its Online Control Method

In order to control the local load roll gap, the cooling water flow rate needs to be changed independently at different positions of the hot roll along the lateral direction. Thus, a new type of partition-cooling beam is designed, as shown in Figure 1a. The red line indicates the width range of the hot strip. According to the product requirements, the cooling beam is divided into five areas along the lateral direction, which includes the middle rapid cooling area, the strip edge cooling area, and two end auxiliary cooling areas. Flow control valves at the pipelines are able to achieve individual control or simultaneous control of the flow rates in each lateral area. The control methods for both upward and downward flow are consistent and synchronized, as shown in Figure 1b. The flow rate corresponding to each nozzle is proportionally adjusted by changing the nozzle aperture or the model setting data; flowmeters and piezometers are used to calculate the opening proportion of water for partition cooling. The number of nozzles is finely designed to match base model and process arrangement. There are three rows of nozzles on both the upper and lower beams, with 10–25 nozzles per row. To ensure the cooling effectiveness of each row of nozzles, the angle between each row of nozzles and the roll surface lateral varies. The angles between the nozzle rows and the roll surface transverse for the upper beam are 5°, 10°, and 10° for the first, second, and third rows, respectively. For the lower beam, the angles are 0°, 5°, and 10° for the first, second, and third rows, respectively. Considering the different distances between each row of nozzles and the roll surface, the lengths of nozzle bases for the first to third rows on the upper beam are 110 mm, 27 mm, and 27 mm, respectively. The lengths of bases for the lower beam are all 110 mm.

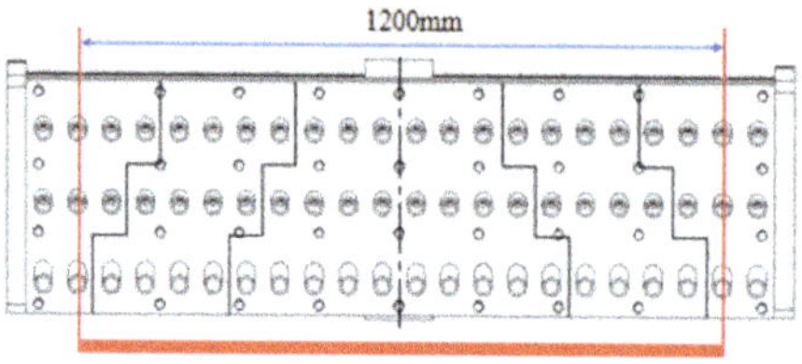

(a)

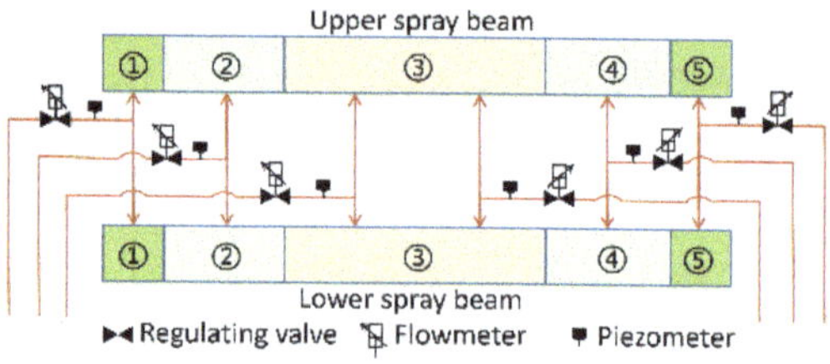

(b)

Figure 1. Partition-cooling system: (a) partition-cooling beam; (b) flow control method.

Due to the high-temperature nature and the characteristics of hot roll, the partition-cooling beam must be different from the classical static cooling method of hot roll as well as the fine control method of cold roll. By this special partition-cooling system of hot roll, online fine adjustments are achieved to regulate the local load of the roll gap. Obviously, it is helpful in minimizing the occurrence of problems such as the edge drop, the edge wave, and the strip misalignment of silicon steel. The specific control process is illustrated in Figure 2. By selecting the appropriate cooling curve from the sample database, the best process is able to be obtained based on current product specifications and online rolling process parameters. Because the roll thermal expansion is calculated in real time, the shape and crown of the hot strip are formed by practical rolling conditions. Subsequently, the coupled roll profile or the load roll gap is also determined based on the different rolling parameters. According to the quality requirements of product (e.g., shape and crown of the strip), online regulation is carried out at any time. If any deviation is detected, corresponding cooling flow adjustments must be made through the target cooling curve. This iterative process continues until the product quality meets the standards. If the adjusted cooling curve is better than the previous ones, it will be stored in the sample database for subsequent production of products with the same specifications. Moreover, continuous updating and optimization of the sample database are carried out to expand the cooling curves in the database. It is necessary to achieve automation control of any local shape and convexity in the hot-rolling process.

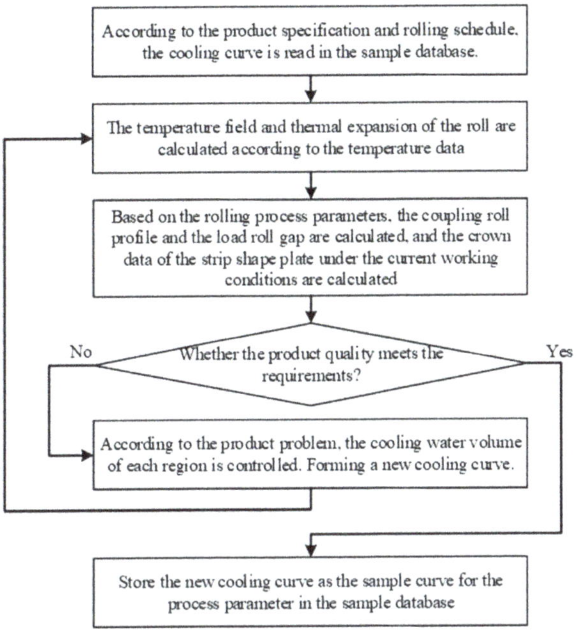

Figure 2. Partition-cooling control process of hot-rolling process.

3. Control Model of Partition-Cooling System

3.1. Dynamic Conditions of Complex Roll Temperature Field

Utilizing the principle of energy conservation, a practical roll's temperature field model is established by formulating the finite difference equation. First, the roll is discretized by dividing areas along the lateral direction. And, then, the actual temperature changes at various grid points are able to be computed. The grid points are categorized as internal points and surface points, as shown in Figure 3 [22].

Figure 3. Grid points and their energy transfer relationship in the roll: (**a**) surface grid points of the roll; (**b**) internal grid points of the roll.

Figure 3 represents the energy relationship of surface points and internal points, respectively. The energy at internal grid points comes from the heat transfer with the surrounding four grid points, while surface grid points, in addition to heat transfer with the surrounding points, also exchange heat with the external environment. In Figure 3, the dashed lines represent the control volumes, where *i* and *j* are the row and column indices of the corresponding grid points. The *i* direction is from the core to the surface of the work roll, and the *j* direction is from the drive side to the operator side. The dashed line represents the controlled volume. The heat increment ΔU according to the law of conservation of energy is expressed as:

$$\Delta U = Q_1 + Q_2 + Q_3 + Q_4 \tag{1}$$

where Q_1, Q_2, Q_3, and Q_4, are the heat received by the target grid point from the grid points $(i, j-1)$, $(i, j+1)$, $(i+1, j)$, and $(i-1, j)$ at the next time step, respectively.

Combining the heat transfer equation of energy at each grid point, the temperature change equations at the corresponding node are established. And, then, the node temperature difference equation for the next time step is obtained. The differential equations for interior grid points of the work roll are expressed as:

$$\begin{aligned} T_{i,j}^{n+1} = T_{i,j}^{n} + \frac{\alpha\Delta t}{(\Delta x)^2}T_{i,j-1}^{n} + \frac{\alpha\Delta t}{(\Delta x)^2}T_{i,j+1}^{n} + 2\frac{\alpha\Delta t}{(\Delta r)^2}\left(1 - \frac{\Delta r}{4r-\Delta r}\right)T_{i+1,j}^{n} - \Big[2\frac{\alpha\Delta t}{(\Delta x)^2} \\ +2\frac{\alpha\Delta t}{(\Delta r)^2}\left(1 - \frac{\Delta r}{4r-\Delta r}\right) + \frac{2h\Delta t}{\rho c\Delta r}\left(\frac{4r}{4r-\Delta r}\right)\Big]T_{i,j}^{n} + \frac{2h\Delta t}{\rho c\Delta r}\left(\frac{4r}{4r-\Delta r}\right)T_{out} \end{aligned} \tag{2}$$

The differential equations at the grid points on the roll's surface are expressed as:

$$\begin{aligned} T_{i,j}^{n+1} = T_{i,j}^{n} + \frac{\alpha\Delta t}{(\Delta r)^2}T_{i-1,j}^{n} + \frac{\alpha\Delta t}{(\Delta x)^2}\left(T_{i,j-1}^{n} + T_{i,j+1}^{n}\right) + \frac{\alpha\Delta t}{(\Delta r)^2}\left(\frac{r+\Delta r}{r}\right)T_{i+1,j}^{n} - \\ \left(2\frac{\alpha\Delta t}{(\Delta r)^2} + \frac{\alpha\Delta t}{(\Delta r)^2}\frac{\Delta r}{r} + 2\frac{\alpha\Delta t}{(\Delta x)^2}\right)T_{i,j}^{n} \end{aligned} \tag{3}$$

where α is the thermal conductivity, r is the roll radius, Δr is the radial unit length of the roll, Δx is the length of roll transverse cell, T^n is the current temperature of the roll, Δt is the time step, h is the convective heat transfer coefficient, T_{out} is the external temperature, ρ is roll density, c is the specific heat capacity of roll, ΔU is the volume variable, and T^{n+1} is the roll temperature at the next moment.

Due to the hot-rolling process being complex, the online thermal roll profile plays an important role in the local shape and crown of the strip. In order to enhance the calculation stability and accuracy of the transient temperature field model, it is necessary to conduct a detailed analysis of these boundary conditions during the hot-rolling process in Figure 4.

Figure 4. Boundary conditions of the work roll: (**a**) lateral partition; (**b**) circumferential cooling conditions.

As shown in Figure 4a, along the lateral direction, the roll is divided into the non-contact low-temperature areas 1 and 5 on both sides, the high-temperature areas 2 and 4 in contact with the hot strip on both sides, and the high-temperature area 3 in the middle of the roll. It is necessary to consider the lateral movement of the rolls and the variation in the strip width during the calculation process. Therefore, the high-temperature area in contact with the strip edge includes a portion of the non-contact low-temperature area. As shown in Figure 4b, the circumferential boundary of the roll is divided into eight different regions, each with specific boundary conditions. This division facilitates the impact of various boundary conditions on the temperature and thermal profile of the roll. Additionally, it allows for the precise setting of equivalent heat transfer conditions. Among them, A is the contact heat transfer in the deformation zone and B and I are the air cooling and the radiation heat transfer at the lower part of the work roll at the exit and entrance, respectively. C and H are the water cooling at the exit and entrance of the work roll, respectively. D and

G are the air cooling at the upper part of the work roll at the exit and entrance, respectively. F is the contact heat transfer between the work roll and the backup roll.

During the high-speed rotation of the work roll, the boundary conditions periodically change. To precisely control the partition water flow, it is necessary to apply the equivalent treatment to the boundary conditions of the work roll. In relatively stable operating conditions, the main factors affecting the work roll's temperature are the strip contact heat transfer coefficient h_{s0} and the basic heat transfer coefficient h_{w0} of the cooling water, represented as [23]:

$$h_{s0} = \frac{\sqrt{R(\Delta h + 2.2 \times 10^{-5} P_r / B_s)}}{2\pi R} \left[15,100 + 230 \times \left(4.168 + 1.712 \times 10^{-6} e^{0.0146 T_s}\right)\right] \quad (4)$$

$$h_{w0} = \kappa L_r L_d + k h_r \quad (5)$$

where R is the work roll's radius, Δh is the reduction, P_r the rolling force, B is the strip width, T_s is the strip temperature, L_r is the cooling water pressure, L_d is the distance between the nozzle and the roll, κ is the correction coefficient, k indicates whether the inlet cooling water is open, and h_r is the heat transfer coefficient of the inlet's cooling water.

According to Figure 4b, dividing the roll into n regions along the circumferential direction, the basic equivalent heat transfer coefficient h_{eq} and basic equivalent temperature T_{eq} are expressed as:

$$h_{eq} = \frac{\sum_{i=1}^{n} h_i \omega_i}{\sum_{i=1}^{n} \omega_i}, T_{eq} = \frac{\sum_{i=1}^{n} h_i T_i \omega_i}{h_{eq} \sum_{i=1}^{n} \omega_i} \quad (6)$$

where h_i is the heat transfer coefficient in the i region, ω_i is the contact area in the i region, and T_i is the temperature in the i region.

In addition, the comprehensive heat generation process must be required in the hot-rolling process by fine-tuning the boundary conditions. For example, besides normal heat exchange, there is also the frictional heat between the roll and the strip, as well as the deformation heat generated by the strip deformation. Assuming that the deformation heat is entirely absorbed by the strip, the frictional heat is distributed between the roll and the strip in inverse proportion to the temperature [24]. Thus, the heat flux density q_d of the deformation heat per unit volume is:

$$t_r = \frac{\sqrt{R(h_0 - h_1)}}{v_r}, \ q_d = \frac{\eta_d \sigma_F \ln \frac{h_0}{h_1}}{\sqrt{3} t_r} \quad (7)$$

The temperature rise caused by frictional heat is as follows:

$$\Delta T_f = \eta_f \frac{f}{c\rho} \frac{l_c}{h_m} \sigma_F \ln \frac{h_0}{h_1} \quad (8)$$

where t_r is the strip's contact time in the deformation zone, v_r is the rolling speed, h_0 is the inlet thickness of the strip, h_1 is the outlet thickness of the strip, σ_F is the deformation resistance of the strip, l_c is the contact arc length of the deformation zone, h_m is the average thickness of the strip in the deformation zone, and η_d and η_f are the heat transfer coefficients of the deformation heat and the friction heat, respectively. According to the above formulas, the online temperature model with high precision and high efficiency is able to be established. Obviously, further optimization and correction of boundary conditions enable accurate prediction of the hot strip.

3.2. Setting Rules of Partition-Cooling System

After installing the partition cooling beams at the exit of the mill, the cooling water flow is able to be regulated in different lateral regions of rolls. And, then, the local thermal roll profile will be changed according to the lateral water flow. However, it is necessary to calculate the online heat transfer coefficient for the corresponding grid points as [23]:

$$h_w = h_{w0} h_q h_p h_\beta h_T h_d \tag{9}$$

where h_q is the flow density coefficient, h_p is the water pressure coefficient, h_β is the nozzle injection angle coefficient, h_T is the roll's temperature coefficient, and h_d is the distance coefficient from nozzle.

Considering the practical situation of mill on site, the installation positions of these cooling beams are often fixed. Therefore, except for the water pressure and the water flow density, other coefficients are treated as constants. Combining with the heat transfer coefficient in Equation (5), further processing of water cooling is shown as:

$$h_w = \chi h_q h_p h_{w0} \tag{10}$$

where χ is a constant determined by factors such as the installation location.

In the process of designing a partitioned-cooling beam, various parameters, such as the nozzle types, the base lengths, and the angles, differ across different regions. These differences lead to variations in the heat transfer coefficients in the cooling water regions. Analyzing in conjunction with production loading, the opening ratio of the nozzle corresponds to the flow rate and the water pressure of the cooling water. Consequently, by adjusting two types of parameters, it is possible to calculate the sectional heat transfer coefficients for the lateral cooling water. Combining practical production data of the 1580 hot mill, the heat transfer coefficient of water cooling is calculated by Equation (5), where κ is 0.25 and k = 53.9. During the rolling process, the flow rate and the pressure of the cooling water serve as a criterion for determining the opening ratio. And, then, the logical relationship is defined as:

$$r = \frac{L_1 \times p_1}{L_{\max} \times p_{\max}} \times 100\% \tag{11}$$

where r is the opening ratio of the partition-cooling beam, L_1 is the practical flow rate of the cooling water, p_1 is the practical pressure of the cooling water, $L_{\max}$ is the maximum water flow rate of the cooling water, and $p_{\max}$ is the maximum pressure of the cooling water.

In the other stable cases, adjusting the flow rate of the cooling water also significantly alters the thermal profile or the load roll gap. Properly, it compensates for the dynamic fluctuations of other roll profiles. The powerful adjustment capability of local roll profile is the most effective method of improving strip edge drop and local crown. When the flow rate is between 0 and 8 L/s, the heat transfer coefficient rapidly increases. This indicates that increasing the water flow effectively regulates the surface temperature of the rolls within this range. When the flow rate is above 8 L/s, the heat transfer coefficient still shows an increasing trend but the rate noticeably slows down. This suggests that, when the water flow reaches a certain level, solely relying on increasing the cooling water flow rate has limited potential to enhance cooling efficiency. In the practical control process, the lateral distribution of the flow rate should be considered as the optimization variable. Clearly, the online thermal roll profile is helpful in minimizing the lateral thickness variations and the edge deviation of the strip.

3.3. *Effective Model of Thermal Roll Profile*

The online thermal roll profile of the current stand has a direct impact on the subsequent stand, so the partition-cooling curve should be designed and selected to meet the different rolling conditions. This achieves a smooth transition in the roll gap of all stands. Actually, the synchronous coupling calculation is conducted to minimize online forecasting deviations. First, the roll's temperature distribution is solved in real time based on the practical operating parameters. Simultaneously, the current thermal expansion is computed to forecast subsequent changes in the thermal roll profile. This information serves as the basis for the control adjustments of the local roll gap.

Assuming the work roll as an infinitely long cylindrical body, the temperature field distribution is symmetric about the cross-section of the roll. Simultaneously, by discretizing the integral part, the thermal expansion of the roll of u_r is obtained [25].

$$u_r = 2(1+v)\frac{\beta_r}{R}\int_0^R (T - T_0)rdr \tag{12}$$

where β_r is the linear expansion coefficient, v is the Poisson's ratio of the roll material, T_0 is the initial temperature or the previous temperature, and T is the current roll temperature.

However, it is necessary to consider the cascading relationship of parameters between front and rear stands [26,27]. It is essential for better predicting the evolution pattern of the thermal roll profile. After obtaining these quasi-static rolling parameters and their variations for each stand, the thermal roll profile changes should be isolated for each stand based on the measured data. And, then, the real-time calculation is performed by using the partition cooling to closely match the local roll gap. Meanwhile, based on the measured crown at the exit stand, the most suitable thermal roll profiles will be retro-calculated for each stand. Referring to the technical route in Figure 2, the original profile C_o at room temperature, the thermal profile C_T, the worn profile C_a, the elastic bending profile C_d, and the anti-symmetric wandering profile caused by the roll shifting C_S are set; the multi-parameter coupled profile C_Z of the target stand is as shown in Equation (13).

$$C_z = \zeta(\lambda_o C_o + \lambda_T C_T + \lambda_a C_a + \lambda_d C_d + \lambda_s C_s) + (1-\zeta)C_{z0} \tag{13}$$

where λ_o, λ_T, λ_a, λ_d, and λ_s are the adaptive weighting factors corresponding to the above roll profile, respectively. C_{z0} is the setting curve of the last similar working condition and ζ is the adaptive adjustment factor.

4. Experimental Results

4.1. Analysis of Partition Water Flow Rates

The core of partition cooling lies in achieving precise control of the thermal profile by adjusting the cooling water quantity in each region. Based on this rule, after the new partition-cooling beam was installed at the exit of the mill, some key cooling experiments were conducted. Then, the effects of adjusting the water flow rates, the strip width, and the entry water flow on the thermal profile were analyzed separately. Taking the most typical S14 steel as an example, the strip width was 1200 mm, the rolling pace was 125 s, the pass quantity was 50 pieces, and the roll diameter was 680 mm. The opening ratios of areas 1 and 5 were 10–100%, corresponding to the water flow range of 1.5–13 m^3/h. For areas 2 and 4, the opening ratios were 10–100%, corresponding to 1.8–16 m^3/h. For area 3, the opening ratios were 10–100%, corresponding to 1.8–16.5 m^3/h. During the analysis of one or multiple areas, the water flows in other areas were set at a fixed opening ratio of 100%. By the changes in thermal profile of the edge and center positions under different cooling conditions, the opening ratios and the thermal profiles are shown in Figure 5.

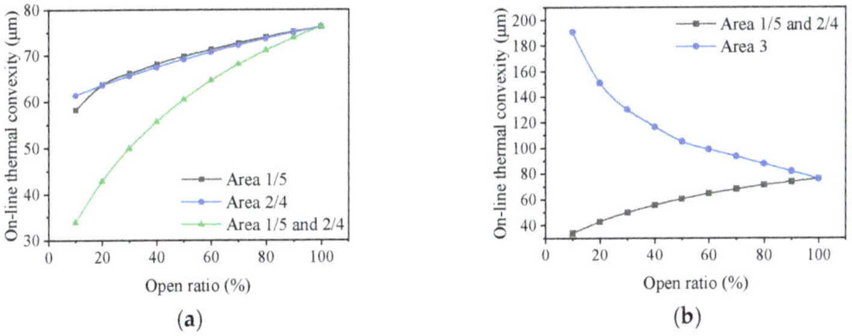

Figure 5. The relationship between the opening ratios and the thermal profiles: (**a**) the independent regulations and the joint regulations; (**b**) opening ratio and thermal convexity in area 3.

Figure 5a shows the regulation process of the water flow in areas 1/5 and 2/4, as well as the joint regulation of the water flow. When individually controlled in areas 1/5, there was a clear nonlinear relationship in thermal profile. Areas 1/5, as the cooling control area of the roll's edge, had a thermal profile range of 50–75 μm corresponding to the opening ratios of 10–100%. The roll surface in this area did not contact with the strip; hence, its influence on the central and local convexity was limited. When individually controlled in areas 2/4, the focus was primarily on controlling the contact area between the strip edge and the rolls. The thermal profiles changed corresponding to the opening ratios of 10–100% from 62–76 μm. Its influence on the overall convexity of the roll was relatively weak but significantly affected the local convexity or the strip edge reduction. Through theoretical calculations, it was observed that the individually controlling methods of the opening ratios of areas 1/5 and 2/4 had little impact on the thermal profile of the roll. The regulation at the edges was more suitable for fine-tuning at the local profile. By jointly adjusting the two areas, the control range of the thermal profile reached 33–76 μm. Clearly, this provided a larger range for the thermal profile or the local roll gap. This more flexible adjustment method was able to meet the rapid online adjustment requirements of any thermal roll profiles. From Figure 5b, area 3 had the greatest impact on the thermal profile. The opening ratios of 10–100% corresponded to the range of 76–191 μm. Based on these above requirements, an appropriate opening ratio of area 3 was selected to ensure a suitable temperature range and meet the rated adjustment range.

Moreover, when the strip width ranged from 800 mm to 1250 mm and the pass quantity was 50, the rolling pace was 125 s. The online thermal profile was analyzed for two kinds of limiting water (10% and 100%). And, then, the partition-cooling range under different widths was obtained, as shown in Figure 6.

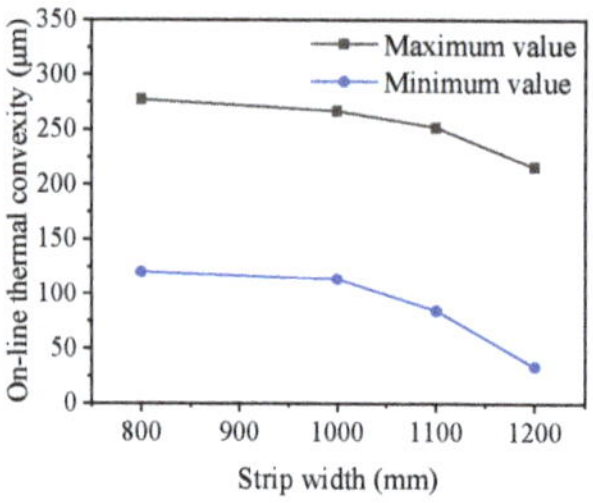

Figure 6. Thermal profile range of different strip widths.

As shown in Figure 6, for these strip widths of 800 mm, 1000 mm, 1100 mm, and 1200 mm, the corresponding ranges of the thermal profile were 120.6 μm to 277.5 μm, 113.9 μm to 267 μm, 84.8 μm to 252 μm, and 33.9 μm to 216 μm, respectively. Obviously, there was a significant variation in the thermal profiles. The partition-cooling capacity also changed with the variation in the strip width. For example, when the strip width was below 1000 mm, the cooling effectiveness of areas 1 and 5 would significantly decrease.

4.2. *Influence of the Inlet Water Flow on the Roll Profile*

The water-cooling process in hot rolling consisted of two parts: the inlet and the outlet. The inlet water cooling used the uniform flow cooling curve and the outlet water cooling realized the temperature control. Whether the inlet water cooling was turned on or not, by the different water volume of the outlet partition cooling, the roll thermal profile and surface temperature changes were analyzed, as shown in Figure 7.

As shown in Figure 7a, under the same distribution of the outlet water flow, when the inlet water was open, the temperatures at the center and edge of the roll were 58 °C and 37 °C, respectively, with a temperature difference of 21 °C. However, when the inlet water was closed, the temperatures at the center and edge of the roll were 72 °C and 41 °C, respectively, with a temperature difference of 31 °C. The inlet cooling overall affected the water coverage area and the cooling capacity of the roll, which directly influenced the thermal stability to a significant extent. As shown in Figure 7b, based on the cooling status of the inlet water, two types of exit water cooling curves were designed to analyze the comprehensive cooling capacity of the inlet and outlet cooling water. It was observed that, when the inlet cooling was open, the control range of the thermal profile was 34–191 μm. And, when the inlet cooling was closed, the adjustment range was 39.5–415 μm, with a minimum difference of 5.5 μm and a maximum difference of 224 μm. This indicated that the combined adjustment of inlet and outlet water flow had a significant impact on thermal profile. As shown in Figure 7c, during the adjustment process, the cooling at the inlet, combined with the cooling in area 3 of the partition cooling, significantly adjusted the overall range of the thermal profile. In the meantime, the other areas provided refined assistance in the adjustment process. When the inlet water is opened and the partition cooling water at the outlet is adjusted to the lowest, the roll temperature is 90 °C, because 90 °C is more sensitive to the cooling water, especially for the partition cooling of roll. When the inlet water is opened and the partition cooling water at the outlet is adjusted to the lowest, the roll temperature is 90 °C. The purpose is to obtain the corresponding relationship of the cooling effect. At the same time, the roll temperature is gradually reduced during the adjustment of the outlet water volume. The corresponding roll temperature is often set to about 60 °C when the outlet water volume is maximum in practical applications. The safe water quantity range was set to ensure that the roll temperature did not exceed the safety threshold of 90 °C. Based on this, the partition cooling control was implemented to assure the thermal profile of hot silicon steel rolling. At the same time, the roll temperature is gradually reduced during the adjustment of the outlet water volume. The corresponding roll temperature is about 60 °C when the outlet water volume is maximum. In practical applications, the safe water quantity range was often set to ensure that the roll temperature did not exceed 90 °C. Based on this, the partition cooling control was implemented for a more complex thermal profile of hot silicon steel rolling.

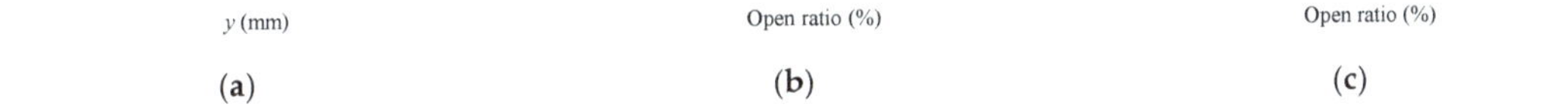

Figure 7. The temperature and the thermal profile: (**a**) offline roll temperature; (**b**) influence of the inlet water flow under different cooling curves; (**c**) relationship between the opening ratio of area 3 and the inlet water flow.

4.3. *Influence of Cooling Curve on Thermal Roll Profile*

Based on theoretical analysis and empirical observations, it was evident that partition cooling was highly effective in adjusting the thermal roll profile. To better meet the specifications for the shape, the convexity, and the local load roll gap, it is necessary to implement partition cooling control to achieve the specific thermal roll profile and the best shape. The diameters of upper roll and lower work were the same, although there was some difference in the practical process, the overall pattern was almost consistent. According to different water distribution curves under rolling parameters in Table 1, the set water opening ratios for each section are indicated in Table 2. The corresponding thermal profiles are illustrated in Figure 8.

Table 1. Rolling parameters.

Steel Grade	Width (mm)	Force (kN)	Rhythm (s)	Throughput	Roll Diameter (mm)
S14	1200	11,882	125	50	680

Table 2. Opening ratio of cooling water.

	Area 1 (%)	Area 2 (%)	Area 3 (%)	Area 4 (%)	Area 5 (%)
Experiment 1	100	100	32	100	100
Experiment 2	100	100	43	100	100
Experiment 3	100	100	56	100	100

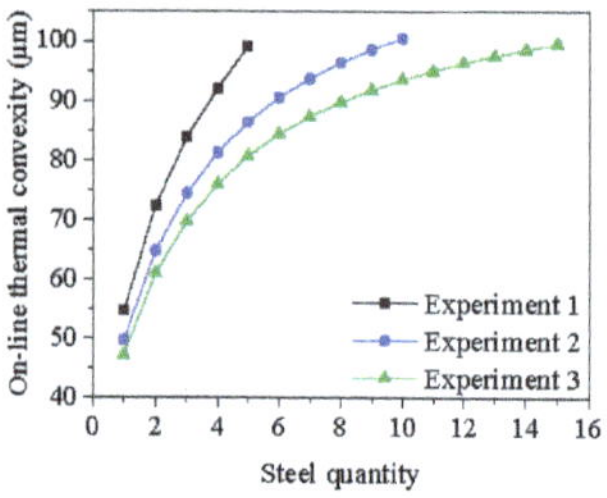

Figure 8. Different cooling curves.

In Figure 8, the stable thermal profile is formed in less time by different the partition-cooling modes. It was very meaningful to minimize fluctuations of rolling process. Based on this important rule, the better cooling curves should be designed to satisfy special requirements of the silicon steel rolling process.

5. Industry Application

The 1580 mill was mainly used for rolling silicon steel of 800~1250 mm. Considering the stability and safety of production, the F5 stand was selected to observe the influence of different partition-cooling curves on the thermal roll profile and the strip crown. The work roll diameter was 680 mm. The new cooling system was divided into upper and lower parts, corresponding to the upper and lower work roll, respectively. The cooling beam of 1249 kg was entirely made of stainless steel material. Each cooling beam needed 132 cooling nozzles. And the control system was equipped with five groups of the control valve, the flow meter, and the pressure gauges. The response time of the valve was within 10 s. The working pressure was in 0.1–1.8 MPa. Accuracy of thermometer was ±0.1 °C. Up to now, this new partition-cooling beam has been applied on the F5 stand for over a year, running stably, and has formed a cooling standard process. And further testing will be conducted on the cooling effectiveness of F6 and F7. By utilizing partition cooling to assist the mechanical control methods, traditional control strategies will be broken through and the local profile will be finely adjusted to achieve good indexes and minimal edge drop of the strip.

The production capacity and strip quality have greatly improved under different water flow conditions for the varied steel grade. Two typical tests were used as cases to verify the industrial application effect of partition cooling. Two sets of experiments are shown in Table 3, and the water quantities of all cooling regions are presented in Table 4. After the rolling process, offline measurements were taken for obtaining the temperature of the roll surface by a contact thermometer. The comparisons between the measured results and the calculated rules are depicted in Figure 9.

Table 3. Experimental rolling parameters.

	Steel Grade	Width (mm)	Exit Thickness (mm)	Rhythm (s)	Force (kN)	Throughput
Experiment 1	S14	1129	2.6	133	10804	56
Experiment 2	S14	1150	2.6	124	11647	60

Table 4. The opening ratio of each area of partition cooling.

	Area 1 (%)	Area 2 (%)	Area 3 (%)	Area 4 (%)	Area 5 (%)
Experiment 1	10	30	90	30	10
Experiment 2	70	80	40	80	70

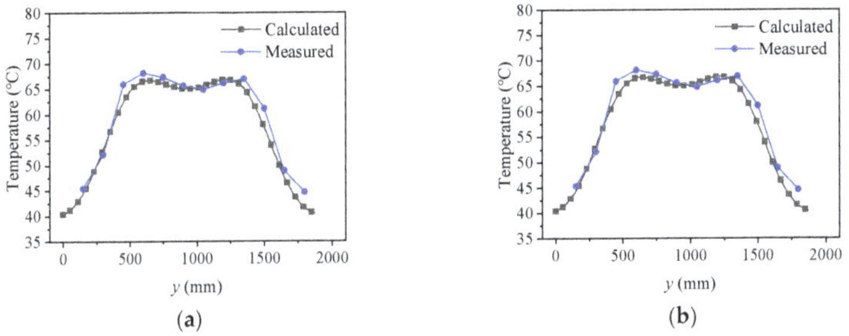

Figure 9. Surface temperatures of the offline roll: (a) Experiment 1; (b) Experiment 2.

As shown in Figure 9, the surface temperature varied obviously with different partition-cooling conditions. In Experiment 1, the measured temperature at the middle position of the roll was 65.7 °C, and the corresponding calculation result was 65 °C. In Experiment 2, the measured temperature at the middle position of the roll was 66.6 °C, and the calculation result was 66.2 °C. Taken overall, the error was within 5%, which demonstrated the new model had a good predictive performance. Before the modification, different temperatures were measured under 10 different operating conditions. After the modification, the same measurements were performed. These temperature differences are shown in Figure 10. It shows that the deviation of the original mode was stable at 7–12 °C but those of the modified mode were within 5 °C.

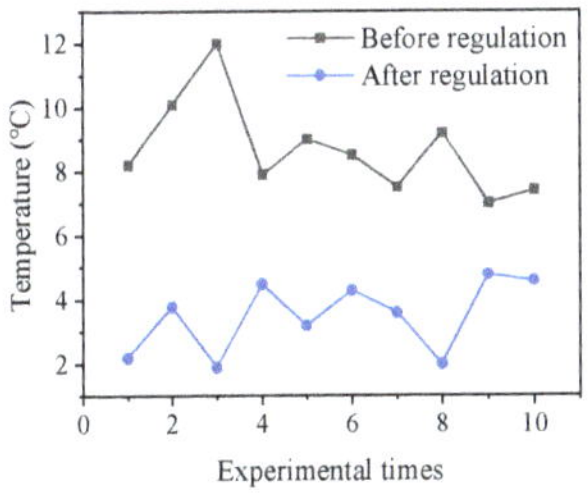

Figure 10. Temperature differences between upper and lower rolls before and after regulation.

Based on different rolling processes, the water flow rates for all areas were calibrated after the modification. The cooling modes were applied to the practical production process, as shown in Table 5. Through the optimization in each area, the temperature difference between the two sides of the roll had improved from 6.1 °C to 0.7 °C, as shown in Figure 11.

Table 5. The opening ratio of each area before and after the adjustment.

	Area 1 (%)	Area 2 (%)	Area 3 (%)	Area 4 (%)	Area 5 (%)
Before adjustment	70	80	100	80	70
After adjustment	56	74	100	86	84

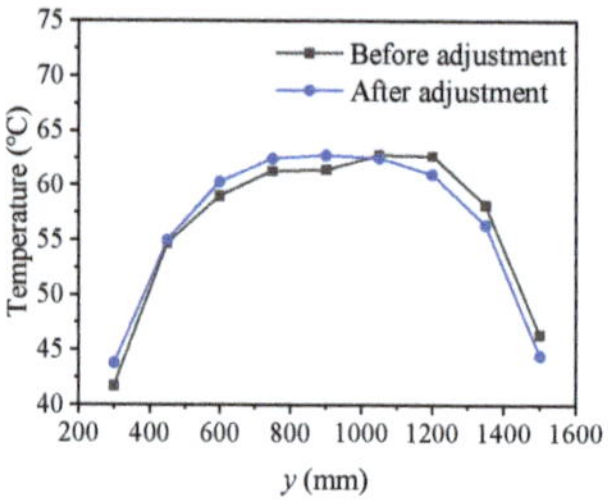

Figure 11. Comparison of roll temperature distribution before and after water adjustment.

By the new partition-cooling system of hot silicon steel, the dynamic control of the roll thermal expansion has been finished. In the practical process, a relatively small amount of water in the central region was able to be set to quickly establish the basic thermal roll profile in the early stages of rolling. It ensured a stable production state. Based on the process requirements, partition cooling was dynamically adjusted to achieve high-quality production. The C40 is the thickness difference between the middle of the strip and 40 mm from the edge of the strip, which is often measured by the thickness measuring system at the outlet of the production line. Now, the hitting rate of C40 has increased by 33%. It significantly reduces the quality fluctuation of the product, as shown in Figure 12.

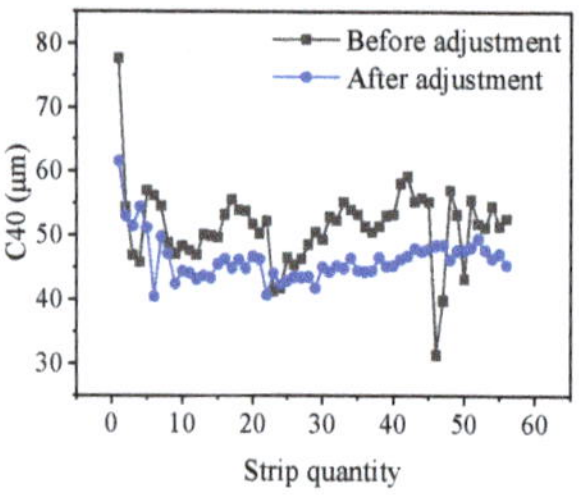

Figure 12. Comparisons of C40.

6. Conclusions

1. A new partition-cooling system of the hot rolling process has been designed to achieve online control of the thermal profile and the precise setting of the lateral roll gap. This enables rapid adjustments of the water flow and the distribution according to the rolling process goals, which provides an effective method for quickly adjusting the local roll gap and the strip shape or the edge drop.
2. An online temperature prediction model has been constructed for the partition-cooling process. Based on the practical production conditions and the partition-cooling modes, most cases were handled to achieve accurate temperature prediction by considering complex conditions in hot continuous rolling. According to the experimental results of partition cooling, the prediction error was within 5%.
3. After application of the partition-cooling system, the temperature difference between the upper and lower rolls was within 5 °C, the lateral edge temperature difference was within 0.7 °C, and the hit rate of C40 increased by 33%.

Author Contributions: Conceptualization, Q.W. and L.Y.; methodology, J.S.; software, Z.Z.; validation, L.D. and J.L.; formal analysis, Q.W.; investigation, L.D and H.W.; resources, Y.J. and H.W; data curation, Q.W.; writing—original draft preparation, J.Y.; writing—review and editing, L.Y.; supervision, J.S.; project administration, Q.W. All authors have read and agreed to the published version of the manuscript.

Funding: This research was funded by Returned Overseas Scholar Foundation of Hebei Province (Grant No. C20210321), Natural Science Foundation of Hebei Province (Grant No. E2021203106), and S&T Program of Hebei (Grant No. 236Z1019G).

Data Availability Statement: The original contributions presented in the study are included in the article, further inquiries can be directed to the corresponding authors.

Conflicts of Interest: Authors Qiuna Wang, Haishen Wang, Lijie Dong, Yanlong Jiao and Jieming Li were employed by the Shougang Company Limited. The remaining authors declare that the research was conducted in the absence of any commercial or financial relationships that could be construed as a potential conflict of interest. The Shougang Company and the funders had no role in the design of the study; in the collection, analyses, or interpretation of data; in the writing of the manuscript, or in the decision to publish the results.

References

1. Özgür, A.; Uygun, Y.; Hütt, M.-T. A Review of Planning and Scheduling Methods for Hot Rolling Mills in Steel Production. *Comput. Ind. Eng.* **2021**, *151*, 106606. [CrossRef]
2. Huang, Q. Research Progress on Key Equipment and Technology of High Quality Steel Plate and Strip Rolling. *J. Mech. Eng.* **2023**, *59*, 34–63.
3. Peng, Y.; Shi, B.; Liu, C.; Xing, J. Review of the Integrated Development of Strip Rolling Equipment process-product Quality Control. *J. Mech. Eng.* **2023**, *59*, 96–118.
4. Müller, M.; Prinz, K.; Steinboeck, A.; Schausberger, F.; Kugi, A. Adaptive Feedforward Thickness Control in Hot Strip Rolling with Oil Lubrication. *Control Eng. Pract.* **2020**, *103*, 104584. [CrossRef]
5. Shi, X.; Feng, D.; Sun, F.; Liu, X.; Zhang, F. Hor Rolling Equipments and Technical Progress for Electrical Steel in China. *Electr. Steel* **2022**, *4*, 6–15.
6. Shao, J.; He, A.; Yang, Q.; Jiang, L.; Yao, C.; Zhou, B. Multi-parameter Coupled Subsection Cooling Regulation Characteristics of Work Rolls in Aluminum Cold Rolling. *Chin. J. Eng.* **2015**, *37*, 1092–1097.
7. Zhang, Y.; Li, X.; Zhao, M.; Qu, F.; Zhang, Y.; Peng, W.; Zhao, D.; Di, H.; Zhang, D. Novel Analytical Heat Source Model for Cold Rolling Based on an Energy Method and Unified Yield Criterion. *Int. J. Adv. Manuf. Technol.* **2022**, *122*, 3725–3738. [CrossRef]
8. Wang, P.; Deng, J.; Li, X.; Hua, C.; Su, L.; Deng, G. A Novel Strategy Based on Machine Learning of Selective Cooling Control of Work Roll for Improvement of Cold Rolled Strip Flatness. *J. Intel. Manuf.* **2023**, *11*. [CrossRef]
9. Gao, S.; Liu, H.; Xi, A.; Yang, X. Closed-loop Control Strategy of Segmented Cooling in Hot Rolling of Aluminum Alloys. *J. Mech. Eng.* **2016**, *52*, 207–212. [CrossRef]
10. Guo, X.; He, A.; Shao, J.; Zhou, B.; Li, Q. Modeling and Simulation of Subsectional Cooling System during Hot Aluminum Rolling. *J. Mech. Eng.* **2013**, *49*, 70–74. [CrossRef]
11. Bohacek, J.; Raudensky, M.; Kotrbacek, P. Remote Cooling of Rolls in Hot Rolling; Applicability to Other Processes. *Metals* **2021**, *11*, 1061. [CrossRef]
12. Wang, X.; Zhang, W.; Ai, Y. Influence of Edge Thermal Spraying on the Temperature Field and Thermal Crown of Work Roll during Cold Rolling of Aluminum Alloy Strip. *Int. J. Adv. Manuf. Technol.* **2023**, *127*, 4331–4338. [CrossRef]
13. Yang, L.; Wang, D.; Yu, B.; Wang, Y. Research on the Quick Simulation Model for the Transient Temperature Field of the Hot Roll. *J. Plast. Eng.* **2010**, *17*, 123–128.
14. Gavalas, E.; Papaefthymiou, S. Thermal Camber and Temperature Evolution on Work Roll during Aluminum Hot Rolling. *Metals* **2020**, *10*, 1434. [CrossRef]
15. Han, G.; Li, H.; Zhang, J.; Kong, N.; Liu, Y.; You, X.; Xie, Y.; Shang, F. Prediction and Analysis of Rolling Process Temperature Field for Silicon Steel in Tandem Cold Rolling. *Int. J. Adv. Manuf. Technol.* **2021**, *115*, 1637–1655. [CrossRef]
16. Wu, H.; Sun, J.; Lu, X.; Peng, W.; Wang, Q.; Zhang, D. Predicting Stress and Flatness in Hot-Rolled Strips during Run-out Table Cooling. *J. Manuf. Process.* **2022**, *84*, 815–831. [CrossRef]
17. Hu, K.; Xue, R.; Shi, Q.; Han, W.; Zhu, F.; Chen, J. FEM Simulation of Thermo-Mechanical Stress and Thermal Fatigue Life Assessment of High-Speed Steel Work Rolls during Hot Strip Rolling Process. *J. Therm. Stress.* **2022**, *45*, 538–558. [CrossRef]
18. Li, Z.; Liu, L.; Yin, B.; Kuang, S.; Wang, J.; Bai, Z. Research on Roll Temperature Field and Hot Roll Crown of Hot Continuous Rolling Mills. *Iron Steel.* **2023**, *10*. [CrossRef]
19. Bao, Q.; Zhang, S.; Guo, J.; Xu, Z.; Zhang, Z. Modeling of Dynamic Data-Driven Approach for the Distributed Steel Rolling Heating Furnace Temperature Field. *Neural Comput. Applic.* **2022**, *34*, 8959–8975. [CrossRef]

20. Li, F.; He, A.; Song, Y.; Wang, Z.; Xu, X.; Zhang, S.; Qiang, Y.; Liu, C. Deep Learning for Predictive Mechanical Properties of Hot-Rolled Strip in Complex Manufacturing Systems. *Int. J. Min. Met. Mater.* **2023**, *30*, 1093–1103. [CrossRef]
21. Meng, L.; Ding, J.; Dong, Z.; Sun, J.; Zhang, D.; Gou, J. Prediction of Roll Wear and Thermal Expansion Based on Informer Network in Hot Rolling Process and Application in the Control of Crown and Thickness. *J. Manuf. Process.* **2023**, *103*, 248–260. [CrossRef]
22. Yang, L.; Jiang, Z.; Zhu, J.; Yu, H. Analysis of Transient Heat Source and Coupling Temperature Field during Cold Strip Rolling. *Int. J. Adv. Manuf. Technol.* **2018**, *95*, 835–846. [CrossRef]
23. Li, Y.; Cao, J.; Qiu, L.; Yang, G.; Kong, N.; He, A.; Zhou, Y. Effect of Strip Edge Temperature Drop of Electrical Steel on Profile and Flatness during Hot Rolling Process. *Adv Mech. Eng.* **2019**, *11*, 753305129. [CrossRef]
24. Chen, C.; Shao, J.; He, A. Research on Online Calculation Methods of Temperature Field of Hot Strip. *J. Mech. Eng.* **2014**, *50*, 135–142. [CrossRef]
25. Wang, L.; Ma, L.; Song, Z. Shape Control Model of Magnesium Alloy Single-stand Four-high Mill. *J. Netshape Form. Eng.* **2022**, *14*, 36–43.
26. Peng, G.; Xu, D.; Zhou, J.; Yang, Q.; Shen, W. A Novel Curve Pattern Recognition Framework for Hot-Rolling Slab Camber. *IEEE Trans. Ind. Inform.* **2023**, *19*, 1270–1278. [CrossRef]
27. Peng, G.; Cheng, Y.; Wang, H.; Shen, W. Industrial Big Data-driven Mechanical Performances Prediction for Hot-rolling Steel Using Lower Upper bound Estimation Method. *J. Manuf. Syst.* **2022**, *65*, 104–114. [CrossRef]

processes

MDPI

Article

Novel Multi-Criteria Group Decision Making Method for Production Scheduling Based on Group AHP and Cloud Model Enhanced TOPSIS

Xuejun Zhang [1], Zhimin Lv [2,*], Yang Liu [3], Xiong Xiao [4] and Dong Xu [1]

1 National Engineering Research Center of Flat Rolling Equipment, University of Science and Technology Beijing, Beijing 100083, China; zxj@ustb.edu.cn (X.Z.); xudong@ustb.edu.cn (D.X.)
2 Collaborative Innovation Center of Steel Technology, University of Science and Technology Beijing, Beijing 100083, China
3 Design and Research Institute Co., Ltd., University of Science and Technology Beijing, Beijing 100083, China; ustb_liuyang@163.com
4 National Engineering Research Center for Advanced Rolling and Intelligent Manufacturing, University of Science and Technology Beijing, Beijing 100083, China; docxxiao@ustb.edu.cn
* Correspondence: lvzhimin@nercar.ustb.edu.cn

Abstract: Optimized production scheduling can greatly improve efficiency and reduce waste in the steel manufacturing industry. With the increasing demands on the economy, the environment, and society, more and more factors need to be considered in the production scheduling process. Currently, only a few methods are developed for the comprehensive evaluation and prioritization of scheduling schemes. This paper proposes a novel MCGDM (multi-criteria group decision making) method for the ranking and selection of production scheduling schemes. First, a novel indicator system involving both qualitative and quantitative indicators is put forward. Diverse statistical methods and evaluation functions are proposed for the evaluation of quantitative indicators. The evaluation method of qualitative indicators is proposed based on heterogeneous data, cloud model theory, and group decision-making techniques. Then, a novel Group AHP model is proposed to determine the weights of all evaluation indicators. Finally, a novel cloud-model-enhanced TOPSIS (technique for order of preference by similarity to ideal solution) method is proposed to rank alternative production scheduling schemes. A practical example is presented to show the implementation details and demonstrate the feasibility of our proposed method. The results and comparative analysis indicate that our hybrid MCGDM method is more reasonable, flexible, practical, and effective in evaluating and ranking production scheduling schemes in an uncertain environment.

Keywords: production scheduling; multi-criteria group decision making; cloud model enhanced TOPSIS; Group AHP; evaluation indicator system

1. Introduction

With the increasing demand for steel products and extending production scale, production planning in a steel mill plant is becoming more complex [1]. Iron and steel enterprises are characterized by long production lines, multiple processes, and complex equipment. The full-flow manufacturing of steel generally includes ironmaking, steelmaking, hot rolling, cold rolling, and finishing processes, which is a complex process with a mixture of continuous and discrete production and manufacturing. The formulation of a steel production scheduling scheme is the key to the stability of steel production and the smooth flow of logistics [2]. Optimal scheduling of steelmaking production contributes to boosting productivity, reducing costs, and achieving sustainable manufacturing for an integrated steel company [3,4]. Therefore, in the field of steel production, it is often necessary to develop and prioritize scheduling schemes.

Steel scheduling is known to be one of the most difficult industrial scheduling problems [5]. Many approaches have been put forward to establish and optimize production planning and scheduling problems, such as the mathematical model method, the artificial intelligence method, and the parallel scheduling method [6]. Mathematical programming is a powerful tool for production planning. However, practical production scheduling problems may be extremely complex, with large numbers of variables, constraints, and dynamic changes. It is difficult to develop an optimization mathematical model that can be solved efficiently. Most models are relatively simple in the setting of optimization objectives and constraints and ignore the indicators that are difficult to quantify and the experts' empirical knowledge that is difficult to express. As a result, many models fail to reflect the actual production conditions, leading to insufficient application feasibility of the solutions. Moreover, most schedule schemes mainly focus on ensuring the continuity of production, increasing productivity, and reducing resource consumption with little consideration of environmental or even social factors such as waste emissions, let alone with comprehensive analysis of production efficiency and environmental and social factors, which results in most scheduling designs lagging behind the strategic development needs of steel plants. In addition, the schemes obtained using multi-objective optimization models are usually Pareto solutions, and it is necessary to select the optimal scheme through a rational comprehensive evaluation. What is more, in a production environment with inevitable dynamic disturbances, scheduling often needs to be adjusted and re-planned, and an effective evaluation system will benefit the optimization of the scheme.

Currently, only a few methods are employed to aid in the preference for an optimal scheduling scheme and almost no practical evaluation system is available for steel production scheduling schemes. With the increasing demands of manufacturing on the economy, the environment, and society, more and more factors need to be considered in the production scheduling process. It is particularly urgent to develop a comprehensive evaluation method for production scheduling under multiple indicators. Multi-criteria decision making (MCDM) methods have shown promising results in comprehensive evaluation and prioritization of scheduling schemes [7–9]. Recently, as a preliminary trial in production scheduling, Thenarasu et al. carried out the performance evaluation of a flexible job shop by employing discrete event simulation and hybrid MCDM approaches [7]. The weights of five indicators were assigned by a fuzzy analytic hierarchy process (AHP). Then, TOPSIS was used to prioritize the jobs. It is found that the MCDM-based rules perform better for large-scale problems by considering multiple realistic indicators. Carlucci proposed a decision model for scheduling in a job-shop manufacturing system that simultaneously deals with the power constraint and the variable speed of machine tools [10]. Gharoun presented a new bi-objective model to deal with an integrated production planning and reliability-based multi-level preventive maintenance scheduling problem using a hybrid multi-attribute decision making (MADM) approach to identify profitable customers [11].

As a branch of operations research, MCDM has been widely applied in decision-making processes and selection, ranking, and evaluation problems in many fields. However, some issues still need to be addressed such as the determination of different indicator weights and the process of various data types. Joshi and Kumar proposed a trapezium cloud model-based TOPSIS method to incorporate uncertainty, hesitancy, and inconsistency using both qualitative and quantitative information [12]. Morente-Molinera et al. proposed an MCDM method that is capable of working in heterogeneous and dynamic environments [13]. Liu et al. developed an integrated failure mode and effect analysis model based on cloud model theory and the hierarchical TOPSIS method to evaluate and rank the risk of failure modes [14]. Lourenzutti and Krohling introduced a generalization of the TOPSIS method, which is capable of processing multiple decision makers and different types of information (crisp, interval, fuzzy, and intuitionistic fuzzy numbers) [15]. Khan et al. developed an MCDM approach for ranking the effectiveness of COVID-19 tests based on the TOPSIS method and complex hesitant fuzzy sets [16]. More recently, Zheng et al.

proposed an MCDM method to evaluate the mental state of personnel by combining cloud model theory, AHP, and TOPSIS [17].

From the previous studies, it can be seen that (1) as a mathematical tool, MCDM is effective in comprehensively evaluating various alternative solutions based on multiple performance indicators. (2) The indicators of the MCDM evaluation system are usually multidimensional and complex, and the performance of the MCDM method largely depends on whether the indicator weights are set reasonably and whether heterogeneous data processing is effective. However, simple MCDM tools cannot solve these problems, and other models and tools must be combined to improve the effectiveness of the MCDM method. (3) AHP integrates the relationships between multiple indicators into a hierarchical structure, thereby determining the weights more accurately. (4) The data types reflected by multi-criteria decision evaluation indicators may be complex and diverse, such as precise numbers, interval numbers, linguistic terms, or expressions. Effective methods are also needed to support the processing of fuzzy information and uncertain data. Cloud model theory and TOPSIS have been widely studied to effectively handle fuzzy information and uncertain data and enhance the scientific nature of evaluation ranking. Undoubtedly, MCDM is appropriate for the evaluation of steel production scheduling schemes, but it has its limitations. To improve the effectiveness of MCDM, it is necessary to explore a Hybrid MCDM model.

This paper proposes a novel MCGDM method for production scheduling based on Group AHP and cloud-model-enhanced TOPSIS. Relative to previous work by other researchers, the major contribution of this study has the following novel features: (1) An evaluation system specifically for evaluating the production scheduling effect of steel plants is proposed, which evaluates and ranks production scheduling schemes, and provides some guidance for the optimization direction of scheduling schemes. (2) Evaluation indicators including both qualitative and quantitative indicators are built from multiple dimensions of economic, environmental, and social aspects with the consideration of key factors that affect the production process and results, which break through the limitations of most evaluation indicators that are limited to a single production process or a single indicator dimension. (3) A novel Group AHP is used for indicator weighting, which greatly enhances the rationality of group decision-making. (4) A novel cloud-model-enhanced TOPSIS is proposed to describe the natural uncertainty and manage various data types, and the final results contain both magnitude and uncertainty information. The proposed method enables a more scientific selection of the optimal scheduling scheme. In addition, it can assist in the optimization process of scheduling schemes by adjusting the scheme and then comparing the evaluation results.

The rest of this paper is organized as follows. Section 2 reviews some preliminaries related to cloud model theory, linguistic terms, and the conversion of heterogeneous data. In Section 3, the evaluation indicator system, Group AHP model, and cloud-model-enhanced TOPSIS are proposed in detail. An application example is presented in Section 4 to show the implementation details and demonstrate the feasibility of the proposed method. Furthermore, the discussion is given in Section 5. Finally, some concluding remarks and suggestions for future research are described in Section 6.

2. Preliminaries

In this section, we first review the cloud model theory. Then, we introduce some basic concepts and representation methods about linguistic terms and linguistic expressions. Finally, the conversion from heterogeneous data to its corresponding normal cloud model (NCM) is presented.

2.1. Cloud Model Theory

The cloud model is a mathematical model of uncertainty proposed based on fuzzy sets and probability theory. It offers a flexible computing tool that can effectively process fuzzy information and uncertain data. Thus far, NCM has been applied widely in various fields.

In the following part of this subsection, we review some basic concepts and operations of the NCM.

Definition 1 ([18]). *Let U be the universe of discourse and T be a qualitative concept in U. Expectation (Ex), Entropy (En), and Hyper-entropy (He) are three numerical characteristics of the qualitative concept T, where $En \geq 0$ and $He \geq 0$. If $x \in U$ is a random instantiation of T, which satisfies $x \sim N(Ex, En'^2)$, $En' \sim N(En, He^2)$, and the certainty degree of x belonging to T satisfies*

$$\mu_T(x) = \exp\left(-\frac{(x - Ex)^2}{2(En')^2}\right), \tag{1}$$

then the distribution of x in U is called as a NCM and each x with its certainty degree $\mu_T(x)$ is called as a cloud droplet of T.

Definition 2 ([19]). *Given any two NCMs $T_1 = (Ex_1, En_1, He_1)$ and $T_2 = (Ex_2, En_2, He_2)$ in the same universe of discourse, the basic operations are defined as follows:*

$$T_1 + T_2 = \left(Ex_1 + Ex_2, \sqrt{En_1^2 + En_2^2}, \sqrt{He_1^2 + He_2^2}\right), \tag{2}$$

$$T_1 - T_2 = \left(Ex_1 - Ex_2, \sqrt{En_1^2 + En_2^2}, \sqrt{He_1^2 + He_2^2}\right), \tag{3}$$

$$T_1 \times T_2 = \left(Ex_1 Ex_2, \sqrt{(En_1 Ex_2)^2 + (En_2 Ex_1)^2}, \sqrt{(He_1 Ex_2)^2 + (He_2 Ex_1)^2}\right), \tag{4}$$

$$T_1 / T_2 = \left(\frac{Ex_1}{Ex_2}, \sqrt{\left(\frac{En_1}{Ex_2}\right)^2 + \left(\frac{Ex_1 En_2}{Ex_2^2}\right)^2}, \sqrt{\left(\frac{He_1}{Ex_2}\right)^2 + \left(\frac{Ex_1 He_2}{Ex_2^2}\right)^2}\right). \tag{5}$$

Definition 3 ([19]). *Let $T_i = (Ex_i, En_i, He_i)$ $(i = 1, 2, \ldots, n)$ be n NCMs in the same universe of discourse. Then cloud synthetic operator f_{CS} is a mapping $T^n \to T$ as*

$$f_{CS}(T_1, T_2, \ldots, T_n) = \left(\frac{1}{n}\sum_{i=1}^{n} Ex_i, \frac{1}{6}\left(\max_i(Ex_i + 3En_i) - \min_j(Ex_j - 3En_j)\right), \sqrt{\sum_{i=1}^{n} He_i^2}\right). \tag{6}$$

Definition 4 ([19]). *Let $T_i = (Ex_i, En_i, He_i)$ $(i = 1, 2, \ldots, n)$ be a collection of NCMs in the same universe of discourse, and $w_i > 0$ is the weight of T_i. Then the cloud-weighted average operator f_{CWA} is a mapping $T^n \to T$ as*

$$f_{CWA}(T_1, T_2, \ldots, T_n) = \sum_{i=1}^{n} w_i T_i / \sum_{i=1}^{n} w_i. \tag{7}$$

According to the derivation by Yang et al. [19], if $w_i > 0$ is an exact number for all $i = 1, 2, \ldots, n$, and $\sum_{i=1}^{n} w_i = 1$, then Equation (7) can be simplified as

$$f_{CWA}(T_1, T_2, \ldots, T_n) = \left(\sum_{i=1}^{n} w_i Ex_i, \sqrt{\sum_{i=1}^{n} (w_i En_i)^2}, \sqrt{\sum_{i=1}^{n} (w_i He_i)^2}\right). \tag{8}$$

Definition 5. *Let $T_1 = (Ex_1, En_1, He_1)$ and $T_2 = (Ex_2, En_2, He_2)$ be two arbitrary NCMs in the same universe of discourse. Then the distance between them is defined as*

$$d(T_1, T_2) = |T_1 - T_2| = \left(|Ex_1 - Ex_2|, \sqrt{En_1^2 + En_2^2}, \sqrt{He_1^2 + He_2^2}\right). \tag{9}$$

Definition 6 ([20]). *Let $T_1 = (Ex_1, En_1, He_1)$ and $T_2 = (Ex_2, En_2, He_2)$ be two arbitrary NCMs in the same universe of discourse. Then the comparison method for NCMs is given as follows:*

(1) If $Ex_1 > Ex_2$, then $T_1 > T_2$;
(2) If $Ex_1 = Ex_2$ and $En_1 < En_2$, then $T_1 > T_2$;
(3) If $Ex_1 = Ex_2$, $En_1 = En_2$, and $He_1 < He_2$, then $T_1 > T_2$;
(4) If $Ex_1 = Ex_2$, $En_1 = En_2$, and $He_1 = He_2$, then $T_1 = T_2$.

2.2. Linguistic Terms

Qualitative assessment results are more suitable to be expressed as natural language. Thus, Zadeh proposed the concept of linguistic variables [21] and the idea of "computing with words" [22]. Linguistic terms are employed to handle cases that are excessively complex or inadequately defined to be adequately represented by quantitative expressions. Linguistic terms are usually modeled by fuzzy sets or NCMs. The NCM is a good tool to deal with the fuzziness and randomness of a linguistic term.

Peng et al. proposed an iterative self-organizing data analysis clustering algorithm combined with DDTW, and a structure dictionary learning-based method for multimode process monitoring; furthermore, a data-driven method was used to estimate the prediction interval of the attributes [23–26]. Shen et al. proposed a new manifold visualization method, SLISEMAP, to reduce the feature dimensions with interpretability [24].

Determining how to convert linguistic terms into NCMs is a challenging problem. Wang and Feng [27] used the golden ratio method to generate NCMs. Wang et al. [28] presented a linguistic scale function model to carry out the conversion between linguistic terms and NCMs. In general, words carry different interpretations for different individuals. Exploring methods to capture and represent such variations is a prominent subject of interest in academic research. For this reason, Yang et al. [29] applied fuzzy statistics and membership function fitting to encode 32 linguistic terms into NCMs based on a dataset collected from 175 individuals. A linguistic term is represented by an NCM as $T(Ex, En, He)$. Recently, nine of these linguistic terms have been applied to large-scale group decision making [30]. The present study also adopts the nine linguistic terms with various scales of uncertainties as listed in Table 1.

Table 1. Nine linguistic terms encoded by NCMs.

Linguistic Terms	NCMs
T_0: none (*n*)	(0.00, 0.100, 0.020)
T_1: very low (*vl*)	(0.197, 0.457, 0.022)
T_2: low	(1.246, 1.062, 0.034)
T_3: slightly low (*sl*)	(3.083, 1.030, 0.033)
T_4: medium (*med*)	(5.058, 1.109, 0.056)
T_5: slightly high (*sh*)	(7.105, 1.027, 0.041)
T_6: high	(8.743, 1.242, 0.041)
T_7: very high (*vh*)	(9.775, 1.054, 0.027)
T_8: maximum (*m*)	(10.00, 0.100, 0.020)

In addition, when conducting qualitative evaluations, a decision maker may hesitate among multiple linguistic terms or pursue a richer linguistic expression, such as "between medium and very high" or "greater than medium" to express his/her opinion. To manage such situations, Huang and Yang [31] introduced the concept of the hesitant cloud linguistic term set.

Definition 7 ([31]). *Let $H = \{T_i | i = 0, 1, \ldots, 2\tau,\ \tau \in N^*\}$ be a finite and totally ordered discrete linguistic term set, where T_i is a linguistic term encoded by an NCM (Ex_i, En_i, He_i). An HCLTS H_S is defined as one ordered finite subset of the consecutive linguistic terms in H.*

The comparative linguistic methods defined by context-free grammars G_H [32] are much closer to human beings' expressions, and thus can be used to elicit linguistic expressions.

Definition 8 ([32]). *Let G_H be a context-free grammar, and $H = \{T_0, \ldots, T_n\}$ be a linguistic term set. The elements of $G_H = (V_N, V_T, I, P)$ are defined as follows:*

V_N = {<primary term>, <composite term>, <unary relation>, <binary relation>, <conjunction>};
V_T = {lower than, greater than, at least, at most, between, and, $T_0, T_1, \ldots, T_n$};
$I \in V_N$;
P = {I ::= <primary term> | <composite term>,
<primary term> ::= $T_0 \mid T_1 \mid \ldots \mid T_n$,
<composite term> ::= <unary relation><primary term> |
<binary relation><primary term> <conjunction><primary term>,
<unary relation> ::= lower than | greater than | at least | at most,
<binary relation> ::= between,
<conjunction> ::= and}.

2.3. Conversion of Heterogeneous Data

In multi-criteria group decision making, diverse types of data representations are appropriate for the evaluations of different indicators, such as exact numbers, interval numbers, linguistic terms, or linguistic expressions. Thus, the evaluation results of different indicators or from different decision makers may have different representation forms. To process these heterogeneous results, the conversion methods of heterogeneous data proposed in [30,33] are used herein.

Both *En* and *He* of an exact number should be zero. Thus, the conversion from an exact number v to its corresponding NCM is defined as

$$v \rightarrow T(v, 0, 0) \tag{10}$$

According to the "3σ rule", approximately 99.7% of values are within the interval $[Ex - 3 \times En, Ex + 3 \times En]$. Thus, the conversion from an interval number $I = [I^L, I^U]$ to its corresponding NCM is defined as

$$I \rightarrow T\left(\frac{I^L + I^U}{2}, \frac{I^U - I^L}{6}, 0\right) \tag{11}$$

The preference formats of decision makers can be represented by the following four structures: utility functions, preference orderings, multiplicative preference relations, and fuzzy preference relations [34–36]. In this study, the preferences of decision makers are expressed as linguistic terms, which can be transformed into fuzzy functions. A decision maker can select one of the nine linguistic terms listed in Table 1 to express his/her evaluation result. Thus, the linguistic term t is converted into its corresponding NCM.

$$t \overset{f_{encoded}}{\rightarrow} T(Ex, En, He) \tag{12}$$

In Equation (12), the $f_{encoded}$ method represents the encoded method based on fuzzy statistics and membership function fitting.

A decision maker may provide an evaluation result using a linguistic expression. Let the linguistic term set H be defined as listed in Table 1. The context-free grammars G_H in Definition 8 are used to generate a linguistic expression. First, each linguistic expression is transformed into an HCLTS by the transformation function f_{G_H}.

$$\begin{cases} f_{G_H}(T_i) = \{T_i | T_i \in H\} \\ f_{G_H}(\text{at most } T_i) = \{T_j | T_j \in H \text{ and } T_j \leq T_i\} \\ f_{G_H}(\text{lower than } T_i) = \{T_j | T_j \in H \text{ and } T_j < T_i\} \\ f_{G_H}(\text{at least } T_i) = \{T_j | T_j \in H \text{ and } T_j \geq T_i\} \\ f_{G_H}(\text{greater than } T_i) = \{T_j | T_j \in H \text{ and } T_j > T_i\} \\ f_{G_H}(\text{between } T_i \text{ and } T_j) = \{T_k | T_k \in H \text{ and } T_i \leq T_k \leq T_j\} \end{cases} . \tag{13}$$

Then, the HCLTS is converted into its corresponding NCM using the cloud synthetic operator f_{CS} in Equation (6).

Therefore, the conversion from a linguistic expression L_t to its corresponding NCM is defined as

$$L_t \overset{f_{G_H}}{\rightarrow} HCLTS \overset{f_{CS}}{\rightarrow} T(Ex, En, He). \quad (14)$$

Example 1. *Let H be the set of the nine linguistic terms listed in Table 1. Suppose a decision maker represents his evaluation result as "between slightly high and very high" using context-free grammar. Then, this linguistic expression is transformed into an HCLTS as* $H_S = \{T_5, T_6, T_7\}$. *Eventually, its corresponding NCM is calculated by (6) as (8.5410, 1.4855, 0.0640).*

Consequently, exact numbers, interval numbers, linguistic terms, and linguistic expressions can all be converted into their corresponding NCMs. Furthermore, the uncertainties associated with heterogeneous data are preserved and propagated throughout the following calculations.

3. Methods

3.1. Indicator System

The comprehensive evaluation of production scheduling is a multi-criteria decision-making problem. An evaluation indicator system can be constructed to measure it. Based on the practices of several large-scale steel enterprises and the literature review on production scheduling in the steel industry in recent years, a total of 14 evaluation indicators are selected from three aspects: the economy, the environment, and society, as listed in Table 2. There are 8 indicators under the economic domain: total processing time (C_1), order production delay time (C_2), number of furnaces in the same casting plan (C_3), change in steel grade between adjacent furnaces (C_4), furnace residual capacity (C_5), balance of production capacity between processes (C_6), bottleneck process planning (C_7), and penalties for jumping production specifications (C_8). There are 3 indicators under the environmental domain: total energy consumption (C_9), consumption of tools (C_{10}), and waste discharge (C_{11}). There are 3 indicators under the social domain: job satisfaction (C_{12}), worker workload balance (C_{13}), and hazard exposure (C_{14}). These indicators can be categorized into two directions: benefit (+) and cost (−). For a benefit indicator, a higher value is better, while the cost indicators are the opposite. In terms of the availability of quantitative data in the evaluation process, these indicators can be categorized into two types: quantitative and qualitative. A quantitative indicator has clear quantitative data, while a qualitative indicator usually requires subjective evaluation by experts.

The total processing time (C_1) is the total time required from the start of production to the end of production, including the waiting time required for the replacement of tools.

$$C_1 = \sum_{j=1}^{g} (Tcc_j + Tch_j). \quad (15)$$

where g is the number of castings; Tcc_j denotes the casting time of the j-th casting; Tch_j denotes the time for replacement of tools after the j-th casting.

The smaller the value of C_1, the better it is.

The order production delay time (C_2) is the number of days that the completion of the production date exceeds the contractual delivery date.

$$X_r = \max(0, P_r - D_r). \quad (16)$$

where P_r and D_r are the completion of production date and the contractual delivery date of the r-th order, respectively.

Table 2. Evaluation indicators and their properties for production scheduling evaluation.

Level I	Level II	Unit	Direction	Type
economy	total processing time (C_1)	minute	−	quantitative
	order production delay time (C_2)	day	−	quantitative
	number of furnaces in the same casting plan (C_3)	amount	+	quantitative
	change in steel grade between adjacent furnaces (C_4)	amount	−	quantitative
	furnace residual capacity (C_5)	ton	−	quantitative
	balance of production capacity between processes (C_6)	ton	−	quantitative
	bottleneck process planning (C_7)	ton	−	quantitative
	penalties for jumping production specifications (C_8)	amount	−	quantitative
environment	total energy consumption (C_9)	score	−	qualitative
	consumption of tools (C_{10})	score	−	qualitative
	waste discharge (C_{11})	score	−	qualitative
society	job satisfaction (C_{12})	score	+	qualitative
	worker workload balance (C_{13})	score	+	qualitative
	hazard exposure (C_{14})	score	−	qualitative

To reflect the complexity of reality, a nonlinear segmented function is used to calculate the penalty value of X_r.

$$f(X_r) = \begin{cases} 0 & X_r \leq T_{\min} \\ \left(\dfrac{X_r - T_{\min}}{T_{\max} - T_{\min}}\right)^2 & T_{\min} < X_r < T_{\max} \\ 1 & X_r \geq T_{\max} \end{cases}. \tag{17}$$

where $T_{\min}$ denotes the tolerable non-punishable overdue date and $T_{\max}$ denotes the maximum allowable overdue date.

The total penalty value of C_2 can be calculated as

$$f(C_2) = \sum_{r=1}^{R} f(X_r). \tag{18}$$

where R is the number of orders.

The smaller the calculated total penalty value $f(C_2)$, the better it is.

The number of furnaces in the same casting plan (C_3) denotes the number of steelmaking furnaces in one casting plan. The number of furnaces in one casting must not exceed the maximum permissible value $N_{\max}$. The evaluation of C_3 can be expressed as

$$f(C_3) = \frac{1}{g}\sum_{j=1}^{g}\left(\frac{N_j}{N_{\max}}\right)^2. \tag{19}$$

where g is the number of castings and N_j denotes the number of furnaces in the j-th casting.

The larger the calculated value of $f(C_3)$, the better it is.

The change in steel grade between adjacent furnaces (C_4) is the total steel grade change between all adjacent furnaces.

$$C_4 = \sum_{j=1}^{g}\sum_{i=2}^{n_j}\left|G_{ij} - G_{(i-1)j}\right|. \tag{20}$$

where g is the number of castings; n_j is the number of furnaces in the j-th casting; G_{ij} denotes the steel grade of the i-th furnace of the j-th casting.

The smaller the value of C_4, the better it is.

The furnace residual capacity (C_5) is the sum of the residual capacity of all furnaces.

$$C_5 = \sum_{j=1}^{g} \sum_{i=1}^{n_j} \left| Mc - W_{ij} \right|. \tag{21}$$

where g is the number of castings; n_j is the number of furnaces in the j-th casting; Mc denotes the maximum permissible capacity; W_{ij} denotes the capacity of the i-th furnace of the j th casting.

The smaller the value of C_5, the better it is.

The balance of production capacity between processes (C_6) is the sum of the absolute value of the difference between the production capacity and the planned production of each unit.

$$C_6 = \sum_{d=1}^{D} \sum_{m=1}^{M} \left| Ca_m - Wa_{md} \right|. \tag{22}$$

where D is the number of days of production; M is the number of units; Ca_m denotes the daily production capacity of the m-th unit; Wa_{md} denotes the planned production of the m-th unit on the d-th day.

The smaller the value of C_6, the better it is.

The bottleneck process planning (C_7) is the sum of the absolute value of the difference between the production capacity and the planned production of the bottleneck process.

$$C_7 = \sum_{d=1}^{D} \left| B - Wb_d \right|. \tag{23}$$

where D is the number of days of production; B denotes the daily production capacity of the bottleneck process; Wb_d denotes the planned production of the bottleneck process on the d-th day.

The smaller the value of C_7, the better it is.

Penalties for jumping production specifications (C_8) include width jumping penalties FW and thickness jumping penalties FH of adjacent productions.

$$C_8 = FW + FH. \tag{24}$$

The FW indicates the sum of width penalty values generated by adjacent slab rolling width jumps throughout the rolling program.

$$FW = \sum_{s=2}^{Q} f_W(W_s - W_{s-1}). \tag{25}$$

where Q is the total number of blocks rolled; W_s and W_{s-1} denote the widths of the s-th and ($s-1$)-th rolled steel, respectively; f_W is a segmented linear function denoted as

$$f_W(x) = \begin{cases} \mu_1 x & 0.5W_{\max} < x \le W_{\max} \\ \mu_2 x & 0.05W_{\max} < x \le 0.5W_{\max} \\ \mu_3 x & 0 < x \le 0.05W_{\max} \\ -\mu_4 x & -0.05W_{\max} < x \le 0 \\ -\mu_5 x & -0.5W_{\max} < x \le -0.05W_{\max} \\ -\mu_6 x & -W_{\max} \le x \le -0.5W_{\max} \end{cases}. \tag{26}$$

where $W_{\max}$ denotes the maximum permissible jumping width; the values of penalty coefficients μ_1, μ_2, μ_3, μ_4, μ_5, and μ_6 are 1.5, 1.2, 1.0, 0.6, 0.5, and 0.4, respectively.

Similarly, the thickness jumping penalties FH can be calculated, and the segmented linear function f_H for thickness jumping penalties is denoted as

$$f_H(x) = \begin{cases} -\eta_1 \frac{H_L}{H_s} x & x < 0, H_s < H_L \\ \eta_2 \frac{H_L}{H_s} x & x \geq 0, H_s < H_L \\ -\eta_3 \frac{H_m}{H_s} x & x < 0, H_s \geq H_L \\ \eta_4 \frac{H_m}{H_s} x & x \geq 0, H_s \geq H_L \end{cases}. \tag{27}$$

where H_m denotes the maximum thickness of steel that can be produced; H_L denotes the critical sensitive thickness of steel; H_s denotes the thickness of the s-th rolled steel; the values of penalty coefficients η_1, η_2, η_3, and η_4 are 200, 20, 18, and 18, respectively.

The smaller the calculated penalty value of C_8, the better it is.

The indicators of total energy consumption (C_9), consumption of tools (C_{10}), and waste discharge (C_{11}) are related to the organization and implementation of the production plan, hot delivery and hot loading, and many other factors. Thus, these indicators are difficult to directly quantify and usually require qualitative empirical evaluations using group decision-making techniques.

The steelmaking process is a process of physicochemical changes in molten iron at high temperatures, and the whole process is carried out at high temperatures. The operation process requires the use of high-voltage electricity, high-pressure water, high-pressure oxygen, large-tonnage trucks, and other high-pressure systems. It will produce a lot of smoke, harmful gases, and noise, polluting the environment and deteriorating labor conditions. Reasonable production scheduling can reduce transportation route intersections and reduce the risk of extreme loads on equipment, etc. For the indicators under the social domain, such as job satisfaction (C_{12}), worker workload balance (C_{13}), and hazard exposure (C_{14}), the relevant conditions and boundaries are difficult to quantitatively define. Thus, these indicators also need to be qualitatively evaluated with the help of experienced experts.

3.2. Weighting Algorithm Based on Group AHP

After constructing the indicator system, it is crucial to determine the weights of these indicators since each indicator holds varying degrees of importance. The AHP proposed by Saaty [37] has been widely applied for determining the weights of different indicators, using pairwise comparisons based on the pyramidal structure of the indicator system. Due to the cognitive limitations or personal biases of a single decision maker, group decision making provides a means for tapping into the "wisdom of the crowd". Group AHP is an extension of the traditional AHP in the context of group decision making, allowing for the participation of multiple decision makers. There are two fundamentally accepted aggregation procedures within Group AHP [38]. One is the aggregation of individual judgments (AIJ) [39,40], and the other is the aggregation of individual priorities (AIP) [41]. Saaty pointed out that if the decision makers are experts, they may wish to aggregate only their final weight vectors (priorities) but not their judgments [42].

Following Saaty's suggestions, the AIP method is used herein. Firstly, a pairwise comparison matrix (PCM) is elicited from each decision maker. Then, the individual weight vector of independent PCM is determined. Lastly, the weight vectors of all decision makers are aggregated to form the group decision result.

Let $\boldsymbol{C} = \{c_1, c_2, \ldots, c_n\}$ be a finite set of n indicators, where c_i denotes the ith indicator. The individual PCM on $\boldsymbol{C}$ is represented as an $n \times n$ matrix $\boldsymbol{A} = (a_{ij})_{n\times n}$, where $a_{ij} = 1/a_{ji}$ and a_{ij} belongs to Saaty's 1–9 fundamental ratio scale representing the relative importance dominance of c_i over c_j. The weight vector can be determined using the eigenvalue method. The normalized eigenvector corresponding to the largest eigenvalue of the matrix $\boldsymbol{A}$ represents the relative weights of n indicators.

A PCM is perfectly consistent if and only if $a_{ij} = a_{ik}a_{kj}$ for $\forall i,j,k$, $1 \leq i,j,k \leq n$. In real-life decision-making situations, it is hard to achieve complete consistency [43]. Moreover, there can be significant differences in the degree of consistency for different PCMs. Thus, the Consistency Ratio (CR) is defined to describe and reflect the quality of a PCM $\boldsymbol{A}$.

$$CR = \frac{CI}{RI}, \tag{28}$$

where CI is the Consistency Index (CI) of $\boldsymbol{A}$ and RI is the average random CI derived from a large number of randomly generated $n \times n$ PCMs (see Table 3).

$$CI = \frac{\lambda_{\max} - n}{n - 1}, \tag{29}$$

where $\lambda_{\max} \geq n$ is the largest eigenvalue of $\boldsymbol{A}$.

Table 3. The RI with different order n.

n	1	2	3	4	5	6	7	8	9	10
RI	0	0	0.52	0.89	1.11	1.25	1.35	1.40	1.45	1.49

A smaller CR indicates a higher quality of PCM. In general, if $CR < 0.10$, then PCM $\boldsymbol{A}$ is acceptably consistent. Otherwise, if CR is greater than 0.10, then the credibility of the PCM is questionable.

Let $\boldsymbol{D} = \{DM_1, DM_2, \ldots, DM_m\}$ be a finite set of m decision makers. The PCM of DM_d is represented as $\boldsymbol{A}^d$, $d = 1, 2, \ldots, m$. The largest eigenvalue of the matrix $\boldsymbol{A}^d$ is calculated as $\lambda^d_{\max}$. Thus, the CI and CR of $\boldsymbol{A}^d$ are calculated as CI^d and CR^d by (29) and (28). The weight vector of DM_d represented as $\boldsymbol{w}^d = [w^d_1, w^d_2, \ldots, w^d_n]$ is the normalized eigenvector corresponding to the largest eigenvalue $\lambda^d_{\max}$ of the matrix $\boldsymbol{A}^d$.

Previously, many papers assigned equal weights to each decision maker in Group AHP models. However, it could be more reasonable to assign different weights to different decision makers according to the quality of their PCMs. Assigning the weight to each decision maker within a group is a key component in the group decision process and plays an important role in obtaining the final group decision result in the Group AHP model. Thus, a method for calculating a decision maker's weight ρ^d using CR^d is proposed in this paper.

$$\rho^d = \frac{\max(0, 0.1 - CR^d)}{\sum\limits_{d=1}^{m} \max(0, 0.1 - CR^d)}, \tag{30}$$

where CR^d ($d = 1, 2, \ldots, m$) is the CR of the matrix $\boldsymbol{A}^d$ from DM_d.

The weight of indicator c_i represented as w_i is computed by the weighted arithmetic mean of m decision makers' weight vectors.

$$w_i = \sum_{d=1}^{m} \rho^d w^d_i, \tag{31}$$

where w^d_i ($i = 1, 2, \ldots, n$, $d = 1, 2, \ldots, m$) is the weight of indicator c_i from the weight vector $\boldsymbol{w}^d$ of DM_d. Eventually, $\boldsymbol{w} = [w_1, w_2, \ldots, w_n]$ is the weight vector of n indicators using Group AHP.

Based on the aforementioned ideas, the algorithm for assigning weights of production scheduling evaluation indicators is described in Algorithm 1.

Algorithm 1: Group AHP for Production Scheduling Evaluation Indicators Weighting.

1: Collect PCMs $\{(A^d)_{n\times n}\}$ on n indicators from m decision makers (d = 1,2, ..., m).
2: ***for*** $d = 1$ ***to*** m ***do***
3: Calculate the largest eigenvalue $\lambda^d_{\max}$ and its normalized eigenvector w^d of A^d.
4: Calculate CR^d of A^d by Equation (28).
5: ***end for***
6: ***for*** $d = 1$ ***to*** m ***do***
7: Calculate each decision maker's weight ρ^d by Equation (30).
8: ***end for***
9: ***for*** $i = 1$ ***to*** n ***do***
10: Calculate the weight of indicator c_i represented as w_i by Equation (31).
11: ***end for***
12: Output the weight vector of n indicators as $w = \{w_1, w_2, \ldots, w_n\}$.

Based on the pyramid structure of the indicator system, PCMs are conducted on indicators belonging to the same parent indicator at each level to obtain the weight vector for these indicators. Therefore, the weights of the bottom-level indicators can be obtained by multiplying the weight vectors layer by layer from the top to the bottom.

3.3. *Cloud Model Enhanced TOPSIS*

3.3.1. Problem Description

The evaluation and selection process of production scheduling schemes is a multi-stakeholder and multi-criteria decision-making problem. Let the alternative production scheduling scheme set be $\boldsymbol{A} = \{A_1, A_2, \ldots, A_s\}$ ($s \geq 2$), the indicator set be $\boldsymbol{C} = \{C_1, C_2, \ldots, C_n\}$, and the decision maker set be $\boldsymbol{D} = \{DM_1, DM_2, \ldots, DM_m\}$. The weights of indicators can be calculated by the Group AHP method as described in Section 3.2. Let $\boldsymbol{X} = [x_{ij}]_{s\times n}$ be the matrix that represents the performance of the scheme set $\boldsymbol{A}$ on the indicator set $\boldsymbol{C}$, as shown in (32).

$$\mathbf{X} = \left[x_{ij}\right]_{s\times n} = \begin{bmatrix} x_{11} & x_{12} & \cdots & x_{1n} \\ x_{21} & x_{22} & \cdots & x_{2n} \\ \vdots & \vdots & \vdots & \vdots \\ x_{s1} & x_{s2} & \cdots & x_{sn} \end{bmatrix} \tag{32}$$

where x_{ij} is the performance evaluation value of scheme A_i on indicator C_j.

Remark 1. *The performance evaluation value x_{ij} can be a precisely quantified numerical value obtained through statistical calculations or some qualitative evaluation data from a group of stakeholders. For a qualitative indicator C_j, the evaluation data $\boldsymbol{x}_{ij} = \left\{x^1_{ij}, x^2_{ij}, \ldots, x^{n_j}_{ij}\right\}$ are derived from multiple stakeholders utilizing group decision making techniques. Additionally, each stakeholder can use an exact number, an interval number, a linguistic term, or a linguistic expression to represent his/her opinion on indicator C_j. Thus, the evaluation data on a qualitative indicator may be multiple heterogeneous data. This is significantly different from reference [8].*

Based on the problem description, the following subsection aims to address the established multi-stakeholder and multi-criteria decision-making problem using a novel hybrid MCGDM method.

3.3.2. Main Steps of the Proposed Method

In this subsection, a novel hybrid MCGDM method based on group decision making principles, cloud model theory, and the TOPSIS method is developed to prioritize alternative production scheduling schemes. The proposed method is composed of three main stages and its flowchart is depicted in Figure 1. Firstly, the indicator system is constructed and the evaluation results of indicators are calculated. Secondly, the weights of indicators are derived based on the Group AHP method. Finally, the ranking of alternative production scheduling schemes is determined by implementing a cloud-model-enhanced TOPSIS method.

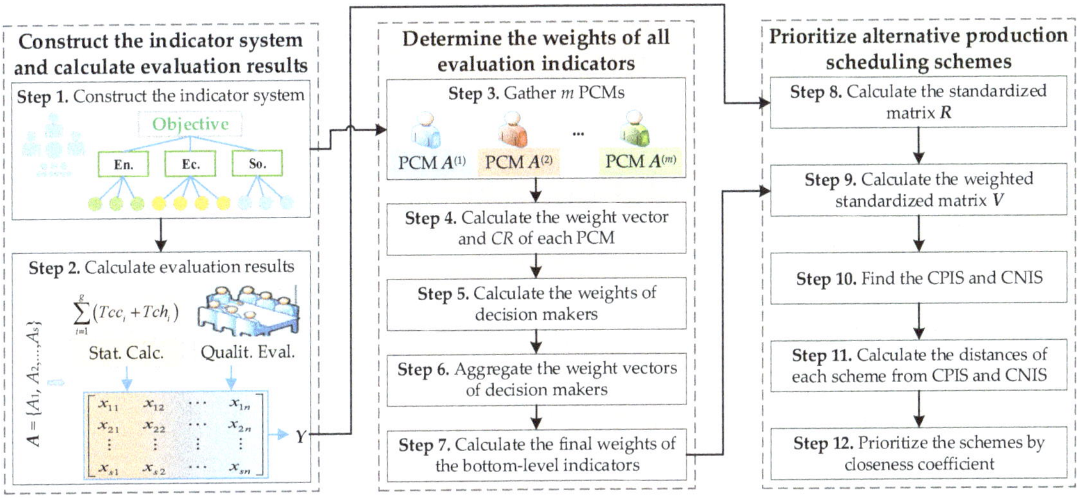

Figure 1. Flowchart of the proposed method for prioritizing production scheduling schemes.

Stage 1: Construct the indicator system and calculate evaluation results.

Step 1. Construct the indicator system for the comprehensive evaluation of alternative production scheduling schemes.

As aforementioned in Section 3.1, the evaluation indicators can be selected from three aspects: the economy, society, and the environment. The board of directors or the decision-making committee of the production planning department determines the final evaluation indicator system according to the practical situation.

Step 2. Calculate evaluation results of all quantitative and qualitative indicators.

For a quantitative indicator, the evaluation result x_{ij} can be calculated using a statistical method or an evaluation function as described in Section 3.1. The results of indicators C_1, C_4, C_5, C_6, and C_7 are directly obtained through statistical calculations, while the results of indicators C_2, C_3, and C_8 are calculated by evaluation functions.

For a qualitative indicator, the evaluation results are typically derived through the involvement of multiple stakeholders utilizing group decision-making techniques. Different indicators can be evaluated by distinct groups of decision makers. For instance, indicators related to society are primarily evaluated by decision makers responsible for social concerns, while indicators related to the environment are primarily evaluated by decision makers responsible for environmental concerns. As described in Section 3.3.1, the evaluation data of a qualitative indicator C_j on A_i can be represented as a set $x_{ij} = \left\{x_{ij}^1, x_{ij}^2, \ldots, x_{ij}^{n_j}\right\}$, where x_{ij}^k provided by the k-th decision maker can be an exact number, an interval number, a linguistic term, or a linguistic expression. Firstly, heterogeneous evaluation data from multiple decision makers are transformed into NCMs using the methods described in Section 2.3. Then, multiple NCMs are aggregated using the cloud-weighted average operator in (8) to calculate the evaluation result of the qualitative indicator. The indicators C_9~C_{14} are evaluated by experts using group decision-making techniques.

In order to represent evaluation results in a uniform manner, evaluation results of quantitative indicators are transformed into NCMs by (10). Therefore, the evaluation result of each quantitative or qualitative indicator is uniformly represented as an NCM. The original performance matrix **X** is transformed into a matrix **Y** represented by NCMs.

$$\boldsymbol{Y} = \left[y_{ij}\right]_{s\times n} = \begin{bmatrix} y_{11} & y_{12} & \cdots & y_{1n} \\ y_{21} & y_{22} & \cdots & y_{2n} \\ \vdots & \vdots & \vdots & \vdots \\ y_{s1} & y_{s2} & \cdots & y_{sn} \end{bmatrix}, \tag{33}$$

where y_{ij} is the evaluation result of indicator C_j on A_i represented by an NCM.

Stage 2: Determine the weights of all evaluation indicators.

Step 3. Gather PCMs from m decision makers.

Based on the indicator system, each decision maker provides a PCM for each level of indicator belonging to the same parent indicator. Saaty's classic 1~9 ratio scale method is used for pairwise comparison of every two indicators belonging to the same parent indicator.

Step 4. Calculate the weight vector and *CR* of each PCM.

The weight vector of each PCM is calculated using the eigenvalue method. The *CR* of each PCM is calculated by (28).

Step 5. Calculate the weights of decision makers.

The weight of each decision maker is calculated by (30) based on the *CR* of decision makers' PCMs.

Step 6. Aggregate the weight vectors of decision makers.

The group result is calculated by aggregating m weight vectors using the weighted arithmetic mean method. Therefore, the weight vector of multiple indicators belonging to the same parent indicator at each level can be calculated.

Step 7. Calculate the final weights of the bottom-level indicators.

After obtaining the relative weights of each level's indicators, the weights of the bottom-level indicators are calculated by multiplying the weights layer by layer from the top to the bottom.

Stage 3: Prioritize alternative production scheduling schemes.

Step 8. Calculate the standardized matrix $\boldsymbol{R} = [r_{ij}]$.

In order to obtain accurate evaluation results, performance scales need to be standardized. This involves converting the values of each indicator to a range between 0 and 1, thereby eliminating the influence of different measurement scales.

For benefit indicators (where a higher value is better):

$$r_{ij} = \frac{y_{ij} - \min\limits_i(y_{ij})}{\max\limits_i(y_{ij}) - \min\limits_i(y_{ij})}. \tag{34}$$

For cost indicators (where a lower value is better):

$$r_{ij} = \frac{\max\limits_i(y_{ij}) - y_{ij}}{\max\limits_i(y_{ij}) - \min\limits_i(y_{ij})}. \tag{35}$$

Step 9. Calculate the weighted standardized matrix $\boldsymbol{V} = [r_{ij}]$.

$$v_{ij} = w_j \times r_{ij}. \tag{36}$$

where w_j is the weight of indicator C_j.

Step 10. Find the cloud-positive ideal solution (CPIS) and the cloud-negative ideal solution (CNIS) in the weighted standardized matrix $\boldsymbol{V}$.

CPIS A^+ denotes maximizing the scores of all indicators, and CNIS A^- denotes minimizing the scores of all indicators.

$$\begin{aligned} A^+ &= \{\max v_{i1}, \max v_{i2}, \ldots, \max v_{in}\} \\ A^- &= \{\min v_{i1}, \min v_{i2}, \ldots, \min v_{in}\} \end{aligned} . \tag{37}$$

Step 11. Calculate the distances of each scheme from CPIS and CNIS.

$$\begin{aligned} d_i^+ &= \sum_{j=1}^{n} d(v_{ij}, v_j^+) \\ d_i^- &= \sum_{j=1}^{n} d(v_{ij}, v_j^-) \end{aligned} . \tag{38}$$

Step 12. Prioritize the schemes by closeness coefficient.
The closeness coefficient CC_i of each scheme is calculated by

$$CC_i = \frac{d_i^-}{d_i^+ + d_i^-}. \tag{39}$$

The larger the value of CC_i, the better the scheme A_i performs. As a result, the ranking order of all schemes is able to be determined according to their closeness coefficients CC_i $(i = 1, 2, \ldots, s)$ in descending order.

4. An Application Example

In this section, an application example of prioritizing production scheduling schemes adapted from the real world is presented to show the implementation details and demonstrate the feasibility of the proposed method. Steel scheduling is the sequencing and planning of steel ladle under conditions that satisfy process constraints as well as equipment requirements to ensure the quality of the product and the efficiency of production. Among them, process constraints mainly include width, hardness, and temperature jumps between neighboring ladles. Figure 2 presents a steel scheduling scheme, which shows the sequence and timing of different ladles being processed on different types of equipment.

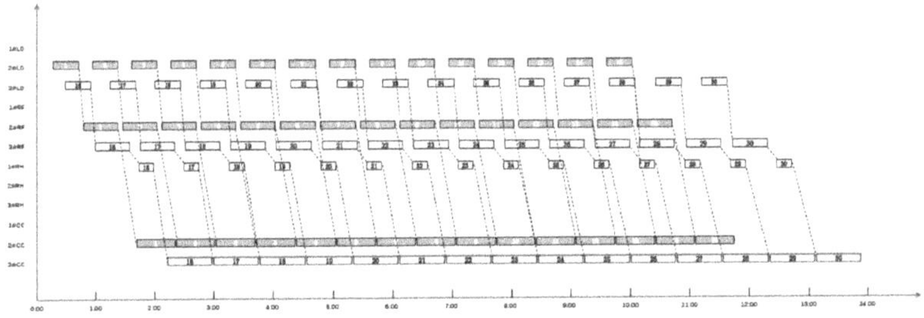

Figure 2. The schematic diagram of a steel scheduling scheme.

4.1. Obtain the Evaluation Matrix

The data are derived and adapted from a practical production scheduling of a steel plant. Five competitive scheduling schemes A_1, A_2, A_3, A_4, and A_5 were obtained using several different mathematical optimization models, simulation methods, and expert or rule-based scheduling methods. Using the methods described in Section 3.1, the performance of 5 schemes on 14 indicators was obtained as shown in Table 4. The performance of quantitative indicators C_1~C_8 was calculated by (15)~(27). The performances of qualitative indicators C_9~C_{11} were evaluated by three decision makers responsible for environmental concerns using heterogeneous data. The performance of qualitative indicators C_{12}~C_{14} was evaluated by four decision makers responsible for social concerns using heterogeneous data.

Table 4. The performance of 5 schemes on 14 indicators.

	C_1	C_2	C_3	C_4	C_5	C_6	C_7	C_8
A_1	1366	1.85	0.85	8	166	3250	785	2356
A_2	1278	0	0.78	7	138	2886	904	1968
A_3	1445	0.86	0.66	11	155	3763	1048	2087
A_4	1210	0	0.82	6	147	2596	816	1789
A_5	1312	0.42	0.72	9	176	3015	953	2523

	C_9	C_{10}	C_{11}
A_1	DM_1: between *med* and *high* DM_2: at least *med* DM_3: 5.5	DM_1: *med* DM_2: at least *med* DM_3: *med*	DM_1: *sl* DM_2: *med* DM_3: *sl*
A_2	DM_1: *low* DM_2: between *low* and *med* DM_3: 3.5	DM_1: *sl* DM_2: *med* DM_3: *med*	DM_1: *low* DM_2: *sl* DM_3: *low*
A_3	DM_1: *high* DM_2: [8.5,9.5] DM_3: 7	DM_1: *med* DM_2: *sh* DM_3: *sh*	DM_1: *med* DM_2: *med* DM_3: *sh*
A_4	DM_1: *sl* DM_2: *med* DM_3: *sl*	DM_1: *low* DM_2: *sl* DM_3: *sl*	DM_1: *low* DM_2: *sl* DM_3: *sl*
A_5	DM_1: at least *sh* DM_2: *med* DM_3: between *med* and *vh*	DM_1: *med* DM_2: *med* DM_3: *med*	DM_1: *sl* DM_2: *sl* DM_3: *med*

	C_{12}	C_{13}	C_{14}
A_1	DM_1: between *med* and *high* DM_2: at least *med* DM_3: *med* DM_4: *med*	DM_1: *med* DM_2: at least *med* DM_3: 6.5 DM_4: at least *med*	DM_1: *low* DM_2: *sl* DM_3: *low* DM_4: 2
A_2	DM_1: 8 DM_2: [7,9] DM_3: greater than *med* DM_4: *med*	DM_1: *med* DM_2: [6,7] DM_3: greater than *med* DM_4: [7,9]	DM_1: *low* DM_2: *low* DM_3: *vl* DM_4: 1
A_3	DM_1: [5,6] DM_2: [4.5,5.5] DM_3: between *low* and *med* DM_4: *low*	DM_1: *sl* DM_2: *low* DM_3: between *low* and *med* DM_4: [6.5,7.5]	DM_1: *sl* DM_2: *low* DM_3: *sl* DM_4: *sl*
A_4	DM_1: *sh* DM_2: *med* DM_3: [6,8] DM_4: *high*	DM_1: *sh* DM_2: *med* DM_3: [6,8] DM_4: *med*	DM_1: *n* DM_2: *vl* DM_3: *n* DM_4: *vl*
A_5	DM_1: at least *sh* DM_2: [7,8] DM_3: *sh* DM_4: between *med* and *vh*	DM_1: at least *sh* DM_2: *med* DM_3: between *med* and *vh* DM_4: [7,8]	DM_1: *low* DM_2: *n* DM_3: *low* DM_4: 1.5

Then, these quantitative and qualitative performance data were transformed into NCMs as shown in Table 5.

Therefore, the original heterogeneous performance matrix $\mathbf{X}$ (Table 4) was transformed into a uniform matrix $\boldsymbol{Y}$ represented by NCMs (Table 5).

Table 5. The performance of 5 schemes on 14 indicators.

	C_1	C_2	C_3	C_4	C_5
A_1	(1366.0000, 0, 0)	(1.8500, 0, 0)	(0.8500, 0, 0)	(8.0000, 0, 0)	(166.0000, 0, 0)
A_2	(1278.0000, 0, 0)	(0.0000, 0, 0)	(0.7800, 0, 0)	(7.0000, 0, 0)	(138.0000, 0, 0)
A_3	(1445.0000, 0, 0)	(0.8600, 0, 0)	(0.6600, 0, 0)	(11.0000, 0, 0)	(155.0000, 0, 0)
A_4	(1210.0000, 0, 0)	(0.0000, 0, 0)	(0.8200, 0, 0)	(6.0000, 0, 0)	(147.0000, 0, 0)
A_5	(1312.0000, 0, 0)	(0.4200, 0, 0)	(0.7200, 0, 0)	(9.0000, 0, 0)	(176.0000, 0, 0)

	C_6	C_7	C_8	C_9	C_{10}
A_1	(3250.0000, 0, 0)	(785.0000, 0, 0)	(2356.0000, 0, 0)	(6.8683, 0.8622, 0.0396)	(6.0841, 0.8129, 0.0393)
A_2	(2886.0000, 0, 0)	(904.0000, 0, 0)	(1968.0000, 0, 0)	(2.6250, 0.6741, 0.0270)	(4.3997, 0.6254, 0.0286)
A_3	(3763.0000, 0, 0)	(1048.0000, 0, 0)	(2087.0000, 0, 0)	(8.2477, 0.4177, 0.0137)	(6.4227, 0.6091, 0.0269)
A_4	(2596.0000, 0, 0)	(816.0000, 0, 0)	(1789.0000, 0, 0)	(3.7413, 0.6103, 0.0243)	(2.4707, 0.6009, 0.0192)
A_5	(3015.0000, 0, 0)	(953.0000, 0, 0)	(2523.0000, 0, 0)	(7.2113, 0.8772, 0.0406)	(5.0580, 0.6403, 0.0323)

	C_{11}	C_{12}	C_{13}	C_{14}
A_1	(3.7413, 0.6103, 0.0243)	(6.3052, 0.7563, 0.0357)	(6.9576, 0.7162, 0.0339)	(1.8938, 0.4553, 0.0146)
A_2	(1.8583, 0.6071, 0.0194)	(7.4909, 0.4709, 0.0218)	(7.1159, 0.4727, 0.0218)	(0.9223, 0.3925, 0.0132)
A_3	(5.7403, 0.6249, 0.0297)	(3.7188, 0.5090, 0.0202)	(3.6145, 0.5689, 0.0218)	(2.6238, 0.5190, 0.0166)
A_4	(2.4707, 0.6009, 0.0192)	(6.9765, 0.4961, 0.0202)	(6.0553, 0.4760, 0.0223)	(0.0985, 0.1654, 0.0105)
A_5	(3.7413, 0.6103, 0.0243)	(7.7953, 0.6508, 0.0289)	(7.2835, 0.6592, 0.0305)	(0.9980, 0.3763, 0.0130)

4.2. Calculate the Weights of Indicators

The weights of indicators can be calculated according to Steps 3 to 7 as described in Section 3.3.2.

For the first level of the indicator system, there are three indicators, i.e., the economy, the environment, and society, that require pairwise comparisons. Six 3 × 3 PCMs were collected from six decision makers as listed in Table 6.

Table 6. Six PCMs for the first level indicators.

1	3	5	1	2	4	1	3	4	1	2	5	1	3	4	1	2	4
1/3	1	2	1/2	1	2	1/3	1	1.5	1/2	1	2	1/3	1	1	1/2	1	2
1/5	1/2	1	1/4	1/2	1	1/4	1/1.5	1	1/5	1/2	1	1/4	1	1	1/4	1/2	1

The weight vectors of six PCMs were calculated as [0.6483, 0.2297, 0.1220], [0.5714, 0.2857, 0.1429], [0.6301, 0.2184, 0.1515], [0.5954, 0.2764, 0.1283], [0.6337, 0.1919, 0.1744], and [0.5714, 0.2857, 0.1429], respectively. The *CR*s were calculated as 0.0036, 0, 0.0015, 0.0053, 0.0088, and 0, respectively. Then, the weights of the six decision makers were calculated as 0.1661, 0.1722, 0.1696, 0.1630, 0.1569, and 0.1722, respectively. Finally, the weight vector of three indicators was calculated as [0.6078, 0.2487, 0.1434] by the weighted arithmetic mean of six weight vectors.

Similarly, five 8 × 8 PCMs were collected from five decision makers for the second-level indicators under economy. The weight vector of eight indicators under economy was calculated as [0.0855, 0.1509, 0.1080, 0.0866, 0.1747, 0.1294, 0.1339, 0.1311]. Three 3 × 3 PCMs were collected from three decision makers for the second-level indicators under environment. The weight vector of three indicators under environment was calculated as [0.5390, 0.2973, 0.1638]. Three 3 × 3 PCMs were collected from three decision makers for the second-level indicators under society. The weight vector of three indicators under society was calculated as [0.3146, 0.2024, 0.4830].

Therefore, the weights of 14 bottom-level indicators were calculated as 0.0520, 0.0917, 0.0656, 0.0526, 0.1062, 0.0786, 0.0814, 0.0797, 0.1340, 0.0739, 0.0407, 0.0451, 0.0290, and 0.0693, respectively. The indicator with the largest weight is C_9, while the smallest is C_{13}.

4.3. Prioritize Alternative Schemes

Firstly, the standardized matrix $\boldsymbol{R}$ was calculated as listed in Table 7 by normalizing the matrix $\boldsymbol{Y}$ in Table 5.

Table 7. The standardized performance matrix of 5 schemes on 14 indicators.

	C_1	C_2	C_3	C_4	C_5
A_1	(0.3362, 0, 0)	(0, 0, 0)	(1.0000, 0, 0)	(0.6000, 0, 0)	(0.2632, 0, 0)
A_2	(0.7106, 0, 0)	(1.0000, 0, 0)	(0.6316, 0, 0)	(0.8000, 0, 0)	(1.0000, 0, 0)
A_3	(0, 0, 0)	(0.5351, 0, 0)	(0, 0, 0)	(0, 0, 0)	(0.5526, 0, 0)
A_4	(1.0000, 0, 0)	(1.0000, 0, 0)	(0.8421, 0, 0)	(1.0000, 0, 0)	(0.7632, 0, 0)
A_5	(0.5660, 0, 0)	(0.7730, 0, 0)	(0.3158, 0, 0)	(0.4000, 0, 0)	(0, 0, 0)
	C_6	C_7	C_8	C_9	C_{10}
A_1	(0.4396, 0, 0)	(1.0000, 0, 0)	(0.2275, 0, 0)	(0.2453, 0.1739, 0.0076)	(0.0857, 0.2577, 0.0121)
A_2	(0.7515, 0, 0)	(0.5475, 0, 0)	(0.7561, 0, 0)	(1.0000, 0.1995, 0.0076)	(0.5119, 0.2471, 0.0108)
A_3	(0, 0, 0)	(0, 0, 0)	(0.5940, 0, 0)	(0, 0.1051, 0.0034)	(0, 0.2180, 0.0096)
A_4	(1.0000, 0, 0)	(0.8821, 0, 0)	(1.0000, 0, 0)	(0.8015, 0.1734, 0.0066)	(1.0000, 0.3062, 0.0118)
A_5	(0.6410, 0, 0)	(0.3612, 0, 0)	(0, 0, 0)	(0.1843, 0.1747, 0.0077)	(0.3453, 0.2358, 0.0110)
	C_{11}	C_{12}	C_{13}	C_{14}	
A_1	(0.5149, 0.2530, 0.0109)	(0.6345, 0.2580, 0.0115)	(0.9112, 0.3300, 0.0144)	(0.2891, 0.2804, 0.0090)	
A_2	(1.0000, 0.3174, 0.0129)	(0.9253, 0.2532, 0.0108)	(0.9543, 0.3032, 0.0129)	(0.6738, 0.2958, 0.0099)	
A_3	(0, 0.2277, 0.0108)	(0, 0.1766, 0.0070)	(0, 0.2193, 0.0084)	(0, 0.2906, 0.0093)	
A_4	(0.8422, 0.2926, 0.0119)	(0.7991, 0.2380, 0.0098)	(0.6652, 0.2565, 0.0109)	(1.0000, 0.3051, 0.0110)	
A_5	(0.5149, 0.2530, 0.0109)	(1.0000, 0.2866, 0.0122)	(1.0000, 0.3356, 0.0145)	(0.6438, 0.2894, 0.0097)	

Then, according to the weights of 14 bottom-level indicators calculated in Section 4.2, the weighted standardized matrix $\boldsymbol{V}$ was calculated as listed in Table 8.

Therefore, as seen in Table 8, the CPIS and CNIS were [(0.0520, 0, 0), (0.0917, 0, 0), (0.0656, 0, 0), (0.0526, 0, 0), (0.1062, 0, 0), (0.0786, 0, 0), (0.0814, 0, 0), (0.0797, 0, 0), (0.1340, 0.0267, 0.0010), (0.0739, 0.0226, 0.0009), (0.0407, 0.0129, 0.0005), (0.0451, 0.0129, 0.0006), (0.0290, 0.0097, 0.0004), (0.0693, 0.0211, 0.0008)] and [(0, 0, 0), (0, 0, 0), (0, 0, 0), (0, 0, 0), (0, 0, 0), (0, 0, 0), (0, 0, 0), (0, 0, 0), (0, 0.0141, 0.0005), (0, 0.0161, 0.0007), (0, 0.0093, 0.0004), (0, 0.0080, 0.0003), (0, 0.0064, 0.0002), (0, 0.0201, 0.0006)], respectively.

Additionally, the distances of five schemes from CPIS and CNIS were calculated as listed in Table 9.

Finally, the closeness coefficients of five schemes were calculated as (0.4120, 0.0613, 0.0025), (0.8111, 0.0860, 0.0033), (0.1551, 0.0473, 0.0018), (0.9031, 0.0916, 0.0035), and (0.4118, 0.0613, 0.0025), respectively. Thus, these five schemes were ranked as $A_4 \succ A_2 \succ A_1 \succ A_5 \succ A_3$ in descending order of their closeness coefficients. It can be seen that A_4 is the most desired scheduling scheme among the five schemes.

Table 8. The weighted standardized performance matrix of 5 schemes on 14 indicators.

	C_1	C_2	C_3	C_4	C_5
A_1	(0.0175, 0, 0)	(0, 0, 0)	(0.0656, 0, 0)	(0.0316, 0, 0)	(0.0279, 0, 0)
A_2	(0.0370, 0, 0)	(0.0917, 0, 0)	(0.0414, 0, 0)	(0.0421, 0, 0)	(0.1062, 0, 0)
A_3	(0, 0, 0)	(0.0491, 0, 0)	(0, 0, 0)	(0, 0, 0)	(0.0587, 0, 0)
A_4	(0.0520, 0, 0)	(0.0917, 0, 0)	(0.0552, 0, 0)	(0.0526, 0, 0)	(0.0810, 0, 0)
A_5	(0.0294, 0, 0)	(0.0709, 0, 0)	(0.0207, 0, 0)	(0.0210, 0, 0)	(0, 0, 0)
	C_6	C_7	C_8	C_9	C_{10}
A_1	(0.0346, 0, 0)	(0.0814, 0, 0)	(0.0181, 0, 0)	(0.0329, 0.0233, 0.0010)	(0.0063, 0.0190, 0.0009)
A_2	(0.0591, 0, 0)	(0.0446, 0, 0)	(0.0603, 0, 0)	(0.1340, 0.0267, 0.0010)	(0.0378, 0.0183, 0.0008)
A_3	(0, 0, 0)	(0, 0, 0)	(0.0473, 0, 0)	(0, 0.0141, 0.0005)	(0, 0.0161, 0.0007)
A_4	(0.0786, 0, 0)	(0.0718, 0, 0)	(0.0797, 0, 0)	(0.1074, 0.0232, 0.0009)	(0.0739, 0.0226, 0.0009)
A_5	(0.0504, 0, 0)	(0.0294, 0, 0)	(0, 0, 0)	(0.0247, 0.0234, 0.0010)	(0.0255, 0.0174, 0.0008)
	C_{11}	C_{12}	C_{13}	C_{14}	
A_1	(0.0210, 0.0103, 0.0004)	(0.0286, 0.0116, 0.0005)	(0.0264, 0.0096, 0.0004)	(0.0200, 0.0194, 0.0006)	
A_2	(0.0407, 0.0129, 0.0005)	(0.0417, 0.0114, 0.0005)	(0.0277, 0.0088, 0.0004)	(0.0467, 0.0205, 0.0007)	
A_3	(0, 0.0093, 0.0004)	(0, 0.0080, 0.0003)	(0, 0.0064, 0.0002)	(0, 0.0201, 0.0006)	
A_4	(0.0343, 0.0119, 0.0005)	(0.0360, 0.0107, 0.0004)	(0.0193, 0.0074, 0.0003)	(0.0693, 0.0211, 0.0008)	
A_5	(0.0210, 0.0103, 0.0004)	(0.0451, 0.0129, 0.0006)	(0.0290, 0.0097, 0.0004)	(0.0446, 0.0201, 0.0007)	

Table 9. The distances of five schemes from CPIS and CNIS.

Schemes	Distance from CPIS (d_i^+)	Distance from CNIS (d_i^-)
A_1	(0.5879, 0.0610, 0.0024)	(0.4119, 0.0517, 0.0021)
A_2	(0.1889, 0.0628, 0.0024)	(0.8109, 0.0538, 0.0021)
A_3	(0.8447, 0.0562, 0.0022)	(0.1551, 0.0459, 0.0017)
A_4	(0.0969, 0.0626, 0.0024)	(0.9029, 0.0535, 0.0020)
A_5	(0.5881, 0.0610, 0.0024)	(0.4117, 0.0517, 0.0021)

5. Discussion

As the final ranking is based on the performance of 14 bottom-level indicators, further analysis is performed to reveal the results on each indicator, as listed in Table 10. According to the standardized performance matrix in Table 7, A_4 and A_2 show the best performance on seven and four indicators, respectively, and A_1 and A_5 both perform the best performance on two indicators. A_3 only has better performance than both A_1 and A_5 on two indicators. In addition, A_4 and A_2 dominate A_3 on all indicators. A_3 does not show the best performance on any indicator but shows the worst performance on 11 indicators. Therefore, it could be concluded that A_3 performs the worst while A_4 performs the best. A_1 optimizes C_3 and C_7, and A_5 optimizes C_{12} and C_{13}, but their performance on other indicator sets is negative. Thus, A_1 and A_5 should be inferior to A_2. These results are in accordance with the final ranking. As a result, the final decision can be made: A_4 will be chosen as the production scheduling scheme.

Based on the analysis above, it is found that one scheduling scheme can only ensure superiority on partial indicators. Thus, it is valuable to address the comprehensive evaluation of scheduling schemes to find the potential all-around optimal solution using MCDM approaches. The proposed cloud-model-enhanced TOPSIS method enables a more scientific selection of the optimal scheduling scheme by considering both qualitative and quantitative indicators, involving heterogeneous data with various uncertainties, and using group decision-making techniques.

Table 10. The rankings of five schemes on each indicator.

Indicator	Ranking	Indicator	Ranking
C_1	$A_4 \succ A_2 \succ A_5 \succ A_1 \succ A_3$	C_8	$A_4 \succ A_2 \succ A_3 \succ A_1 \succ A_5$
C_2	$A_2 \sim A_4 \succ A_5 \succ A_3 \succ A_1$	C_9	$A_2 \succ A_4 \succ A_1 \succ A_5 \succ A_3$
C_3	$A_1 \succ A_4 \succ A_2 \succ A_5 \succ A_3$	C_{10}	$A_4 \succ A_2 \succ A_5 \succ A_1 \succ A_3$
C_4	$A_4 \succ A_2 \succ A_1 \succ A_5 \succ A_3$	C_{11}	$A_2 \succ A_4 \succ A_1 \sim A_5 \succ A_3$
C_5	$A_2 \succ A_4 \succ A_3 \succ A_1 \succ A_5$	C_{12}	$A_5 \succ A_2 \succ A_4 \succ A_1 \succ A_3$
C_6	$A_4 \succ A_2 \succ A_5 \succ A_1 \succ A_3$	C_{13}	$A_5 \succ A_2 \succ A_1 \succ A_4 \succ A_3$
C_7	$A_1 \succ A_4 \succ A_2 \succ A_5 \succ A_3$	C_{14}	$A_4 \succ A_2 \succ A_5 \succ A_1 \succ A_3$

When solving MCDM problems, various studies usually use one or two of the following methods: AHP (including Grey AHP, GAHP, and FAHP), Cloud model, or TOPSIS, but rarely integrate these three together. This paper integrates the three in our proposed decision method. Then a further comparative analysis is carried out to demonstrate the notable features of the method. The comparisons of our proposed method with two recent methods are listed in Table 11.

Table 11. The comparative analysis with recent methods.

Methods	Performance Calculations	Data Types	Weight Assignments	Ranking Algorithms
Thenarasu et al. (2022) [1]	Statistical calculations	Exact numbers	Fuzzy AHP	TOPSIS
Zhang et al. (2022) [8]	Statistical calculations and cumulative prospect theory	Exact numbers	Fuzzy ANP	VIKOR
Proposed method	Quantitative indicators: Statistical calculations or/and evaluation functions Qualitative indicators: Group decision making	Exact numbers Interval numbers Linguistic terms Linguistic expressions	Group AHP	Cloud model enhanced TOPSIS

From the above table, it can be seen that the other two methods mainly consider quantitative indicators; the performances of quantitative indicators calculated by plain statistical methods usually cannot reflect their real benefits. In Ref. [2], cumulative prospect theory is used to reflect the psychological sensitivities of decision-makers to gains and losses, but the same value function and parameters used make it difficult to express the diversity of different indicators. In this paper, the performances of quantitative indicators are calculated by diverse statistical methods or/and evaluation functions to accurately reflect their real benefits. In addition, the performances of qualitative indicators are evaluated using group decision-making techniques under uncertainty in our proposed method. As shown in Table 11, different from the other two methods, which focus on dealing with exact numbers, our proposed method performs well in managing heterogeneous data involving exact numbers, interval numbers, linguistic terms, and linguistic expressions. When assigning weights to indicators, we adopt group decision-making techniques and the proposed Group AHP model that allows the participation of multiple decision makers to determine the weights of indicators. Finally, the proposed cloud-model-enhanced TOPSIS is able to manage and propagate uncertain information during the evaluation process. The final result contains both magnitude and uncertainties. However, previous methods are unable to manage uncertainties, and the final outcome is deterministic. For example, suppose the closeness coefficients of two schemes A_1 and A_2 are calculated by our proposed method as (0.8120, 0.5610, 0.0725) and (0.8119, 0.0601, 0.0005), respectively. Although the expectation of the closeness coefficient of A_1 is slightly higher than that of A_2, the uncertainty of A_1 is significantly greater than A_2. Therefore, for those who are concerned with uncertainty avoidance, A_2 should be preferred.

From the application example and the above comparative analysis, it can be concluded that a more rational and effective ranking of scheduling schemes can be achieved using the developed novel MCDM method based on Group AHP and cloud-model-enhanced TOPSIS. Compared with previous methods, the advantages of our proposed method are summarized as follows:

(1) We propose a systematic evaluation system for steel production scheduling, which is more reasonable, flexible, practical, and effective in evaluating and ranking steel production scheduling schemes in an uncertain environment. It can also provide some guidance for the optimization direction of scheduling schemes according to indicator performance.
(2) The evaluation indicators include both qualitative and quantitative indicators, which are built from the entire production scheduling process, simultaneously considering key factors that affect the production process and results from multiple dimensions of economic, environmental, and social aspects, breaking through the limitations of most evaluation indicators that are limited to a single production process or a single indicator dimension.
(3) We present diverse statistical methods and evaluation functions to evaluate quantitative indicators to accurately reflect their real benefits, and develop a hybrid decision method integrating Group AHP, a cloud model, and the TOPSIS method to evaluate qualitative indicators. The hybrid method has good data capabilities of managing uncertainties of heterogeneous data, achieving better multi-indicator trade-offs, enhancing group decision-making rationality, and obtaining a more effective full and distinguishing ranking of schemes; hence, enhancing the effectiveness of evaluation.
(4) The proposed hybrid MCGDM method is transparent with well-defined problem descriptions and implementation procedures and is evaluated on a real practical steel production scheduling dataset.

6. Conclusions

In this study, a novel hybrid MCGDM method based on group decision-making principles, cloud model theory, and the TOPSIS method is developed to prioritize alternative production scheduling schemes. A novel indicator system involving both qualitative and quantitative indicators is put forward for the comprehensive evaluation of production scheduling schemes. Diverse statistical methods or/and evaluation functions are proposed for the evaluation of quantitative indicators. Heterogeneous data, cloud model theory, and group decision-making techniques are utilized for the evaluation of qualitative indicators. Moreover, a novel Group AHP model is proposed to determine the weights of all evaluation indicators. Then, a novel cloud-model-enhanced TOPSIS method is proposed to rank production scheduling schemes. The proposed hybrid MCGDM method provides a structured and systematic framework for the comprehensive evaluation of production scheduling schemes and overcomes many weaknesses of previous methods. Finally, a practical example is presented to demonstrate the feasibility and effectiveness of our proposed method. The results and comparative analysis indicate that the hybrid MCGDM method developed in this study is more reasonable, flexible, practical, and effective in evaluating and ranking production scheduling schemes under an uncertain environment.

Even though there are a large number of articles on the development of production scheduling schemes, the literature on the selection/ranking of alternative schemes is scarce. Furthermore, the existing literature does not consider both qualitative and quantitative indicators and involves heterogeneous data for comprehensive evaluation and prioritization of production scheduling schemes. This study fills this gap.

Despite all the advantages, the proposed hybrid MCGDM method has some drawbacks that suggest directions for future research. Although a novel Group AHP model is proposed, only subjective judgments are utilized to determine the weights of indicators. Nevertheless, a comprehensive weighting method considering both subjective and objective weights is meaningful for further research, such as the Goal-based Participatory weighting scheme [44]. The consensus among different experts' opinions should also be considered to maintain the robustness of the method [45]. In addition, it should be pointed out that the proposed hybrid MCGDM method has the potential for prioritizing jobs in large-scale flexible job shop scheduling problems. Future research will be concerned with more practical applications using the proposed method presented in this article.

Author Contributions: Conceptualization, X.Z. and Z.L.; methodology, X.Z., Z.L. and Y.L.; investigation, X.Z. and X.X.; data curation, X.Z. and Y.L.; writing—original draft preparation, X.Z. and Y.L.; writing—review and editing, X.Z. and Z.L.; visualization, X.Z., X.X. and D.X.; project administration, X.Z. and Z.L. All authors have read and agreed to the published version of the manuscript.

Funding: This work was supported in part by the National Key Research and Development Plan under grant number 2022YFB3304905 and 2020YFB1713600.

Data Availability Statement: The data presented in this study are available upon request from the corresponding author.

Conflicts of Interest: Authors X.Z., Z.L., Y.L., X.X. and D.X. were employed by the University of Science and Technology Beijing. All the authors declare that the research was conducted in the absence of any commercial or financial relationships that could be construed as a potential conflict of interest.

References

1. Lin, J.; Liu, M.; Hao, J.; Jiang, S. A multi-objective optimization approach for integrated production planning under interval uncertainties in the steel industry. *Comput. Oper. Res.* **2016**, *72*, 189–203. [CrossRef]
2. Li, Q.; Wang, X.; Zhang, X. A Scheduling Method Based on NSGA2 for Steelmaking and Continuous Casting Production Process. *IFAC—PapersOnLine* **2018**, *51*, 174–179. [CrossRef]
3. Jiang, S.-L.; Liu, M.; Lin, J.-H.; Zhong, H.-X. A prediction-based online soft scheduling algorithm for the real-world steelmaking-continuous casting production. *Knowl.-Based Syst.* **2016**, *111*, 159–172. [CrossRef]
4. Jiang, S.-L.; Peng, G.; Bogle, I.D.L.; Zheng, Z. Two-stage robust optimization approach for flexible oxygen distribution under uncertainty in integrated iron and steel plants. *Appl. Energy* **2021**, *306*, 118022. [CrossRef]
5. Pacciarelli, D.; Pranzo, M. Production scheduling in a steelmaking-continuous casting plant. *Comput. Chem. Eng.* **2004**, *28*, 2823–2835. [CrossRef]
6. Long, J.-Y.; Zheng, Z.; Gao, X.-Q.; Gong, Y.-M. Production Planning System for the Whole Steelmaking Process of Panzhihua Iron and Steel Corporation. *J. Iron Steel Res. Int.* **2014**, *21*, 44–50. [CrossRef]
7. Thenarasu, M.; Rameshkumar, K.; Rousseau, J.; Anbuudayasankar, S. Development and analysis of priority decision rules using MCDM approach for a flexible job shop scheduling: A simulation study. *Simul. Model. Pract. Theory* **2021**, *114*, 102416. [CrossRef]
8. Zhang, P.; Qiao, F.; Wang, J.; Sutherland, J.W. Novel Multi-Criteria Sustainable Evaluation for Production Scheduling Based on Fuzzy Analytic Network Process and Cumulative Prospect Theory-Enhanced VIKOR. *IEEE Robot. Autom. Lett.* **2022**, *7*, 9969–9976. [CrossRef]
9. Nakhaeinejad, M.; Nahavandi, N. An interactive algorithm for multi-objective flow shop scheduling with fuzzy processing time through resolution method and TOPSIS. *Int. J. Adv. Manuf. Technol.* **2012**, *66*, 1047–1064. [CrossRef]
10. Carlucci, D.; Renna, P.; Materi, S. A Job-Shop Scheduling Decision-Making Model for Sustainable Production Planning with Power Constraint. *IEEE Trans. Eng. Manag.* **2021**, *70*, 1923–1932. [CrossRef]
11. Gharoun, H.; Hamid, M.; Torabi, S.A. An integrated approach to joint production planning and reliability-based multi-level preventive maintenance scheduling optimisation for a deteriorating system considering due-date satisfaction. *Int. J. Syst. Sci.—Oper. Logist.* **2022**, *9*, 489–511. [CrossRef]
12. Joshi, D.K.; Kumar, S. Trapezium cloud TOPSIS method with interval-valued intuitionistic hesitant fuzzy linguistic information. *Granul. Comput.* **2017**, *3*, 139–152. [CrossRef]
13. Morente-Molinera, J.; Wu, X.; Morfeq, A.; Al-Hmouz, R.; Herrera-Viedma, E. A novel multi-criteria group decision-making method for heterogeneous and dynamic contexts using multi-granular fuzzy linguistic modelling and consensus measures. *Inf. Fusion* **2019**, *53*, 240–250. [CrossRef]
14. Liu, H.-C.; Wang, L.-E.; Li, Z.; Hu, Y.-P. Improving Risk Evaluation in FMEA with Cloud Model and Hierarchical TOPSIS Method. *IEEE Trans. Fuzzy Syst.* **2018**, *27*, 84–95. [CrossRef]
15. Lourenzutti, R.; Krohling, R.A. A generalized TOPSIS method for group decision making with heterogeneous information in a dynamic environment. *Inf. Sci.* **2016**, *330*, 1–18. [CrossRef]
16. Khan, M.S.A.; Anjum, F.; Ullah, I.; Senapati, T.; Moslem, S. Priority Degrees and Distance Measures of Complex Hesitant Fuzzy Sets with Application to Multi-Criteria Decision Making. *IEEE Access* **2022**, *11*, 13647–13666. [CrossRef]
17. Zheng, X.; Hao, T.; Wang, H.; Xu, K. Quantification study of mental load state based on AHP-TOPSIS integration extended with cloud model: Methodological and experimental research. *Complex Intell. Syst.* **2023**, *9*, 5501–5525. [CrossRef]

18. Li, D.Y.; Du, Y. *Artificial Intelligence with Uncertainty*; Chapman & Hall/CRC Press: Boca Raton, FL, USA, 2007.
19. Yang, X.J.; Zeng, L.; Zhang, R. Cloud Delphi method. *Int. J. Uncertain. Fuzziness Knowl.-Based Syst.* **2012**, *20*, 77–97. [CrossRef]
20. Yang, X.J.; Zeng, L.; Luo, F.; Wang, S.X. Cloud hierarchical analysis. *J. Inf. Comput. Sci.* **2010**, *7*, 2468–2477.
21. Zadeh, L.A. The concept of a linguistic variable and its application to approximate reasoning—I. *Inf. Sci.* **1975**, *8*, 199–249. [CrossRef]
22. Zadeh, L. Fuzzy logic = computing with words. *IEEE Trans. Fuzzy Syst.* **1996**, *4*, 103–111. [CrossRef]
23. Peng, G.; Xu, D.; Zhou, J.; Yang, Q.; Shen, W. A Novel Curve Pattern Recognition Framework for Hot-Rolling Slab Camber. *IEEE Trans. Ind. Inform.* **2022**, *19*, 1270–1278. [CrossRef]
24. Peng, G.; Cheng, Y.; Zhang, Y.; Shao, J.; Wang, H.; Shen, W. Industrial big data-driven mechanical performance prediction for hot-rolling steel using lower upper bound estimation method. *J. Manuf. Syst.* **2023**, *65*, 104–114. [CrossRef]
25. Peng, G.; Cheng, Y.; Wang, H.; Shen, W. Industrial IoT-enabled prediction interval estimation of mechanical performances for hot-rolling steel. *IEEE Trans. Instrum. Meas.* **2022**, *71*, 1–10. [CrossRef]
26. Huang, K.; Wu, Y.; Yang, C.; Peng, G.; Shen, W. Structure Dictionary Learning-Based Multimode Process Monitoring and its Application to Aluminum Electrolysis Process. *IEEE Trans. Autom. Sci. Eng.* **2020**, *17*, 1989–2003. [CrossRef]
27. Wang, H.L.; Feng, Y.Q. On multiple attribute group decision making with linguistic assessment information based on cloud model. *Control Decis.* **2005**, *20*, 679–685.
28. Wang, P.; Xu, X.; Huang, S.; Cai, C. A Linguistic Large Group Decision Making Method Based on the Cloud Model. *IEEE Trans. Fuzzy Syst.* **2018**, *26*, 3314–3326. [CrossRef]
29. Yang, X.; Yan, L.; Peng, H.; Gao, X. Encoding words into Cloud models from interval-valued data via fuzzy statistics and membership function fitting. *Knowl.-Based Syst.* **2014**, *55*, 114–124. [CrossRef]
30. Yang, X.; Xu, Z.; Xu, J. Large-scale group Delphi method with heterogeneous decision information and dynamic weights. *Expert Syst. Appl.* **2023**, *213*, 118782. [CrossRef]
31. Huang, H.-C.; Yang, X. Representation of The Pairwise Comparisons in AHP Using Hesitant Cloud Linguistic Term Sets. *Fundam. Informaticae* **2016**, *144*, 349–362. [CrossRef]
32. Rodríguez, R.M.; Martínez, L.; Herrera, F. A group decision making model dealing with comparative linguistic expressions based on hesitant fuzzy linguistic term sets. *Inf. Sci.* **2013**, *241*, 28–42. [CrossRef]
33. Yang, X.J.; Xu, Z.F.; He, R.M.; Xue, F.X. Credibility assessment of complex simulation models using cloud models to represent and aggregate diverse evaluation results. In *Intelligent Computing Methodologies*; Huang, D.S., Huang, Z.K., Hussain, A., Eds.; ICIC 2019; Lecture Notes in Computer Science; Springer: Berlin/Heidelberg, Germany, 2019; Volume 11645, pp. 306–317.
34. Zhang, B.; Dong, Y.; Herrera-Viedma, E. Group Decision Making with Heterogeneous Preference Structures: An Automatic Mechanism to Support Consensus Reaching. *Group Decis. Negot.* **2019**, *28*, 585–617. [CrossRef]
35. Chiclana, F.; Herrera, F.; Herrera-Viedma, E. Integrating multiplicative preference relations in a multipurpose decision-making model based on fuzzy preference relations. *Fuzzy Sets Syst.* **2001**, *122*, 277–291. [CrossRef]
36. Xu, Z. Multiple-Attribute Group Decision Making with Different Formats of Preference Information on Attributes. *IEEE Trans. Syst. Man Cybern. Part B Cybern.* **2007**, *37*, 1500–1511. [CrossRef]
37. Saaty, T.L. *The Analytic Hierarchy Process*; McGraw-Hill: New York, NY, USA, 1980.
38. Ossadnik, W.; Schinke, S.; Kaspar, R.H. Group Aggregation Techniques for Analytic Hierarchy Process and Analytic Network Process: A Comparative Analysis. *Group Decis. Negot.* **2016**, *25*, 421–457. [CrossRef]
39. Dong, Q.; Cooper, O. A peer-to-peer dynamic adaptive consensus reaching model for the group AHP decision making. *Eur. J. Oper. Res.* **2016**, *250*, 521–530. [CrossRef]
40. Grošelj, P.; Dolinar, G. Group AHP framework based on geometric standard deviation and interval group pairwise comparisons. *Inf. Sci.* **2023**, *626*, 370–389. [CrossRef]
41. Yang, X.; Yan, L.; Zeng, L. How to handle uncertainties in AHP: The Cloud Delphi hierarchical analysis. *Inf. Sci.* **2013**, *222*, 384–404. [CrossRef]
42. Saaty, T.L. Decision making with the analytic hierarchy process. *Int. J. Serv. Sci.* **2008**, *1*, 83–98. [CrossRef]
43. Pant, S.; Kumar, A.; Ram, M.; Klochkov, Y.; Sharma, H.K. Consistency Indices in Analytic Hierarchy Process: A Review. *Mathematics* **2022**, *10*, 1206. [CrossRef]
44. Correa Machado, A.M.; Ekel, P.I.; Libório, M.P. Goal-based participatory weighting scheme: Balancing objectivity and subjectivity in the construction of composite indicators. *Qual. Quant.* **2023**, *57*, 4387–4407. [CrossRef]
45. Herrera-Viedma, E.; Cabrerizo, F.J.; Kacprzyk, J.; Pedrycz, W. A review of soft consensus models in a fuzzy environment. *Inf. Fusion* **2014**, *17*, 4–13. [CrossRef]

MDPI AG
Grosspeteranlage 5
4052 Basel
Switzerland
Tel.: +41 61 683 77 34

Processes Editorial Office
E-mail: processes@mdpi.com
www.mdpi.com/journal/processes

www.ingramcontent.com/pod-product-compliance
Lightning Source LLC
LaVergne TN
LVHW070540071126
851858LV00011BA/247
* 9 7 8 3 7 2 5 8 8 1 3 1 4 *